Records arithmetick, or, The ground of arts teaching the perfect work and practice of arithmetic both in whole numbers and fractions after a more easie and exact form then in former time hath been set forth (1658)

Robert Hartwell

Records arithmetick, or, The ground of arts teaching the perfect work and practice of arithmetic both in whole numbers and fractions after a more easie and exact form then in former time hath been set forth
Record, Robert, 1510?-1558.
Dee, John, 1527-1608.
Mellis, John, fl. 1588.
Imperfect: pages stained.
[21], 535 p.
London : Printed by James Flesher and are to be sold by Joseph Crawford ..., 1658.
Wing / R645
English
Reproduction of the original in the University of Illinois (Urbana-Champaign Campus)

Early English Books Online (EEBO) Editions

Imagine holding history in your hands.

Now you can. Digitally preserved and previously accessible only through libraries as Early English Books Online, this rare material is now available in single print editions. Thousands of books written between 1475 and 1700 and ranging from religion to astronomy, medicine to music, can be delivered to your doorstep in individual volumes of high-quality historical reproductions.

We have been compiling these historic treasures for more than 70 years. Long before such a thing as "digital" even existed, ProQuest founder Eugene Power began the noble task of preserving the British Museum's collection on microfilm. He then sought out other rare and endangered titles, providing unparalleled access to these works and collaborating with the world's top academic institutions to make them widely available for the first time. This project furthers that original vision.

These texts have now made the full journey -- from their original printing-press versions available only in rare-book rooms to online library access to new single volumes made possible by the partnership between artifact preservation and modern printing technology. A portion of the proceeds from every book sold supports the libraries and institutions that made this collection possible, and that still work to preserve these invaluable treasures passed down through time.

This is history, traveling through time since the dawn of printing to your own personal library.

Initial Proquest EEBO Print Editions collections include:

Early Literature

This comprehensive collection begins with the famous Elizabethan Era that saw such literary giants as Chaucer, Shakespeare and Marlowe, as well as the introduction of the sonnet. Traveling through Jacobean and Restoration literature, the highlight of this series is the Pollard and Redgrave 1475-1640 selection of the rarest works from the English Renaissance.

Early Documents of World History

This collection combines early English perspectives on world history with documentation of Parliament records, royal decrees and military documents that reveal the delicate balance of Church and State in early English government. For social historians, almanacs and calendars offer insight into daily life of common citizens. This exhaustively complete series presents a thorough picture of history through the English Civil War.

Historical Almanacs

Historically, almanacs served a variety of purposes from the more practical, such as planting and harvesting crops and plotting nautical routes, to predicting the future through the movements of the stars. This collection provides a wide range of consecutive years of "almanacks" and calendars that depict a vast array of everyday life as it was several hundred years ago.

Early History of Astronomy & Space

Humankind has studied the skies for centuries, seeking to find our place in the universe. Some of the most important discoveries in the field of astronomy were made in these texts recorded by ancient stargazers, but almost as impactful were the perspectives of those who considered their discoveries to be heresy. Any independent astronomer will find this an invaluable collection of titles arguing the truth of the cosmic system.

Early History of Industry & Science

Acting as a kind of historical Wall Street, this collection of industry manuals and records explores the thriving industries of construction; textile, especially wool and linen; salt; livestock; and many more.

Early English Wit, Poetry & Satire

The power of literary device was never more in its prime than during this period of history, where a wide array of political and religious satire mocked the status quo and poetry called humankind to transcend the rigors of daily life through love, God or principle. This series comments on historical patterns of the human condition that are still visible today.

Early English Drama & Theatre

This collection needs no introduction, combining the works of some of the greatest canonical writers of all time, including many plays composed for royalty such as Queen Elizabeth I and King Edward VI. In addition, this series includes history and criticism of drama, as well as examinations of technique.

Early History of Travel & Geography

Offering a fascinating view into the perception of the world during the sixteenth and seventeenth centuries, this collection includes accounts of Columbus's discovery of the Americas and encompasses most of the Age of Discovery, during which Europeans and their descendants intensively explored and mapped the world. This series is a wealth of information from some the most groundbreaking explorers.

Early Fables & Fairy Tales

This series includes many translations, some illustrated, of some of the most well-known mythologies of today, including Aesop's Fables and English fairy tales, as well as many Greek, Latin and even Oriental parables and criticism and interpretation on the subject.

Early Documents of Language & Linguistics

The evolution of English and foreign languages is documented in these original texts studying and recording early philology from the study of a variety of languages including Greek, Latin and Chinese, as well as multilingual volumes, to current slang and obscure words. Translations from Latin, Hebrew and Aramaic, grammar treatises and even dictionaries and guides to translation make this collection rich in cultures from around the world.

Early History of the Law

With extensive collections of land tenure and business law "forms" in Great Britain, this is a comprehensive resource for all kinds of early English legal precedents from feudal to constitutional law, Jewish and Jesuit law, laws about public finance to food supply and forestry, and even "immoral conditions." An abundance of law dictionaries, philosophy and history and criticism completes this series.

Early History of Kings, Queens and Royalty

This collection includes debates on the divine right of kings, royal statutes and proclamations, and political ballads and songs as related to a number of English kings and queens, with notable concentrations on foreign rulers King Louis IX and King Louis XIV of France, and King Philip II of Spain. Writings on ancient rulers and royal tradition focus on Scottish and Roman kings, Cleopatra and the Biblical kings Nebuchadnezzar and Solomon.

Early History of Love, Marriage & Sex

Human relationships intrigued and baffled thinkers and writers well before the postmodern age of psychology and self-help. Now readers can access the insights and intricacies of Anglo-Saxon interactions in sex and love, marriage and politics, and the truth that lies somewhere in between action and thought.

Early History of Medicine, Health & Disease

This series includes fascinating studies on the human brain from as early as the 16th century, as well as early studies on the physiological effects of tobacco use. Anatomy texts, medical treatises and wound treatment are also discussed, revealing the exponential development of medical theory and practice over more than two hundred years.

Early History of Logic, Science and Math

The "hard sciences" developed exponentially during the 16th and 17th centuries, both relying upon centuries of tradition and adding to the foundation of modern application, as is evidenced by this extensive collection. This is a rich collection of practical mathematics as applied to business, carpentry and geography as well as explorations of mathematical instruments and arithmetic; logic and logicians such as Aristotle and Socrates; and a number of scientific disciplines from natural history to physics.

Early History of Military, War and Weaponry

Any professional or amateur student of war will thrill at the untold riches in this collection of war theory and practice in the early Western World. The Age of Discovery and Enlightenment was also a time of great political and religious unrest, revealed in accounts of conflicts such as the Wars of the Roses.

Early History of Food

This collection combines the commercial aspects of food handling, preservation and supply to the more specific aspects of canning and preserving, meat carving, brewing beer and even candy-making with fruits and flowers, with a large resource of cookery and recipe books. Not to be forgotten is a "the great eater of Kent," a study in food habits.

Early History of Religion

From the beginning of recorded history we have looked to the heavens for inspiration and guidance. In these early religious documents, sermons, and pamphlets, we see the spiritual impact on the lives of both royalty and the commoner. We also get insights into a clergy that was growing ever more powerful as a political force. This is one of the world's largest collections of religious works of this type, revealing much about our interpretation of the modern church and spirituality.

Early Social Customs

Social customs, human interaction and leisure are the driving force of any culture. These unique and quirky works give us a glimpse of interesting aspects of day-to-day life as it existed in an earlier time. With books on games, sports, traditions, festivals, and hobbies it is one of the most fascinating collections in the series.

old books. new life.

The BiblioLife Network

This project was made possible in part by the BiblioLife Network (BLN), a project aimed at addressing some of the huge challenges facing book preservationists around the world. The BLN includes libraries, library networks, archives, subject matter experts, online communities and library service providers. We believe every book ever published should be available as a high-quality print reproduction; printed on-demand anywhere in the world. This insures the ongoing accessibility of the content and helps generate sustainable revenue for the libraries and organizations that work to preserve these important materials.

The following book is in the "public domain" and represents an authentic reproduction of the text as printed by the original publisher. While we have attempted to accurately maintain the integrity of the original work, there are sometimes problems with the original work or the micro-film from which the books were digitized. This can result in minor errors in reproduction. Possible imperfections include missing and blurred pages, poor pictures, markings and other reproduction issues beyond our control. Because this work is culturally important, we have made it available as part of our commitment to protecting, preserving, and promoting the world's literature.

GUIDE TO FOLD-OUTS MAPS and OVERSIZED IMAGES

The book you are reading was digitized from microfilm captured over the past thirty to forty years. Years after the creation of the original microfilm, the book was converted to digital files and made available in an online database.

In an online database, page images do not need to conform to the size restrictions found in a printed book. When converting these images back into a printed bound book, the page sizes are standardized in ways that maintain the detail of the original. For large images, such as fold-out maps, the original page image is split into two or more pages

Guidelines used to determine how to split the page image follows:

- Some images are split vertically; large images require vertical and horizontal splits.
- For horizontal splits, the content is split left to right.
- For vertical splits, the content is split from top to bottom.
- For both vertical and horizontal splits, the image is processed from top left to bottom right.

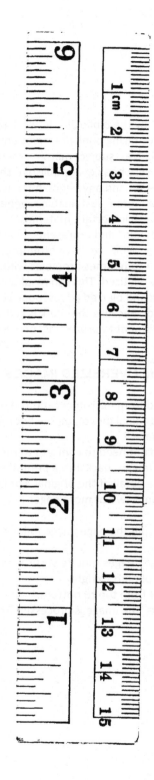

MICROFILMED — 1984

Records Arithmetick:

OR,

THE GROUND
OF ARTS;

TEACHING

The perfect work and practice of Arithmetick,
both in whole Numbers and Fractions, after a more
easie and exact form then in former time hath been set forth :
Made by M. *Robert Record*, D. in Physick.

Afterward, augmented by M. John Dee.

And since enlarged with a third part of Rules of Practise, abridged into a briefer method then hitherto hath been
published, with divers necessary Rules incident to the
Trade of Merchandise : with Tables of the valuation of all Coyns, as they are currant
at this present time,

By JOHN MELLIS.

And now diligently perused, corrected, illustrated and enlarged;
with an *Appendix* of figurative Numbers, and the extraction of
their Roots, according to the method of *Christian Urstitius* : with
Tables of Board and Timber measure; and new tables of Interest, after 10. 8. and 6. per 100 ; with the true value of Annuities
to be bought or sold present, Respited or in Reversion : the first
calculated by *R. C.* but corrected, and the latter diligently calculated by *Ro. Hartwell Philomathemat.*

Scientia non habet inimicum nisi ignorantem.
Fide ———————— sed ———————— Vide.

Sum ex libris Na: Clutterbuck

LONDON.

Printed by *James Flesher*, and are to be sold by *Joseph
Cranford*, at the signe of the Kings head in St. *Pauls*
Church-yard. 1 6 5 8.

That which my friend hath well begun
    For very love to Common-weal,
Need not all whole to be now done,
But now encrease I doe reveal.

Something herein I once redreſt,
And now again for thy behoof,
Of zeal I doe, and at requeſt,
Both mend and adde, fit for all proof.

Of numbers uſe, the endleſſe might,
No wit nor language can expreſſe :
Apply and try both day and night,
And then this truth thou wilt confeſſe.

J. Dee.

The Books Verdiᴄt.

To pleaſe or diſpleaſe ſure I am,
    But not of one ſort to every man :
To pleaſe the beſt ſort would I fain,
    The froward diſpleaſe shall I certain;
Yet wiſh I will though not with hope,
    All eares or mouths to pleaſe or ſtop.

To the moſt mightie Prince,

Edward the 6th by the grace of God,

King of *England, France,* and
Ireland, &c.

THe Excellency of mans nature being ſuch, as it is by Gods divine favour (moſt mighty Prince) not only created in highneſſe of degree far above all other corporall things, but by perfection, reaſon, and ſearch of wit, much approaching toward the image of God, as not onely the holy Scriptures doe teſtifie, but alſo thoſe naturall Philoſophers, which exactly did conſider the nature of man, and namely the far reach and infinite compaſſe of the words of the mind, were inforced to confeſſe, that man ſcarcely was able to know himſelf. And if he would duely ponder the nature of himſelf, he would finde it ſo ſtrange, that it might ſeem to him a very miracle : And thereof ſprang that ſaying; *Magnum miraculum eſt homo, maximum miraculum ſapiens homo.* For undoubtedly, as man is one of the greateſt miracles that ever God wrought, ſo a wiſe man is plainly the greateſt.

And therefore was it that ſome did account the head of a man the greateſt miracle in the world, becauſe not onely of the ſtrange workmanſhip that is in it, but much more of the efficacy of reaſon, wit, memory, imagination, & ſuch other powers, and works of the minde, which can more eaſily conceive any thing in a manner then underſtand it ſelf. Amongſt all the creatures of God, it findeth none more difficult to be perceived then the ſame powers of it ſelf; whereby it doth conceive and judge : as it may be well conjectured by the diverſity of opinions, that the wiſeſt Philoſophers did utter touching the ſpirit of man, and the ſubſtance of it : whereof I now intend to make no reherſal ; but who ſo liſteth to read

A 2

thereof

thereof, may finde it largly set forth, not onely in *Aristotle* his bookes *de Anima*, but also in *Galen* his booke called *Historia Philosophica*: and again in *Plutarch's* work, *De Philosophorum placitis*, whose words are also repeated of *Eusebius* in the xv. book, Τῆς ἐυαγγελικῆς προσραςκευῆς, unto whom I remit them that have desired to understand intricate difficulty of knowing our own selves, as touching our best part, and that part whereby we deserve to bear the name of men.

This matter seemed so obscure and difficult in knowledge; that *Galen* who for his excellent wisdome and judgement in naturall works, is called of many men, a Miracle in Nature, yet in searching the nature and substance of the spirit of man, he not only confesseth himself ignorant, but counteth it plain temerity to attempt to finde it. So farre above the hope of mans knowledge is that part whereby man doth know and judge of things. And although the ignorant sort (which hate all things that they know not) doe litle esteem the profoundnesse of a mans spirit and reason, the chief power and faculty of it: yet as there is a kind of fear and obedience of unreasonable beasts unto man, by the working power of God, so is there in those small reasoned persons a certain kind of reverence toward wisdome and reason, which they do shew oftentimes, and by power of perswasion, are inforced to obey reason, will they nill they. And hereby it came to passe, that the rudenesse of the first age of man was brought unto some more civill trade, as it is well declared by *Cicero*, in the beginning of his first book, *De Inventione Rhetorica*, where he saith thus, *Nam fuit quoddam tempus quum in agris homines passim bestiarum more vagabantur, & sibi victu ferino vitam propagabant, nec ratione animi quicquam, sed pleraq; viribus corporis administrabant. Nondum divinæ religionis, non humani ratio colebatur. Nemo legitimas viderat nuptias, non certos quisquam inspexerat liberos; non jus æquabile quid utilitatis haberet, acceperat: ita propter errorem atq; inscitiam cæca ac temeraria dominatrix animi cupiditas, ad se explendum viribus corporis abutebatur perniciosissimis satellitibus. Quo tempore quidam, magnus videlicet vir & sapiens, cognovit quæ materia esset, & quanta ad maximas res opportunitas in animis inesset hominum, si quis eam possit elicere, & præcipiendo meliorem reddere. Qui dispersos homines in agris & in tectis sylvestribus abditos, ratione quadam compulit in unum locum, & congregavit · & eos in unamquamque rem inducens utilem atque honestam, primo propter insolentiam reclamantes, deinde propter rationem atque orationem*
studiosius

studiosus audientes, ex feris & immanibus, mites reddidit & mansuetos.

This long repetition of *Tullyes* words will seem tedious to them that love but little, and care much lesse for the knowledge of reason, but unto your Majesty (I dare say) it is a dilectable rememberance, and unto me it seemed so pleasant, that I could scarce stay my pen from writing all that mine eyes did so greedily read.

This sentence of *Cicero* am I loath to translate into English, partly for that unto your Majesty it needeth no translation, but especially knowing how far the grace of *Tullyes* eloquence doth excell any English mans tongue, and much more exceedeth the basenesse of my barbarous style : yet for the fruit of my sentence, I had rather unto my meer English Country men utter the rudenesse of my translation, then to defraud them the benefit of so good a lesson, trusting they will so learne to love reason, that they will also gladly and greedily embrace all good Sciences, that may help to the just furniture of the same, when they consider that informed reason was the onely instrument, or at least the chiefest means to bring men into civil regiment, from barbarous manners, and beastly conditions. " For the time was (saith *Tully*) that men " wandered abroad in the fields up and down like beasts, and used " no better order in feeding then they : so that by reasons rule " they wrought nothing, but most of their doings did they atchive " by fore of strength. At this time there was no just regard of " religion towards God, nor of duty towards man. No man had " seen right use of marriage, neither did any man know their own " children from other ; nor no man had felt the commodity of " just Laws : so that through error and ignorance, willfull lust, " like a blind and heady ruler, abused bodily strength as a most " mortall minister for the satisfying of his desire. At that time " was there one which not onely in power , but also in wis-" dome was great, and he considered how that in the mind of " men was both apt instruments, and great occasion to the due " acomplishment of most weighty affaires, if a man could apply " them to use, and by teaching of rules frame them to better trade. " This man with perswasion of reason gathered into one place " the people that were wandring about the fields, and lay lurk-" ing in wild cottages, and woods, and bringing them in one com-" mon society, did trade them to all such things, as either were " profitable or honest, although not without repining at the first,

by

" by reafon then they had not been fo accuftomed before : Yet at
" length through reafon and perfwafion of words they obeyed
" him more diligently, and fo of a wild and cruel people he made
" them curteous and gentile.

Thus hath *Tully* fet forth the efficacy of reafon and perfwafion,
how it was able to convert wild people to a mildneffe, and to
change their furious cruelneffe into gentle courtefie : were it not
now a great reproach in this our time (when knowledge reigneth
fo large) that men fhould fhew themfelves leffe obfequious to reafon ? Unleffe it may be thought that now every man having fuf-
ficient knowledge of himfelfe, needeth not to harken to the perfwafion of others.

Indeed he that thinketh himfelfe wife, will not efteem the reafon of any other, be he never fo wife ; fo that of fuch a one it may
well be faid : He that thinketh himfelfe wifer then he is, may juftly be counted a double fool. Wherefore fuch men are not to be
permitted in open audience to talk, but muft be put to filence,
and be made to give ear to reafon ; which reafon confifteth not
in the multitude of words heaped rafhly together, and applied for
one purpofe, but reafon is the expreffing of a juft matter with witty perfwafions, furnifhed with learned knowledge : fuch knowledge had *Mofes*, being expert in all learning of the *Ægyptians*, as
the Scriptures declare, and therefore was able to perfwade the ftub-
born people of the Jews, although not without pain. Such knowledge, and fuch reafons did *Druys* fhew which was the firft Law-
maker of all the Weft part of Europe. Like reafon and wifdome
did *Xamolxis* amongftthe Goths. *Lycurgus* unto the Lacedemoni-
ans. *Zeleucus* to the Locrians. *Solon* to the Athenienfes, and
Dunwallo Molmutius two thoufand years paft amongft the old
Brittains of this Realm. And hereby came it to paffe, that their
Laws continued long till more perfect reafon altered many of
them, and wilfull power oppreffed moft of them.

> *Druys* was
> fon to King
> *Sarron*, and
> fucceeded
> him in his
> kingdome.

At the beginning when thefe wife men perceived how hard it
was to bring the rude people to underftand reafon, they judged the
beft means to attain this honeft purpofe, to depend of learning in
every kinde : for by learning (as *Ovid* faith) *Pectora mitefcunt,*
afperitáfque fugit ; Stout ftomachs doe wax mild, and fharp fierce-
neffe is exiled. Therefore as *Berofus* doth teftifie, *Sarron* that was
the third King over all this Weft part of Europe, for to bring the
people from beaftly rage to manly reafon : did erect fchools
of liberall Arts, which tooke fo good fucceffe, that his name
continued

continued in that fort famous above two thoufand years after : for *Diodorus Siculus* which was in the time of *Julius Cæſar*, maketh mention of the learned men of Goths or Celtes, and nameth them *Sarronides*, that is to ſay, *Sarron his Schollers and followers.*

Amongſt theſe Arts that then were taught ſome did inform the tongue, and make them able both to utter aptly their minde, and alſo to perſwade ; as Grammar, Logick, and Rhetorick, although not ſo curiouſly as in this time; ſome other did appertain to the juſt order of partition of lands, the true uſing of Weights, Meaſures and reckonings of all ſorts of bargains, and for order of building and ſundry other uſes ; thoſe were Arithmetick and Geometry. Again, to incourage men to the honour of God, they taught Aſtronomy, whereby the wonderfull works of God were ſo manifeſtly ſet forth, that no mans tongue, nor pen can in like ſort expreſſe his infinite Power, his unſpeakable wiſdome, and his exceeding goodneſſe toward man, whereby he doth bountifully provide for man all neceſſaries, not onely to live, but alſo to live pleaſantly. And ſo was their confidence in Gods providence ſtrongly ſtayed, knowing his goodneſſe to be ſuch, that he would help man as he could, & his power to be ſo great, that he would do nothing but that that was beſt. Beſides theſe Sciences they taught alſo Muſick, which moſt commonly they did apply partly to religious ſervices, to draw men to delight therein, and partly to ſongs made of the manners of men, in praiſe of vertue and diſcommendation of vice, whereby it came to paſſe, that no man would diſpleaſe them, nor doe any thing evill that may come to their hearing : for their ſongs made evill men more abhorred in that time then any excōmunication doth in this time. The poſterity of theſe Muſitians continue yet both in Wales and Ireland, called *Bardes* unto this day, by the ancient name of *Bardus* the firſt founder.

This Bardus Druydius the 5. King of the Celtes, reigned 60. years, and died 1832 years before Chriſt.

And as theſe Sciences did encreaſe, ſo did vertue encreaſe thereby. Again as thoſe Sciences did decay ſo vertue loſt her eſtimation, and conſequently was little in uſe : whereof to make a full declaration were a thing meet for a Prince for to hear, but it would require a peculiar Treatiſe. Wherefore at this preſent I count it ſufficient, lightly to have touched this matter in generall words, and to ſay no more of the particularity thereof, but onely touching one of thoſe Sciences, that is Arithmetick, by which not onely juſt partition of Lands was made, but alſo touching buying and ſelling, all Aſſiſes, Weights, and Meaſures were deviſ-

ed

ed, and all reckonings and accounts driven; yea by proportion of it were the true orders of justice limited, as *Aristotle* in his Ethicks doth declare, and the degrees of estates in the common-wealth established; although that proportion be called Geometricall and not Arithmeticall, yet doth that proportion appertain to the art of Arithmetick, and in Arithmetick is taught the progression of such proportions, and all things thereto belonging. Wherefore I may well say, that seeing Arithmetick is so many wayes needfull unto the first planting of a Common-wealth, it must needs be as much required to the preservation of it also: for by the same means is any Common-wealth continued, by which it was erected and established. And if I shall in small matters in appearance, but indeed very weighty, put one example or two. What shall we say for the Statutes of this Realm, which be the onely stay of good order, in manner, now? As touching the measuring of ground by length and breadth, there is a good and an ancient Statute made by art of Arithmetick; and now it shall be to little use, if by the same art it be not practised and tried. For the assise of bread and drink, the two most common and most necessary things for sustentation of man, there was a goodly ordinance in the Law made, which by ignorance hath so grown out of knowledge, and use, that few men do understand it, and therefore the Statute books wonderfully corrupted, and the Commons cruelly oppressed: notwithstanding some men have written that it is too doubtfull a matter to execute those assises by those Statutes, by reason they depend of the standard of the coyne, which is much chang'd from the state of that time, when those Statutes were made. Thus shall every man read (that listeth) in the abridgement of the Statutes, in the title of Weights and Measures, in the seventh number of the English Book, where he should have translated a good ordinance which is set forth in the French Book: but no marvell if the Abridgment doth omit it seeing the great Book of Statutes doth omit the same Statute, as it hath done many other very good laws. And this is the fruit of ignorance to reject and condemn all that it understandeth not, although they use some cloaks for it, but such cloaks as being allowed, migh serve to repell all good laws; which God forbid.

Again, there is an ancient order for assise of fire, wood and coals, which was renued not many years past; and now how avarice and ignorance doth canvase that Statute, it is too pitifull to talk of, and more miserable to feel.

Further

Furthermore, for the Statute of Coynage, and the standard thereof, if the people understood rightly the Statute, they should not nor would (as they often do) gather an excuse for their folly thereby : but as I said, these Statutes by wisdome and good knowledg of Arithmetick were made, & by the same must they be continued. And let ignorance no more meddle with the use of them, then it did with the making of them. Oh in how miserable case is that Realm, where the ministers and interpreters of the law are destitute of all good Sciences, which be the keyes of the lawes ! How can they either make good lawes, or maintain them that lack the true knowledge whereby to judge them? And happy may that Realm be accounted, where the Prince himself is studious of learning, and desireth to understand equity in all laws. Therefore most happy are we the loving subjects of your Majesty, which may see in your Highnesse not onely such towardnesse, but also such knowledge of diverse arts, as seldom hath been seen in any Prince of such years, whereby we are enforced to conceive this hope certainly, that he which in those years seeketh knowledg when knowledge is least esteemed, and of such an age can discern them to be enemies both to his royall Person and to his Realms, which labour to withdraw him from knowledge to excessive pastime, and from reasonable study to idle or noysome pleasures, he must needs when he cometh to more mature years, be a most prudent Prince, a most just Governour, and a right Judge, not only of his subjects commonly, but also of the ministers of his Laws, yea, and of the Laws themselves : and to be able to conceive the true equity and exact understanding of all his Laws and Statutes, to the comfort of his good subjects, and the confusion and reproach of them which labour to obscure or pervert the equity of the same Laws and Statutes. How some of these statutes may be applyed to use, as well in this our time, as in any other time, I have peculiarly declared in this Book, and some other I have omitted for just considerations, till I may offer them first unto your Majesty to weigh them as to your Highnesse shall seem good : for many things in them are not to be published without your Hignesse knowledge and approbation ; namely because in them is declared all the rates of alloyes for all standards from one ounce upward, with other mysteries of Mint matters, & also most part of the varieties of coyns that have been currant in this your Majesties Realm by the space almost of six hundred years last past, and many of them that were currant in the time that the Romans ruled here.

All

All which with the ancient deſcription of England and Ireland, and my ſimple cenſure of the ſame, I have almoſt compleated to be exhibited to your Highneſſe : In the mean ſeaſon moſt humbly beſeeching your Majeſty to accept this ſimple Treatiſe, not worthy to be preſented to ſo high a Prince, but that my lowly requeſt to your Majeſty is, that this amongſt other of my Bookes may paſſe under the protection of your Highneſſe, whom I beſeech God moſt earneſtly and daily, according to my duty, to advance in all honor, and Princely Regality, and to increaſe in all knowledge, juſtice and godly policy, Amen.

Your Majeſties moſt obedient
ſubject and ſervant,

Robert Record.

To

To the loving READERS

The Preface of Mr Robert Record.

SOre oft times have I lamented with my self the unfortunate condition of England, seeing so many great Clerks to arise in sundry other parts of the world, and so few to appear in this our Nation: whereas for pregnancy of naturall wit (I thinke) few Nations doe excelle Englishmen: But I cannot impute the cause to any other thing, then to the contempt, or misreguard of learning. For as Englishmen are inferior to no men in mother wit, so they passe all men in vain pleasures, to which they may attain with great pain and labour: and are slack to any never so great commodity, if there hang of it any painfull study, or travelsome labour.

Howbeit, yet all men are not of that sort, though the most part be, the more pity it is, but of them that are so glad, not onely with painfull study, and studious pain to attain learning, but also with as great study and pain to communicate their learning to others, & make all England(if it might be)partakers of the same; the most part are such that unneath they can support their own necessary charges, so that they are not able to beare any charges, in doing of that good that else they desire to do.

But a greater cause of lamentation is this that when learned men have taken pains to do things for the aid

of

of the unlearned, scarce they shall be allowed for their well doing, but derided and scorned, and so utterly discouraged to take in hand any like enterprise again. So that if any be found (as here are some) that doe favour learning and learned wits, and can be contented to further knowledge, yea onely with their word; such persons though they be rare, yet shall they encourage learned men to enterprise something at the least that England may rejoyce of. And I have good hope that England will (after she hath taken some sure taste of learning) not onely bring forth more favorers of it, but also such learned men, that she shall be able to compare with any Realm in the world. But in the mean season where so few regarders of learning are, how greatly they are to be esteemed that doe favour and further it, my pen will not suffice at full to declare.

Therefore, gentile Reader, whereas I doe upon most just occasion judge, yea and know assuredly, that there be some men in this Realm, which both love and also much desire to further good learning, and am not well able to write their condign praise for the same, I think it better with silence to overpasse it, then either say too little of it, or provoke against them the malice of such other, which doe nothing themselves that is praisworthy, and therefore cannot abide to heare the praise of any other mans good indeed.

And considering their great favour unto learning, though I my selfe be not worthy to be reckoned in the number of great learned men, yet am I bold to put my selfe in Presse, with such ability as God hath lent me, though not with so great cunning as many men, yet with as great affection as any man to helpe my country men, and will not cease daily, (as much as my small ability will

will suffer me) to endite some such thing, that shall be to the instruction, though not of learned men, yet at the least of the vulgar sort, whose argument alwaies shall be such as it shall delight all learned wits though they doe not learne any great thing out of it.

But to speak of this present booke of Arithmetick, I dare not, nor will not set it forth with any words, but remit it to the judgement of all gentile Readers, and namely such as love good learning, beseeching them so to esteem it, as it doth seem worthy. And so either to accept the thing for it selfe, either at the least to allow my good endeavour. But I perceive I need not use any perswasions unto them, whose gentile nature and favourable mind is ready to receive thankfully and interpret to the best of all such enterprises attempted for so good an end, though the thing do not alwayes satisfie mens expectation. This considered, did bolden me to publish abroad this little Booke of the Art of numbering, which if you shall receive favourably, you shall encourage me to gratifie you hereafter with some greater thing.

And as I judge some men of so loving a mind to their native Countrey, that they would much rejoyce to see it prosper in good learning, and witty Arts: so I hope well of all the rest of Englishmen, that they will not be unmindfull of his due praise, by whose means they are helped and further'd in any thing. Neither ought they to esteem this thing of so little value, as many men of little discretion oftentimes doe. For who so setteth small price by the witty device and knowledge of numbering, he little considereth it to be the chief point. (in manner) whereby men differ from all brute beasts: for as in all other things (almost) beasts are partakers with us,

so

The Preface to the Reader.

so in numbering we differ cleane from them, and in manner peculiarly, sith that in many things they excell us again.

The Fox in crafty wit exceedeth most men,
A Dog in smelling hath no man his peer.
To foresight of weather if you look then,
Many beasts excell men ; this is cleer.
The wittinesse of Elephants doth letters attain,
But what cunning doth there in the Bee remain ?
The Emmet foreseeing the hardnesse of winter,
Provideth victuals in the time of summer.
The Nightingale, the Linet, the Thrush, the Lark,
In Musicall harmony passe many a Clark.
The Hedghog of Astronomy seemeth to know,
And stoppeth his cave where the wind will blow.
The Spider in weaving such art doth show,
No man can him mend, nor follow I trow.
When a house will fall, the Mice right quick
Flee thence before; can man doe the like ?

Many things else of the wittinesse of Beasts and Birds might I here say, save that another time of them I intend to write, wherein they excell in manner all men as it is daily seen : but in number was there never beast found so cunning, that could know or discerne one thing from many, by daily experience you may well consider, when a Bitch hath many whelps, or a Hen many chickens : and likewise of other whatsoever they be, take from them all their young saving only one, and you shall perceive plainly that they miss none, though they will resist you in taking them away, and will seek them again if they may know where they be, but else they will

never

The Preface to the Reader.

never miſſe them truly; but take away that one that is left, and then will they cry and complain; and reſtore to them that one, then are they pleaſed again. So that of number, this may I juſtly ſay, it is the onely thing almoſt that ſeparateth man from beſts. He therefore that ſhall contemn number, declareth himſelfe as bruitiſh as a beaſt and unworthy to be count in the Fellowſhip of men. But I truſt there is no man ſo foul overſeen, though many right ſmally doe it regard.

Therefore will I now ſtay to write againſt ſuch, and return again to this my Book, which I have written in the form of a Dialogue, becauſe I judge that to be the eaſieſt way of inſtruction, when the Scholer may aſke every doubt orderly, and the Maſter may anſwer to his queſtion plainly.

Why the Author wrote in Dialogue wiſe.

Howbeit I think not the contrary, but as it is eaſier to make another mans worke then to make the like; ſo there will be ſome that will finde fault becauſe I writ in a Dialogue: but as I conjecture thoſe ſhall be ſuch as do not, cannot, or will not perceive the reaſon of right teaching, and therefore are unmeet to be anſwered unto, for ſuch men with no reaſon will be ſatisfied.

And if any man object, that other books have been written of Arithmetick already ſo ſufficiently that I needed not now to put pen to the book, except I will condemn other mens writings: To them I anſwer, That as I condemn no mans diligence, ſo I know that no one man can ſatisfie every man: and therefore like as many doe eſteeme greatly other bookes, ſo I doubt not but ſome will like this my booke above any other Engliſh Arithmetick hitherto written; and namely ſuch as ſhall lack inſtructors, for whoſe ſake I have ſo plainly ſet forth the Examples, as no booke that I have ſeen hath done hitherto

herto: which thing shall be great ease to the rude Readers.

Therefore (gentile Reader) though this Booke can be but small aid to the learned fort, yet unto the simple ignorant (which needeth most helpe) it may be a good furtherance and mean unto knowledge.

And though unto the King his Majesty privately I doe it dedicate, yet I doubt not (such is his clemency) but that he can be content, yea, and much desirous, that all his loving Subjects shall take the use of it, and imploy the same to their most profit. Which thing if I perceive that they thankfully doe, and receive with as good will as it was written, then will I shortly with no lesse kindnesse set forth such introductions into Geometry and Cosmography, as I have at times promised, and as hitherto in English hath not been enterprised, wherewith I dare say all honest hearts will be pleased, and all studious wits greatly delighted.

I will say no more, but let every man judge as he shall see cause. And thus for this time I will stay my Pen, committing you all to that true fountain of perfect number, which wrought the whole world by number and measure: he is Trinity in Unity, and Glory, Amen.

Here

Here followeth a Table of the whole *Contents of this Book.*

The Contents of the first Dialogue.

THe declaration of the profit of Arithmetick.   Page. 1
Numeration with an easie and large Table.   p. 9

Addition.   p. 22

Subtraction.   p. 38

Multiplication.   p. 57

Division.   p. 73

Reduction with diverse declarations of Coynes, Weights, and Measures of sundry formes newly added, with a new Table, containing most part of the gold Coyn throughout Christendome, with the true Weight and valuation of them now in currant English money.   p. 100

Progression both Arithmetical and Geometrical, with diverse sundry questions touching the same.   p. 117

The Golden Rule, or Rule of Proportion called the Rule of Three direct.   p. 145

The Backer Rule of Three, with diverse questions thereunto belonging, newly added and augmented   p. 150

The double Rule of Proportion direct.   p. 156

The Rule of Proportion composed of five numbers.   p. 162

The Backer Rule, or the second part of the Rule of Proportion, composed.   p. 165

The Rule of Fellowship without time limited.   p. 168

The Rule of Fellowship with time limited.   p. 174

The second Dialogue containeth.

The first 5 kinds of Arithmetick wrought by Counters.   p. 179

The common kinds of casting by Counters, after the Merchants fashion and Auditors also.   p. 213

The Contents of the second part touching Fractions.   Page 216

Numeration in Fractions.   p. 218

B   The

The order of working Fractions. p. 223

Reduction of diverse Fractions into one denomination in 3 varieties. p. 224

Reduction of Fractions of fractions. p. 230

Reduction of improper Fractions. p. 232

Reduction of Fractions to the smallest Denomination, with easie Rules how to convert them thereunto. p. 236

Reduction of a Fraction and how it may be turned into any other Fraction, or into what Denomination you list. p. 240

Addition of Fractions. p. 243

Subtraction of Fractions. p. 244

Multiplication of Fractions. p. 247

Duplation of Fractions. p. 252

Division of Fractions. p. 253

Mediation of Fractions. p. 258

The Golden Rule direct in Fractions. p. 259

The Backer, or reverse Rule in Fractions. p. 262

The Statute of Assise of Bread and Ale recognized, and applyed to this time with new Tables thereunto annexed. p. 264

The Statute of measuring of Ground with a Table thereof faithfully calculated and corrected. p. 275

The Rule of fellowship, or society, with the reasons of the rules and proofs of their work. p. 280

To finde 3 numbers in any proportion. p. 291

The Rule of Alligation, with diverse questions, and the proofs of their works, with many varieties of such solutions. p. 295

The Rule of Falshood, or false Position, with diverse questions, and their proofs. p. 310

The Contents of the third part.

The 1 Chapter entreateth of Rules of brevity and practise, after a briefer method then hitherto hath been published in the English tongue: p. 345

The

The Contents.

The 2 Chapter entreateth of brief Reduction of diverse measures, as Ells, Yards, Braces, &c. Rules of Practise. p. 377

The 3 Chapter entreateth of the Rule of Three in broken numbers, after the trade of Merchants something differing from M. Records order which is comprehended in 3 Rules. p. 381

The 4 Chapter entreateth of losse and gain in the trade of Merchandise. p. 406

The 5 Chapter entreateth of losse and gain in the trade of Merchandise upon time, &c. with necessary questions therein wrought by the double Rule of Three or the Rule of Proportion composed of 5 numbers. p. 412

The 6 Chapter entreateth of Rules of payment and one of the necessariest rules that appertaineth to buying and selling, &c. p. 415

The 7 Chapter entreateth of buying and selling in the trade of Merchandise, wherein is taken part ready money, and diverse dayes of payment given for the rest, and what is wonne and lost in the hundred pound forbearance for 12 moneths. p. 417

The 8 Chapter entreateth of Tare and allowances of Merchandise, sold by weight, and of their losses and gains therein, &c. p. 421

The 9 Chapter entreateth of lengths and breadths of Arras, and other Clothes with diverse questions thereunto belonging. p. 442

The 10 Chapter entreateth of the reducing of pawnes of Geans into English yards. 428

The 11 Chapter entreateth of Rules of Loan and Interest, with divers questions incident thereunto. p. 429

The 12 Chapter entreateth of the making of Factors. p. 434

The 13 Chapter entreateth of Rules of barter or exchanging of Merchandise, wherein is taken part ware, and part ready money, with their proofs, and diverse other necessary questions thereunto belonging. p. 438

The 14 Chapter entreateth of exchanging of money from one place to another, with diverse necessary questions incident thereunto. p. 448

B 2   The

The Contents.

The 15 Chapter entreateth of six sundry forms of practises for Reduction of English, Flemish, and French money, and how each of them may easily be brought to money Sterling. p. 455

The 16 Chapter containeth a brief note of the ordinary coynes of most places of Christendome for trafficke, and the manner of their exchanging from one City or Town to another, which known, the Italians call Parie : whereby they finde the gain or losse upon the Exchange. p. 461

The 17 Chapter containeth also a declaration of the diversity of the weights and Measures of most places in Christendome for trafficke, proportioned in equality one to another, as also unto our English measure and weight, whereby the ingenious Practitioner may easily reduce the weight and Measure of each Countrey into another. p. 465

The 18 Chapter entreateth of divers sports and pastimes, done by Number. p. 473

An Appendix of figurate Numbers, with the extraction of roots. p. 479

A table of Board and timber measure by Robert Hartwell. p. 502

Certain tables of interest at 10 per 100. p. 506

New tables of interest at 8 per 100. p. 516

A Col-

A Collection of such Tables as are
Contained in this Treatise.

A Large Table of Numeration.   *Page* 19

A Table of Multiplication.   *p.* 57

A Table of all gold Coyns in this Realme, with the moſt uſuall gold Coyns throughout Chriſtendome, with their ſeverall weights of pence and grains, and what they are worth in currant money Engliſh.   *p.* 103

The valuation of Coyns this preſent year.   *p.* 104

Certain Tables or notes of the contents of Ale, Bear, Wine, Butter, Sope, Salmon, Eeles,&c. both what ſuch veſſels ought to contain by the Statute, and what thoſe veſſels empty ought to weigh.   *p.* 113

A Table of the quantity of dry Meaſures, as Pecks, Buſhels, Quarters, Weys, &c.   *p.* 115

A Table of the proportion of meaſures, touching length and breadth, to wit, from the inch to the foot, and ſo to the yard, the Ell, with their parts, the Perch, the Rod, the Furlong, the mile, &c.   *ibid.*

A Table made by Progreſſion Arithmeticall which containeth a double Table of Multiplication.   *p.* 119

A Table or Demonſtration of a figure or meaſure for the perfect underſtanding of a Fraction of Fractions.   *p.* 231

A Table of the contents of the Statute, for the aſſiſe of the weight of Bread, from one ſhilling the quarter to 20 ſhillings faithfully corrected.   *p.* 265

Two large Tables containing the aſſiſe of Bread from 3 ſhillings the quarter of wheat, to 40 ſhillings 6 pence.   *p.* 270 & 273

A neceſſary Table of the Statute of meaſuring of ground, upon the breadth given, what length it ought to contain : faithfully corrected according to the equity of the Statute, wherein the Author declareth how neceſſary this worthy art of *Arithmetick* is unto Gentilemen, Students of the Law, and ſuch other as are deſirous of infallible truth.   *p.* 278

B 3   A

The Contents,

A Table of the aliquot parts of a pound, or 20 shillings. *p.* 365

Two Tables, the first of the agreement of the Weights, the other of the measures of most places of *Europe* for traffique, whereby through the aid of the Rule of Three, the ingenuous may easily reduce one measure to a perfect valuation of other Countreyes measure or weight, and likewise theirs to ours. *p.* 461

A Table of Board and timber measure, exactly calculated by R.H. and the use thereof. *p.* 502

Cerrain tables of interest calculated by R. C. at 10 *per* 100. *p.* 494

Certain Tables of interest calculated by R. H. at 8 *per* 100. *p.* 506

A Table of interest at 6 per 100 calculated by T. W. *p.* 530

Before the Introduction of Arithmetick it were very good to have some understanding and knowledge of these Figures and Notes.

j	1	one	ɤɤ	20	twenty
ii	2	two	ɤl	40	fourty
iii	3	three	l	50	fifty
iiii	4	foure	lɤ	60	sixty
ʋ	5	five	lɤɤ	70	seventy
ʋi	6	six	ɤc	90	ninety
ʋii	7	seven	C	100	a hundred
ʋiii	8	eight	CC	200	2 hundred
iɤ	9	nine	D	500	5 hundred
ɤ	10	ten	DC	600	6 hundred
ɤi	11	eleven	M	1000	a thousand
ʋii	12	twelve	MD	1500	a thou. 5 hund.
&c.	&c.	&c.	&c.	&c.	&c.

A

A Dialogue between the Master and the Scholer: teaching the *Art* and use of *Arithmetick* with *Pen*.

The Scholar speaketh.

SIR, such is your authority in mine estimation, that I am content to consent to your saying, and to receive it as truth, though I see none other reason that doth lead me thereunto: whereas else in mine owne conceit it appeareth but vaine, to bestow any time privately in learning of that thing that every Childe may and doth learne at all times and houres, when he doth any thing himselfe alone, and much more when he talketh or reasoneth with others.

M. Lo, this is the fashion and chance of all them that seek to defend their blinde ignorance, that when they think they have made strong reason for themselves, then have they proved quite contrary. For if numbring be so common (as you grant it to be) that no man can doe any thing alone, and much lesse talk or bargain with other, but he shall still have to doe with number: this proveth not number to be contemtible and vile, but rather right excel-

lent

lent and of high reputation, sith it is the ground of all mens affairs, in that without it no tale can be told, no communication without it can be continued, no bargaining without it can duly be ended, or no businesse that man hath justly compleated. These commodities if there were none other, are sufficient to approve the worthinesse of number. But there are other innumerable, far passing all these, which declare number to exceed all praise. Wherefore in all great works are Clerks so much desired? Wherefore are Auditors so richly fed? What causeth Geometricians so highly to be inhaunced? Why are Astronomers so greatly advanced? Because that by number such things they finde, which else would far excell mans minde.

Scholar. Verily, Sir, if it be so, that these men by numbering, their cunning do attain, at whose great works most men doe wonder, then I see well I was much deceived, and numbring is a more cunning thing then I tooke it to be.

Master. If number were so vile a thing as you did esteem it, then need it not to be used so much in mens communication. Exclude number, and answer to this question : How many years old are you ?

Scholar. Mum.

Master. How many dayes in a weeke ? How many weekes in a yeare ? What lands hath your Father ? How many men doth bee keep ? How long is it since you came from him to me ?

Scholar. Mum.

Master. So that if number want, you answer all by mummes : How many miles to London ?

Scholar

Scholar. A peak full of plums.

Master. Why, thus you may see, what rule number beareth, and that if number be lacking it maketh men dumb, so that to most questions they must answer mum.

Scholar. This is the cause sir, that I judged it so vile, because it is so common in talking every while: Nor plenty is not dainty, as the common saying is.

Master. No, nor store is no sore. perceive you this? The more common that the thing is being needfully required, the better is the thing, and the more to be desired. But in numbering as some of it is light and plain, so the most part is difficult and not easie to attain. The easier part serveth all men in common, and the other requireth some learning. Wherefore as without numbring a man can doe almost nothing, so with the help of it you may attaine to all things.

Scholar. Yea sir, why then it were best to learn the Art of numbring, first of all other learning, & then a man need learn no more if all other come with it.

M. Nay, not so: but if it be first learned, then shall a man be able (I mean) to learn, perceive & attain to other Sciences; which without it he could never get.

Scholar. I perceive by your former words, that Astronomy and Geometry depend much on the help of numbring: but that other Sciences, as Musick, Physick, Law, Grammer, and such like, have any help of Arithmetick, I perceive not.

Master. I may perceive your great Clerklinesse by the ordering of your Sciences: but I will let that passe now, because it toucheth not the matter that I

I intend, and I will shew you how Arithmetick doth profit in all these, somewhat grossly, according to your small understanding; omitting other reasons more substantiall.

Musick.

First (as you reckon them) musicke hath not onely great help of Arithmeticke, but is made, and hath its perfectnesse of it: for all musicke standeth by number & proportion: And in Physicke, besde the calculation of criticall dayes, with other things which I omit. how can any man judge the pulse rightly, that is ignorant of the proportion of numbers.

Physick.

And as for the Law, it is plain, that the man that is ignorant of Arithmeticke, is neither meet to be a Judge, neither an Advocate, nor yet a Proctor. For how can hee well understand another mans cause, appertaining to distribution of goods, or other debts or of summes of money, if he be ignorant of Arithmetick? This oftentimes causeth right to bee hindred, when the Judge either delighteth not to heare of a matter that hee perceiveth not, or cannot judge for lacke of understanding: this commeth by ignorance of Arithmeticke.

Law.

Now, as for Grammer, mee thinketh you should not doubt in what it needeth number, sith you have learned that Nouns of all sorts, pronouns, Verbs and Participles are disind diversly by numbers: besides the variety of Nouns of Number, and Adverbs. And if you take away number from Grammer, then is all the quantitie of Syllables lost. And many other wayes doth number help Grammer. Whereby were all kindes of Meeters found and made? was it not by number?

Grammer.

But

But how needfull Arithmeticke is to all parts of Philosophy, they may soon see, that doe reade either Aristotle, Plato, or any other Philosophers writings. For all their examples almost, and their probations depend of Arithmetick. It is the saying of Aristotle, that hee that is ignorant of Arithmeticke, is meet for no Science. And Plato his Master wrote a little sentence over his School-house doore, Let none enter in hither (quoth he) that is ignorant of Geometry. Seeing hee would have all his Scholars expert in Geometry, much rather hee would have the same in Arithmeticke, without which Geometry cannot stand. *Philoso-phy.*

And how needfull Arithmeticke is to Divinity, it appeareth, seeing so many Doctors gather so great mysteries out of number, and so much do write of it. And if I should go about to write all the commodities of Arithmeticke in civill acts, as in governance of Common-weales in time of peace, and in due provision and order of Armies in time of war, for numbering of the Host, summing of their wages, provision of Victuals, viewing of Artillerie, with other Armour; beside the cunningest point of all, for casting of ground, for encamping of men, with such other like : And how many waies also Arithmeticke is conducible for all private Weales, of Lords and all Possessioners, of Merchants, and all other occupiers, and generally for all estates of men, besides Auditors, Treasurers, Receivers, Stewards, Bailiffes, and such like, whose Offices without Arithmeticke are nothing : If I should (I say) particularly repeat all such commodities of the noble Science of Arithmeticke, it were *Divinity* *Armie.*

enough

enough to make a verie great booke.

Scholar. No, no, sir, you shall not need : For I doubt not but this, that you have said, were enough to perswade any man to think this Art to be right excellent and good : and so necessary for man, that (as I thinke now) so much as a man lacketh of it, so much he lacketh of his sense and wit.

Master. What, are you so farre changed since, by hearing these few commodities in generall ? by likelihood you would be farre changed if you knew all the particular Commodities.

Scholar. I beseech you Sir, reserve those Commodities that rest yet behinde unto their place more convenient : and if ye will be so good as to utter at this time this excellent treasure, so that I may be somewhat inriched thereby, if ever I shall be able, I will requite your pain.

Master. I am very glad of your request, and will doe it speedily, sith that to learn it you be so ready.

The duty of a Scholar.

Scholar. And I to your authority my wit doe subdue, whatsoever you say, I take it for true.

Master. That is too much, and meet for no man to be believed in all things, without shewing of reason. Though I might of my Scholar some credence require, yet except I shew reason, I doe it not desire. But now sith you are so earnestly set this Art to attaine, best it is to omit no time, lest some other passion coole this great heat, and then you leave off before you see the end.

Perseverance in study.

Scholar. Though many there be so unconstant of mind, that flitter and turn with every winde, which often begin, and never come to the end, I am one of this sort as I trust you partly know. For p my good will what I once begin, till I have it ally ended, I would never blin.

Master. So have I found you hitherto indeed, nd I trust you will increase rather, then goe acke. For better it were never to assay, then to b2inke and fly in the mid way: But I trust you ill not doe so; therefore tell me briefly: What all you the Science that you desire so greatly.

Scholar. Why sir, you know.

Master. That maketh no matter, I would heare ohether you know, and therefore I aske you. For reat rebuke it were to have studied a Science, nd yet cannot tell how it is named.

Scholar. Some call it Arsemetrick, and some ugrime.

Master. And what doe these names betoken ?

S. That if it please you, of you would I learn.

Master. Both names are corruptly written : *Aριθμι* Arsemetrick for Arithmeticke, as the Greeks call it, *τική.* nd Augrime for Algorisme, as the Arabians found t : which doth betoken the Science of Numbring : o2 Arithmos in Greeke is called Number : and of it ommeth Arihmeticke, the Art of Numbering : bo that Arithmeticke is a Science o2 Art teaching he manner and use of Numbering : This Art may ee w2ought diversly, with Pen, o2 with Coun- ers. But I will first shew you the wo2king with he pen, and then the other in o2der.

Scholar. This will I remember. But how many hings are to bee learned to attaine this Art fully?

Master. There are reckoned commonly seven arts o2 wo2ks of it.

{Numeration

Numeration, Addition, Subtraction, Multiplication, Division, Progression, and Extraction, of roots: to these some men adde Duplation, Triplation, and Mediation. But as for these three last they are contained under the other seven. For Duplation, and Triplation are contained under Multiplication, as it shall appeare in their place: And Mediation is contained under Division, as I will declare in his place also.

Scholar. Yet then there remain the first seven kinds of Numbering.

Master. So there doth: howbeit if I shall speake exactly of the parts of Numbering, I must make but five of them: for Progression is a compound Operation of Addition, Multiplication and Division. And so is the Extractions of roots. But it is no harme to name them as kinds severall, seeing they appeare to have some severall working. For it forceth not so much to contend for the number of them, as for the due knowledge and practising of them.

Scholar. When you will that I shall name them, as seven kindes distinct: But now I desire you to instruct me in the use of each of them.

Master. So I will, but it must be done in order: for you may not learn the last so soon as the first, but you must learne them in that order, as I did rehearse them, if you will learne them speedily and well.

Scholar. Even as you please. Then to begin: Numeration is the first in order, what shall I do with it?

Master

Maſter. Firſt, you muſt know what the thing is, and then after learn the uſe of the ſame.

Numeration.

Umeration *is that* Arithmeticall *skill, whereby we may duely value, ex-preſſe, and read any* Number or Summe *propounded : or elſe in apt* Figures *and* Places *ſet downe any* Number *known or named.*

Scholar. Why? then me thinketh you put a difference between the Value and the Figures.

Maſter. Yea ſo doe I. For the Value is one thing, and the Figures are another thing, and that com-eth partly by the diverſity of Figures, but chiefly in the places wherein they be ſet.

Scholar. Then muſt I know here 3 things, the Value, the Figure, and the Place.

Maſter. Even ſo. But yet adde Order to them as the fourth. And firſt mark, that there are but tenne Figures that are uſed in Arithmetick ; and of thoſe tenne, one doth ſigniſie nothing , which is made like an o, and is privately called a Cypher, A Cypher. though all the other ſometime be likewiſe named, The other nine are called ſignifying Figures, and be thus figured.

B

Figures.

1. 2. 3. 4. 5. 6. 7. 8. 9.

And this is their value.

i. ii. iii. iiii. v. vi. vii. viii. ix.

But here you must mark, that every Figure hath two values: One alwayes certain, that it signifieth properly, which it hath of his form, and the other uncertain, which he taketh of his place.

A place.

A place is called the seate or roome that a Figure standeth in. And looke how many Figures are written in one summe, so many places hath that whole number. And that must be called the first place, that is next to the right hand, and so reckoning by order towards the left hand, so that that place is last that is next to the left hand. As for example. If there stood before you six men in a row, side by side, and you should tell them as they stand in order, beginning with the man that were next to your right hand; then he that were next him should be called your second, and so forth to the furthest from your right hand, which is the sixth and the last.

Scholar. I perceive you well: so might I reckon Letters or any other thing. As if I should write 8 Letters after this order, a, b, c, d, e, f, g, h. then must I say, h, is the first, g, the second, f, the third, e, the fourth, d, the fifth, c, the sixth, b, the seventh, a, the eight.

Master

Master. That is well done. And after the same sort use hereafter, that what I declare by one example, do you expresse by another : and so shall I perceive whether you understand it or no. And so passe over nothing, till you perceive it well, and be expert therein.

Schol. I pray you how many of these places be there in all ?

Master. There is no certain number of them, but they are sometimes moe, and sometimes fewer, according to the sum that is expressed. For so many as the figures are, so many are the places : and the last place is so called, not because it is the last of all other, but it is the last of that present summe, and it may bée the middle place in another summe.

Scholar. Me séemeth I perceive this very well, as touching the order of reckoning of the places : but as for the number of them, you say there is no certainty. Now there resteth to declare the value of the figures by the diversity of places, which you called the value uncertain.

Value uncertain.

Master. But first let me hear whether you know perfectly the certain value.

Value certain.

Scholar. Yes sir, as you wrote them, so I marked them.

Master. How write you then five ?

Scholar. By this figure 5.

Master. And how six ?

Scholar. Thus 6.

Master. Write these thrée numbers, each by it self, as I speake them, vii, iiii, iii.

Scholar. 7. 4. 3.

C   Master.

Mafter. How write you thefe foure other, ii.i.ix.viii.

Scholar. Thus (I trow) 2. 1. 6. 8.

Mafter. Nay there you miffe : loke on mine example again.

Scholar. Sir true it is, I was to blame, I take 6 for 9, but I will beware hereafter.

Mafter. Now then take heed thofe certain values every figure reprefenteth when it is alone written without other figures joyned to him. And alfo when it is in the firft place, though mary other doe follow : as for example, this figure 9 is ix. ftanding now alone.

Scholar. How is he alone, and ftandeth in the middle of fo many letters ?

Mafter. The letters are none of his fellows. For if you were in France in the middle of a thoufand Frenchmen. if there were no English man with you, you would reckon your felf to be alone.

Scholar. So it is. Then 9 without more figures of Arithmetick betokeneth ix. whatfoever other letters be about it.

Mafter. Even fo, and fo doth it, if it be in the firft place joyned with other, how many foever do follow, as in this example. 3679. You fee 9 in the firft place, and doth betoken nine as it were alone.

Scholar. I perceive that, and doth not 7 that ftandeth in the fecond place (between 9 and 6 in the third place) betoken vii. and fo 3 in the fourth place betoken three ?

Mafter. Thefe figures be as you have faid, but their values are not fo. For as in the firft place every figure betokeneth his own value certain onely, fo

in

in the second place every figure betokeneth his own value certain tenne times: as in the example, 7 in the second place is seven times ten, and is lxx. And in the third place, every Figure betokeneth his own value an hundred times, so the 6 in that place beto= keneth vi C. and in the fourth place every figure betokeneth his owne value a M. tim s, as in the aforesaid number 3 in the fourth place standeth for 3 M. and in the fifth place every figure standeth for his own value x M. times, and in the sixth place a C M. times, and in the seventh place a M M. times, and in the eighth place xMM, so that every place exceedeth the former ten times.

Scholar. As thus: if I make this number at all A generall Rule. adventures, 91359684. here are eight places. In the first place is 4 and betokeneth but foure: in the second place is 8 and betokeneth ten times 8 that is 80, in the third place is 6 and betokeneth six hun- dred: in the fourth place 9 is nine thousand, and 5 in the fifth place is xM times 5, that is fifty M. So 3 in the sixth place is a CM times 3 that is, CCCM. Then 1 in the seventh place, one MM, and 9 in the eighth place, ten thousand thousand times 9, that is xMM. But now I cannot easily nor quickly read it in order.

Master. That shall you practise by this meanes. First, put a pricke over the fourth figure, and so over the seventh. And (if you have so many) over the tenth, thirteenth, sixteenth, and so forth, still leaving two figures between each two prickes. And those two roomes between the pricks are called Ternaries. Ternaries

Then begin at the last pricke, and see how many

C 2   figures

figures are betwéen him and the end, which cannot
paſſe thrée, reckoning himſelfe for one : then pronounce them as if they were written alone from
the reſt, and at the end of their value, ſo many
times thouſands as your numbers have pricks.

After that come to the next thrée figures, and
ſound them as if they were apart from the reſt,
and adde to their value ſo many times thouſands,
as there are pricks betwéen them, and the firſt place
of your whole number. And ſo do by every other
thrée figures following, if you have more. As in example, 91359684 this was your number.

Put a prick over 9 in the fourth place, and over
1 in the ſeventh place, and then no more (for your
places come not to ten)as thus : 91359684.

Now go to the laſt prick over 1, and take it and
the figure 9 that followeth it, & value them alone.

Scholar. 91, that is xci.

Maſter. So it is, then adde for the number of
your pricks twice M.

Scholar. that is, xci. thouſand thouſand.

Maſter. So it is. Then take the thrée other
figures from one to the next Prick, and value them.

Scholar. 359, that is, CCC. lix.

Maſter. Now adde for the one prick, that is betwéen them and the firſt place, M.

Scholar. CCC. lix. thouſand.

Maſter. Then come to the other 3 figures that
remain.

Schol. 684, that is, vi C. lxxxiiii.

Maſter. Now have you valued all. And at the
end of the laſt number you ſhall adde nothing, becauſe there remaineth no pricke nor number after
it,

it, yet proue it in another number, as thus,
23086 4089105340.

Schol: 23086 4089105340. I haue pricked them
as you taught me, but I am in doubt whether I
haue done well or no, because of the Cyphers, for
I remember you told me that they do signifie nothing, and therefore I doubt whether I should
reckon them for a Figure in setting of the pricks;
and againe, I know not wherefore they serue.

Master. That will I tell you now. Indeed they
are of no value themselues, but they serue to make
up the number of places, and so make the Figure
following them to bee in a farther place, and therefore to signifie the more value : as in this example, *The use of*
90 the Cypher is of no value, but yet he occupieth *Cyphers.*
the first place, and causeth 9 to be in the second place,
and so to signifie ten times 9, that is xc. so that
two Cyphers thrust the Figure following them into
the third place, and so forth.

Scholar. Then I perceiue in the example aboue,
I haue pricked well enough, for though that Cypher that is pricked signifies nothing, yet must
he haue the prick, because he came in the thirteenth
place. Then will I proue to number that summe.
First, there is 230, M, M, M, M, and then followeth
864, M, M, M. And what shall I now do? There
is a Cypher in the third place, and no Figure after
him, but they that I haue reckoned.

Master. He did serue for them that you haue already reckoned, to make them in a place further
then they should bée, if he were away, and therefore now ye shall let him goe. And so doe alwayes
when he occupieth that place next before any

C 3   pricke,

pricke, which is the last of that Ternary, and a Cypher in the last place doth nothing.

Scholar. Then shall I say but 89, M, M.

Master. So, but go forth.

Scholar. 105 thousand. Now are all my pricks spent, and yet remain 340, so that I must value them, CCC.xl. onely.

Master. Now can you reckon after this sort: and remember that every such room, so parted, is called a Ternary, or Trinity, for you have numbred or valued the summe most truly and by the aid of the pricks each denomination is distinct most plainly.

Trinity.

Scholar. What call you Denomination?

Denomination.

Master. It is the last value or name added to any summe. As when I say, an hundred two and twenty pounds: Pounds is the Denomination. And likewise in saying, 25 men: Men is the Denomination, and so of other: But in this place (that I spake of before) the last number of every Ternary, is the Denomination of it. As for the first Ternary, the Denomination is Vnites, & of the second Ternary, the Denomination is thousands, & of the third Ternary, thousand thousands, or Millions; of the fourth, thousand thousand thousands or thousand millions: & so forth.

Scholar. And what shall I call the value of the three figures that may be pronounced before the Denomination, as in saying, 203000000. that is, two hundred three millions: I perceive by your words, that millions is the Denomination: but what shall I call CCiii. joyned before the millions?

Numerator.

Summe or value.

Master. That is called the Numerator, or Valuer, and the whole summe that resulteth of them both, is called the Summe, Value or Number.

Scholar.

Scholar. Now is there any thing elſe to be learn=
ed in Numeration ? oz elſe have I learned it fully?

Maſter. I might ſhew you here who were the
firſt Inventors of this Art, and the reaſon of all
theſe things that I have taught you, but that I
will reſerve till ye have learned over all the pza=
ctiſe of this Art, leſt I ſhould trouble you with over
many things at the firſt.

But yet this you muſt mark, that there are three
kindes of Numbers, one called Digits, another Articles, and the third mixt numbers.

A Digit is any number under ten, as theſe :
1.2.3.4.5.6.7.8.9.
And 10 with all other that may be divided into ten
parts firſt, & nothing remain, are called articles, ſuch
as are 10. 20.30.40. 50. &c. 100. 200. &c. 1000. &c.

And that number is called mixt, that containeth
articles, oz at the leaſt one article, and a Digit, as
12.16.19.21.38.107.1005.& ſo fozth & foz the moze
eaſe of underſtanding & remembzance, mark this.
The Digit number is never wzitten with moze
then one figure, but the article and the mixt number
are ever wzitten with moze then one figure. And
thus they differ, that the article hath everm=ze
this Cypher o in the firſt place: and the mixt number hath ever there ſome Digit.

Scholar. By theſe laſt wozds I perceive it much
better then I did befoze, and now (I think) I will
never miſſe to know thoſe three aſonder.

Maſter. If you remember now all that I have
ſaid, you have learned ſufficiently this firſt kind
of Arithmeticke called Numeration. Howbeit I will
exhozt you now to remember both this that I have
C 4　　　　ſaid,

Three
kinds of
numbers.
Digits.

Articles.

Mixt.

ſaid, and all that I ſhall ſay, and to exerciſe your
ſelf in the practiſe of it; for rules without practiſe,
are but a light knowledge, and practiſe it is that
maketh men perfect and prompt in all things.

And as you have learned to gather and expreſſe
the value of a ſumme propounded and ſet downe
before you, ſo muſt you practiſe to marke, note, and
write down with apt figures and in due places, any
number onely named, or recited to you, or if your
ſelf imagined; as for a proof. How note you, or
write downe this ſumme, five thouſand two hundred fifty and ſeven?

S. This troubleth me now, whether I ſhould begin at the firſt, or at the laſt. For reaſon (me thinketh) ſhould cauſe me to begin at the firſt, & yet if I
write it as you ſpeake it, I muſt begin at the laſt.

Maſter, When you know your places perfectly,
you may begin where you liſt; but the more eaſe for
your hand is to beginne with the laſt, that is to
ſay, as I did ſpeak them. yet for the more ſurety,
a while you may begin at the firſt, repeating my
words backwords thus: ſeven, fifty, two hundred,
five thouſand: or elſe ſounding them all by their
digit or value, as thus: ſeven, five, two, five; for
that way is eaſieſt: But then muſt you look well
whether there be any cypher in your ſumme, that he
may be ſet in his place: as if the laſt valuer of your
ſumme (as you ſpeak it) be above 9, then is there a
cypher in the firſt place. And if it be an hundred,
or above, then is there two cyphers, one in the firſt
place. and another in the ſecond, and ſo forth.

But becauſe this thing is ſuch that cannot be ſet
forth without many words; I think beſt here now

at the end of Numeration, to adde a Table easie and ready for the first exercise of it.

Lo this is the Table.

Vnites.	Tennes.	Hundreds.	Thousands.	X. of Thousands.	C. of Thousands.	Millions.	X. of Millions.	C. of Millions.	M. of Millions.	X. M. of Millions.	The denominators of the place or value uncertain.	The order of places.
9	9	9	9	9	9	9	9	9	9	9	Nine.	
8	8	8	8	8	8	8	8	8	8	8	Eight.	
7	7	7	7	7	7	7	7	7	7	7	Seven.	
6	6	6	6	6	6	6	6	6	6	6	Six.	
5	5	5	5	5	5	5	5	5	5	5	Five.	
4	4	4	4	4	4	4	4	4	4	4	Foure.	
3	3	3	3	3	3	3	3	3	3	3	Three.	
2	2	2	2	2	2	2	2	2	2	2	two.	
1	1	1	1	1	1	1	1	1	1	1	One.	
0	0	0	0	0	0	0	0	0	0	0	Cyph.	
First.	Second.	Third.	Fourth.	Fifth.	Sixt.	Seventh.	Eighth.	Ninth.	Tenth.	Eleventh.		

The right side or hand. The names of Digits, values certain, or valuers.

The left hand or side.

This Table (as ye may see) hath eleven Places, and in each of them are set all the Digits, whose certain value is written on the right hand of the Table, and the value uncertain on the left hand ; so that by this Table you may learne both how to expresse

expresse any Number that you lift (if that it exceed not eleven places) that is to say, XC. thousand Millions, and so may you by help of it, value all summes proposed under the said number.

For example: take the summe that I proposed before which was five thousand two hundred fifty and seven. And if you will expresse it, take the first number (as I speak it) which is five M, whose valuer or certain value is 5, and his uncertain value, or Denomination is M. First, you shall seek at the right hand of the valuer 5. Then seek along under the title of Denomination toward the left hand till you finde thousands, and under it, right at the foote of the Table, is the number of the place, that is in the fourth, wherein you must write your Digit, or valuer 5.

Afterward come to the second part of the number two hundred, whose valuer is 2, and his Denomination C. Seek two at the right hand of the Table, and go along under the Denomination toward the left hand, till you come under C. then look to the foot of the Table, and there you shall see the number of the place, that is to say, the third, wherein you must set your Digit 2.

Then do so by your other two numbers that remain, and you shall finde 5 in the second place for your fifty, and 7 in the first place for your seven. And thus you may do with other numbers.

Scholar. Master, I thank you heartily. I perceive you seek to instruct me most plainly and briefly, and not to hide your knowledge with subtile words, as many do. For this rule is so plain, that I can desire it no plainer. And though it seem some-

what

what long, yet I perceive it to be a sure way.

Master. So it is, and though it be long, yet it is neither too long, neither too plaine for young learners that lack practise : for this Table is in stead of a Teacher to them that lack one. But now I trust I have said enough of Numeration : which after you have well practised, then may you learn forth.

Scholar. Yet I pray you in one thing to tell me your judgement. Why do men reckon the order of the places backward, from the right hand to the left ?

Why numbers are written backward.

Master. In that thing all men do agree that the Chaldees, which first invented this Art, did set these Figures as they set all their Letters, for they write backward, as you terme it, and so do they read. And that may appear in all Hebrew, Chaldee and Arabick Books ; for they be not onely written from the right hand to the left and so must be read, but also the right end of the book is the beginning of it, whereas the Greeks, Latins, and all Nations of Europe, doe write and reade from the left hand toward the right : and all their Books beginne at the left side.

Scholar. That reason hath satisfied me.

Master. It neither satisfieth me, neither lyketh me well, because I see that the Chaldees and Hebrewes doe not so use their own Numbers, as at another time I will declare. But this plain reason may best satisfie you presently, that seeing in pronouncing of Numbers wee keep the order of our own reading, from the left hand to the right : and again, wee doe ever name the greater numbers

before

before the smaller : It was reason that the lesser places, containing the lesser numbers, should be set on the right hand, and the greater places containing the greater Numbers to proceed toward the left hand.

Scholar. This reason is to me so plain, that it seemeth now against reason to make a doubt of that order. So that now for Numeration I am satisfied : hoping that practise shall make me fully ready and expert in it. And in the mean season I desire to learn the other kinds of Arithmetick.

Master. That is well said : but what should you next learn ? can you tell ?

Scholar. I remember you said, that Addition was next.

Master. Even so and what that is, must you first know.

Addition.

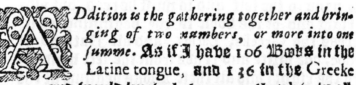

Addition is the gathering together and bringing of two numbers, or more into one summe. As if I have 106 Books in the Latine tongue, and 136 in the Greeke tongue, and would know how many they bee in all, I must write these two numbers one over another, writing the greatest number highest, so that the first figure of the one being under the first figure of the other, and the second under the second, and so forth in order.

When you have so done, draw under them a right line, then will they stand thus .

Now begin at the first places toward the right hand allwayes, and put toge-

160
136

ther

ther the two first figures of these two numbers, and loke what commeth of them write

under them, right under the line.   160

As in saying 6 and 0 is 6, write 6   136

under 6, as thus :   ———

   6

   And then go to the second figures, and doe like-

wise : as saying 3 and 6 is 9,   160

write 9 under 6 and 3, as here you   136

sée.   ———

   96

   And likewise doe you with the figures that be in

the third place, saying, 1 and 1 be 2 ,   160

write 2 under them, and then will   136

your whole summe appear thus :   ———

   296

   So that now you sée that 160, and 136, doe make

in all 296.

Scholar. What ? this is very easie to doe, me

thinketh I can do it even since.

   There came through Cheapside two droves of

Cattell : in the first was 848 sheep, and in the se-

cond was 186 other Beasts.

   Those two summs I must write as you taught

me, thus : Then if I put the two first Fi-   848

gures together, saying, 6 and 8, they make   186

14. That must I write under 6 and 8, thus: ———

   14

Master. Not so: and here you are twice deceived.

First in going about to adde together two summes

of sundry things, which you ought not to doe except

you sék onely the number of them, and care not for

the things : For the summe that should result to

that Addition, should be a summe neither of sheep nor

of other beasts, but a confused summe of both. How-

beit sometimes ye shall have summes of divers De-

nominations to be added, of which I will tell you

   anon :

anon : but first I will shew you where you were deceſved in another point, and that was in wꝛiting 14, which came of 6 and 8 under 6 and 8, which is unpoſſible: foꝛ how can two figures of two places be wꝛitten under one figure and one place ?

Scholar. Truth it is, but yet I did ſo underſtand you.

Maſter. I ſaid indeed, that you ſhould wꝛite that under them that did reſult of them both together: which ſaying is alwayes true, if that ſumme doe not exceed a Digit: But if it be a mixt number, then muſt you wꝛite the Digit of it under your figures, as you have ſaid befoꝛe : and if it be an Article, then wꝛite o under them, and in both ſoꝛts you ſhall kéep the Article in your minde ; and therefoꝛe when you have added your ſecond Figures, which occupy the place of tens, you ſhall put that one thereto, which you kept in your minde ; foꝛ though it were ten indéed, yet in that place it is but as one, becauſe that every one of that place is tenne, foꝛ that it is the place of tens. And in like manner, if you have in the ſecond place ſo great a number that it amounteth above 9, then wꝛite the Digit, & reſerve the Article in your minde ever adding it to the next place following, and ſo of all other places, how many ſoever you have. And if you have a mixt number when

A place. you have added your laſt Figures, then wꝛite the Digit under the laſt Figures, and the Article in the next place beyond them : ſo ſhall your number reſulting of Addition, have one place moꝛe then the numbers which you ſhall adde together.

Scholar. Now do I perceive you, and the reaſon of this is, (as I underſtand) becauſe that no

one

one place can contain aboue 9, which is the greateſt Figure that is, and then all tens oʒ Articles muſt
be put to the next place following : foʒ euery place
(as I may ſée,) pcæbeth the other place next befoʒe
him by 10.

Now, if it pleaſe you, I will return to my example of Cattell. But I remember you ſaid I might
not adde ſummes of ſundʒy things together, and
that I may ſée by reaſon.

Maſter. Crüth it is, if you ſéek the due ſumme
of any thing, but if you only ſéek a bare ſumme, and
haue no reſpect to the thing, then were it better to
name the ſumme onely without any thing : as in
ſaying 848 without naming ſheep oʒ any thing
elſe, And likewiſe 186, naming nothing.
Now let me ſée how you can addethoſe two ſummes.

Scholar. I muſt firſt ſet them ſo that the two firſt
figures ſtand one ouer another, and the other each
one ouer his fellow of the ſame place ; then ſhall I
dʒaw a line under them both. And ſo likewiſe of o-
ther figures, ſetting alwayes the greateſt number
higheſt, thus as followeth.

Then muſt I adde 6 to 8, which make 14,   848
that is a mixt number, therefoʒe muſt I take   186
the digit which is 4, and wʒite it under 6,   ____
and 8, kéeping the Article 1 in my minde thus :   4

Next that, I doe come to the ſecond figures, adding them together, ſaying, 8 and 4 make 12, to
the which I put the one reſerued in my minde, and
that maketh 13 of which number I wʒite the   848
Digit 3 under 8 and 4, and kéep the Article in   186
my minde, thus :   ____
                                                   34
Then come I to the third figures, ſaying, 1
and

and 8 make 9, and 1 in my mind maketh 10. Sir, shall I write the Cypher under 1 and 8 ?

Master. Yea.

Scholar. Then of 10 I write the Cypher under 1 and 8, and keep the Article in my mind.

Master. What needeth that, seeing there follow no more figures ?

Scholar. Sir, I had forgotten, but I will remember better hereafter. Then seeing I am come to the last Figures I must write the Cypher under them, and the Article in a further place after the Cypher thus :

$$\begin{array}{r} 848 \\ 186 \\ \hline 1034 \end{array}$$

Master. So now you see, that of 848, and 186 added together, there amounteth 1034.

Scholar. Now I thinke I am perfect in Addition.

Master. That will I prove by this example. There are two Armies of Souldiers: in the one are 106860, and in the other 9400. How many are there in both Armies say you ?

Scholar. First, I set them one over another, beginning with the first number on the right hand, thus :

$$\begin{array}{r} 106800 \\ 9400 \end{array}$$

But the nether number will not match the over number.

Master. That forceth not.

Scholar. Then do I adde o to o, and there amounteth o, that must I write under the first place thus :

$$\begin{array}{r} 106800 \\ 9400 \\ \hline 0 \end{array}$$

Master. Well said.

Scholar. Then likewise in the second place I adde o to o, & there ariseth o, which I write under the second place thus :

$$\begin{array}{r} 106800 \\ 9400 \\ \hline 00 \end{array}$$

Then

Then I come to the third place, saying, 4 and 8 make 12, of which I write the Digit 2, and kéep the Article 1 in my mind, thus :

106800
9400
200

Then I adde 9 to 6, which make 15, to that I adde the Article 1 that was in my minde, and it is 16, I write 6 under 6 and 9, and kéep 1 in mind, thus :

106800
9400
6200

Master. Why doe you not write both figures, séeing you are come to the last couple of numbers.

Scholar. Nay, reason sheweth me, that I must adde that Article that is in my mind unto the next Figure of the over summe, though there be no more in the nether summe.

Master. That is well considered : then do so.

Scholar. Then say I, o in the over summe, and 1 in my minde maketh 1 : that write I under o. Then followeth there yet one more in the over summe, which hath none to be added to it, for there is none in the nether summe, nor yet in my mind : therefore I thinke I must write that even as it is.

Master. Yea.

Scholar. Then doth my whole summe appear thus :

106800
9400
116200

Master. If you mark this you have learned perfectly the common Addition of all summes which are of one Deno-mination : so that ye observe this also, that in Ad-dition you must have two numbers at the least : or elle how can you say that you doe adde? And ever let the greatest number bée written highest, for that is the best way; though it be not necessary.

And forget not this, that (if you have many numbers

to adde together) you shall habe oftentimes an
Article of a greater value then 10, sometimes 20,
sometimes 30, sometimes moze, yea (peradventure)
100. Therefoze as you did with the Article 10, so
do with them, referving them in your minde, and
adding to the number next following so many,
as their valuer oz value certain is : that is to say, 2
foz 20, 3 foz 30, 5 foz 50, 10 foz 100, 12 foz 120,
and so forth of other like. So that if the Article be
130, then muſt you ſet down the o and kéep 10 in
minde, to be carried to the next row of Figures oz
place, if any such happen to come. Foz your better
underſtanding take this example foz all.　　4889
I would adde theſe thirteen ſummes into　　4599
one, which I ſet after this manner : then　　2290
doe I beginne and gather the ſumme of　　3699
the firſt row of figures, which come to　　2299
107, (foz I take 9 there tenne times,　　4099
and that is 90) then 9 and 8 is 17, that　　1099
is in all 107, of which ſumme I wzite the　　3298
7 under the firſt row of Figures, and then　　299
foz that 100 is tenne tennes, I kéepe　　699
tenne in mind, which tenne I muſt adde　　499
unto the next row of Figures, which are　　899
in the ſecond place :　——————————　389
which ſecond row of figures (when they are added
together with that tenne that I had in my minde)
make in all 125, of which ſumme wzite the
Digit 5 under the ſecond row, and then (foz that 120
containeth twelve tennes) I kéep twelve in minde
to be added to the third place oz row of figures ;
which being added together, make in all 60, the
Cypher o I ſet down under the row of figures in
the third place :　　　　　　　　　　　　And

And the Figure 6 I kéep in minde to be added to the row of Figures in the fourth place, which (when they ware added together) make 29. The Figure or Digit 9 I set downe under the fourth place. And because it is my laſt worke, I ſet downe the 2 alſo that I have in my minde to the 9 in the fiſth place; ſo theſe ſummes doe make in all 29057.

¶ But (for your more eaſe in worke) when you have an Addition of ſo many ſummes to be added together, you were beſt part that ſumme into two or thrée parts, and worke them ſeverally, and ſo put their Additions together, and this were the beſt thing you could doe when over many ſummes fall to be added.

Scholar. This ſéemeth ſomewhat hard, by the reaſon of ſo many numbers together.

4889
4599
2290
3699
2299
4099
1099
3298
299
699
499
899
389
29057

Howbeit, I thinke (if I do often prove, even with the ſame example, either by working of it alone, or elſe by parting it as you ſaid even now) that I ſhall be able to do ſo ſhortly with any other ſumme.

Maſter. So ſhall you. For it is often practice that maketh a man quick and ripe in all things: but becauſe, as well in great ſummes as in ſmall there may chance to be ſome errour, I will teach you how you ſhall prove whether you have done well or no.

Scholar. That were a great help and eaſe.

Maſter. Begin firſt with the higheſt number, and then to all the other orderly, and adde them together, not having regard to their places, but as though

The proof of Addition.

D 2

though they were all Unites : and still (as your
number encreaseth above 9) cast away 9. Then go
forth, ever casting away 9 as often as it amounteth
thereto : and so doe till you have gone over all the
numbers that you intended first to adde ; and whatsoever remaineth after such Addition and casting
away of 9, write it in some void place by the end
of a line, for the better remembrance : and thus is
the first place of your worke proved. Then secondly, put together the Figures that result of the Addition under the line, still casting away 9 also. And
then that that remaineth write at the other end of
the line ; and if those two Figures be like, than
have you well done, but if they be unlike, then have
you missed. As for example, in this present summe.
The first Figure of the over line is 9, let him goe,
then 8 and 8 is 16, take away 9, there resteth 7,
and adde that 7 to 4 that followeth, and it maketh
11 from which if you take 9, there resteth 2. Then
come to the next row, whose first and second numbers are 9, therefore overpasse them both, and
take the 5 to the 2 which did remain in the first
row, that maketh 7, put thereto the 4 following
and that maketh 11, thence take 9, and there remaineth 2. Next unto that, goe to the third line
whose two first numbers you may let passe, becaus
they are nines : then take the two Figures of 2
which (with the other two that remained, in the
second row) make 6. Then goe to the fourth row
whose two first numbers let goe, and take the 6 to
the 6 that remaineth, and that maketh 12 : take
away 9, and there resteth 3, which with the 3 that
is next, maketh 6. And so goe through all the
                                                        oth

other numbers, and you shall finde that there remaineth 5, after you have cast away 9, as often as you can finde it: therefore write 5, at the end of the line in a void place thus:

5 ——————

Then gather all the Figures of the Totall summe, which is under the lowest line, and cast away 9 as often as you can find it; as thus, 7 and 5 make 12, take away 9, there resteth 3. to that if you adde the 2 that is last, (for you may omit the 9) then doth it make 5, which 5 you must write at the other end of the line that you made in the void place, thus :

5 —————— —————— 5

And then you sée that these two Figures be like, whereby you may know that you have done wel', and so you may prove in any other.

Scholar. (If it please you) I will prove in another summe.

Master. With a good will.

Schol. Then will I take one of your former examples, which was this.

First in the highest line 8 and 6 make 14, then 9 taken away, there remains 5, to which I adde the 1 that followeth, and that maketh 6: then come I to the second line, where I finde first 4, which with 6 maketh 10, from that I take 9, and there resteth one, the next Figure is 9, and therefore I let him alone, so finde I 1 remaining, which I set at the end of a line, thus :

$$\begin{aligned} &106800\\ &9400\\ &116200 \end{aligned}$$

1 —————— ——————

Then I come to the Totall summe, and there I

D 3   finde

finde that all the Figures put together make 10, from which I take 9, and there resteth 1 also, which I put at the other end of the line, thus:

I——————I

And becaufe they be like, I know that I have well added.

Mafter. So you know now both how to adde two fummes oz moze together, and alfo how to pzove whether you have done well oz no : and now I will teach you how to adde fummes of divers Denominations together : which thing can never be but when the one Denomination is fuch that it containeth the other certaine times. And yet you fhall adde them to the other, not after this fozt (as you did them that were of one Denomination) but after fuch a fozt as I will now fhew you, that is to fay :

If you have a fumme of divers Denominations, then looke that you fet every Denomination by himfelfe, with fome note oz figure of his Denomination, as they are wont to be wzitten. Then wzite your other fummes fo under that firft, that every one bee fet under the other of the fame Denominations : As foz Example, if your Denominations bee pounds, fhillings, and pence, wzite pounds under pounds, fhillings under fhillings, and pence under pence : and not fhillings under pence, noz pence under pounds.

Scholar. Now that you have fpoken it, me thinketh it needeth not to warne me of it, foz it were againft reafon fo to confound fummes : but yet if you had not fpoken of it, peradventure I fhou'd have been deceived in it.

Mafter. If you doe fay it is plain, I will fpeak

Addition of numbers of divers Denominations.

m

no moze of it, but with an example make the matter
to appear evidently.

First, one man oweth me 22 l. 6 s. 8 d. another
oweth me 5 l. 16 s. 6 d. and another oweth me
4 l. 3 s. I would know what this is all together :
Therefoze must I first set down
my great summe, and then the o=
ther, every one under his Deno=
mination agræing to the greatest
summe, as here you sée with a line
under them.

li.	s.	d.
22	6	8
5	16	6
4	3	0

Then must I begin at the smallest numbers
(which must alwayes be set next to the right) and
adde them together : and if the summe will make 1
oz 2 oz 3 of the next Denomination, then must I
kéep it in my minde till I come to that place, and
under that first place must I note the residue (if
there remain any of the same Denomination) but if
there remaine none, then néed I to wzite under it
nothing. And this is all that you must marke in
this Addition : foz all other things are like to the
manner of Addition befoze mentioned : *Therefore the
chiefest point of this Addition is, to know the values of
common Coines, and rated summes,* As how many
shillings be in a pound, how many pence in a shil=
ling, of which (and of other like things) I will in=
struct you hereafter in teaching of Reduction : But
now I may not disturbe your wit from the thing
that we are about.

Therefoze let us returne to that former example
which I purposed of the Debtors : which summes
when I had set ozderly, they stod thus with a line
under them.

D 4   Then

Then to adde them into one summe, I must begin at the right hand where the smallest Denomination is, and adde them together, first saying, 6 and 8 make 14. Now seeing these 14 are pence, which contain one shilling and 2 pence: the 2 pence I set down under the line of pence, and the one shilling I keep in my minde to carry to the next row being the place of shillings.

li.	s.	d.
22	6	8
5	16	6
4	3	0
		2

Then doe I adde the shillings together, saying, 1 in my minde and 3 make 4, and 6 make 10, and 6 make 16, and 1 in the second place which standeth for 10, make 26, which is 1 pound 6 s. The 6 s, I set down under the place of shillings, as appeareth in the example. And the 1 pound I keep to carry to the pounds.

li:	s.	d.
22	6	8
5	16	6
4	3	0
	6	2

Then come I to the pounds, adding them all together, saying, 1 that I keep and 4 make 5, and 5 make 10, and 2 make 12. The Figure or Digit 2 I set downe right under that place or row of pounds where I gather them, and the Article 1 I keep to carry to the next place, saying, 1 in my minde & 2 is 3, which 3 I set downe directly under the 2. And then appeareth my whole summe thus.

li.	s.	d.
22	6	8
5	16	6
4	3	0
32	6	2

And thus must you doe with any such like summes whatsoever, whether they be money, weight or measure, which (if you practise divers summes) you

you shall be well acquainted with the feat of Addition.

But now can you tell how to prove this Addition, or such other like of divers Denominations, and to try whether you have well done or no?

Scholar. I would I could.

Master. That shall you doe by this means : You must make a Crosse which shall have as many lines as you have sundry Denominations in your Addition : As if you have but two Denominations then you may take it thus : that the over part and nether part may serve for one Denomination. And if you have three Denominations (as pounds, shillings, and pence) then must you make three lines, thus : The upright line may serve for pounds, and the highest thwart line for shillings, and the lowest for pence : as for example, the summe which we last wrought.

Proof of Addition of divers Denominations.

li.	s.	d.			
22	6	8	(	5	6
5	16	6			
4	3	0			
			2		2
				5	

For the proof of which, because it containeth 3 Denominations, I must make a crosse of 3 lines, as in the example before. Then I reckon first at the right hand the pence, 6 & 8 make 14, from which I take 12 for the next Denomination, that is to say, a shilling, and there resteth 2, which I must write

write at one end of the nether thwart line.

After that I gather the summe of the shillings 3, 16, 6, which maketh 25, to whom I put 1 that I took of the pence and that maketh 26, from those I take 20. the quantity of the next greater Denomination, that is to say, a pound, and there resteth 6, which I write at the end of the highest thwart line.

Thirdly, I adde together the pounds, 4, 5, and 2, which make 11. to them I adde the one that came of shillings, and they make 12, from whence I cast 9, and there resteth 3, that thrée I joyne to the 2 in the next place, and they make 5, which 5 I set at the Crosse also. And thus is my first part of my worke proved.

That done I come to the total summe under the line, and examine it, beginning at the pence, where I finde but 2, and cannot take 9 from him: therefore I set him at the other end of the nether thwart line: then I come to the shillings, where I finde onely 6, which (because it is lesse then nine) I set it at the other end of the line of the shillings, that is, the overmost thwart line.

Last of all, of the 32 li. I take thrée times 9. which is 27, and there remaineth 5, which I write under the upright line: either else I may reckon them simply without any respect of their valuation or place: saying, 2 and 3 make 5, which because it is lesse then nine, I set under the upright line as before. Then I consider every number, comparing it to the number that is against it: and because I finde them to be every one like his match I know that I have well done.

Scholar. This Crosse I perceive doth serve for thes

theſe 3 Denominations, pounds, ſhillings, pence : but what if I had l. s. d. ob. and qu.

Maſter. Theſe lines, as I have ſaid, doe ſerve for 3 Denominations, ſuch as they be, as here 3 doe ſerve for pounds, ſhillings, and pence : but if you have no pounds in your ſumme, then may they ſerve for ſhillings, pence, and half penies : yea, for d. ob. and q. or in weight for C. q. and l. or in meaſure for Elles, Quarters, and Nailes, if you have no greater Denomination : ſo that you remember that the upright line ſerveth for the greateſt Denomination, and the higheſt thwart line for the next, and the loweſt for the leaſt.

And ſo if you have foure Denominations, you muſt make your croſſe with ſo many lines : And if your ſumme be of more Denominations, make ſo many lines in your croſſe. And thus will I make an end of Addition, ſaving that here (for the better underſtanding of this Rule) I have ſet you down certain examples both of money, weight, and meaſures with their works and proofs.

Examples of Addition.

li.	s.	d.	li.	s.	d.
23	10	4	130	17	10
45	6	8	28	6	8
37	2	9	13	13	4
25	13	6	120	0	0
131	13	3	292	17	10

The

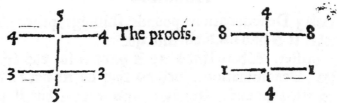

The proofs.

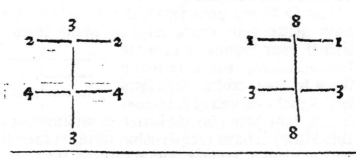

C.	q.	li.	yards	q.	ñayles.
34	1	3	17	3	3
12	2	2	35	2	1
7	3	4	26	1	3
13	0	13	54	2	0
67	2	23	134	1	3

Subtraction.

THen have I learned the two first kindes of Arithmetick: now (as I remember) doth follow Subtraction, whose name (me thinketh) doth sound contrary to Addition.

Master. So it is indeed: for, as Addition increaseth one grosse summe, by bringing many into one: so contrariwise, Subtraction diminisheth a grosse summe by withdrawing of other from it. S
tha

that Subtraction *or* Rebating *is nothing elſe but an Art to withdraw and abate one ſumme from another, that the remainer may appear.*

Scholar. What doe you call the Remainer ?

Maſter. That you may perceive by the name.

Scholar. So me thinketh : but yet it is good to aske the truth of all ſuch things, leſt in truſting to mine owne conjecture, I be deceived.

Maſter. So it is the ſureſt way. And, as I ſæ cauſe, I will ſtill declare things unto you ſo plainly, that you ſhall not næd to doubt. Howbeit, if I doe overpaſſe it ſometimes, (as the manner of men is to forget the ſmall knowledge of them to whom they ſpeake)then doe you put mæ in remembrance your ſelfe, and that way is ſureſt.

And as for this word that you laſt asked me, take you this deſcription : The Remainer is a ſum=me left after one Subtraction made, which declareth the exceſſe or difference of the two other numbers, as if I would abate or ſubtract 14 out of 18, there ſhould remain 4, which is called the Remainer, and is the difference betwæne thoſe two numbers 14 and 18.

Scholar. I perceive then what Subtraction is : now reſteth to know the order how to work it.

Maſter. That ſhall you doe by this meanes. Firſt, you muſt conſider, that if you ſhould goe about to rebate, you muſt have two ſundry ſummes propoſed : the firſt, which is your groſſe ſumme, (or ſumme totall) and it muſt be ſet higheſt : and then the rebatement (or ſumme to be withdrawn) which muſt be ſet under the firſt, (whether it be in one parcell or in many) and that in ſuch ſort, that
the

the firſt figures be one juſt over another, and ſo the
ſecond and third, and all other following, as you
did in Addition: then ſhall you draw under them a
line, and ſo are your ſummes duely ſet to begin
your working.

Then begin you at the right hand (as you did in
Addition) & withdraw the nether number out of the
higher, and if there remain any thing, write that
right under them beneath the line: and if there remain nothing (by reaſon that the two Figures
were equall) then write under them a Cypher of
nought: And ſo doe you with all the other Figures,
evermore abating the lower out of the higher, and
write under them the Remainder ſtill, till you come
to the end : And ſo will there appear under the line
what remaineth of your groſſe ſumme after you
have deducted the other ſumme from it, as in this
example.

I received of your Father 48 s. of which I have
laid out for you 36 s. now would I know what
doth remaine. And therefore I ſet my number thus
in order. Firſt, I write the greateſt ſumme, and
under him the leſſer, ſo that the Figures at the right ſide be even one   48
under another, and ſo the other, thus.   36

Then doe I rebate 6 out of 8, and
there reſteth 2, which I write under   48
them right beneath the line thus.   36
                                              2

Then I goe to the ſecond Figures,
and do rebate 3 out of 4, where there
remaineth 1, which I write under   48
them right, and then the whole ſum   36
and operation appeareth thus.   12

Whereby

Whereby it appeareth, that if I withdraw 36 out of 48 there remaineth 12.

Scholar. Now I will prove in a greater summe, and I will subtract 2367924 out of 3468946, those summes I set in order thus :

$$3468946$$
$$\underline{2367924}$$

Then doe I begin at the right side, and deduct 4 out of 6, and there resteth 2, which I write under them. Then go I to the second figures, and withdraw 2 out of 4, and there remaineth 2, which I set under them also, then I take 9 out of 9 and there resteth 0, which I write under them (for you say, that if the Figures be equall, so that nothing doth remain, I must write the Cypher 0 under them.)

Master. It was well remembred: now go forth.

Scholar. Then I come to the fourth place, and draw 7 out of 8, and there remaineth 1. which I write under them also. Then in the fifth place, I take 6 out of 6, & there resteth 0, (for it I write under them the Cypher 0.) Then in the sixth place 3 rebated from 4 there remaineth 1, which I write under them, and likewise in the seventh and last place, 2 taken from 3 there is left 1, which I write under them : so have I done my whole working, and my summes doe appeare thus. Where-by I see, that (if I doe rebate 2367924 out of 3468946) there remaineth 1101022.

$$3468946$$
$$2367924$$
$$\overline{}$$
$$1101022$$

Master. This is well done. And that you may be sure to perceive fully the Art of Subtraction, let mee see how you can subtract 52984732 out of 8250003456.

Scholar.

Scholar. First, I set downe the greatest summe, and after that I will write under it the lesser number, beginning at the right   8250003456. side, and then my Figures   52984732 will stand thus :

Then take I 2 from 6 and the rest is 4, which I write under them. Then do I withdraw 3 from 5, and there remaines 2, which I write under them. Then take I 7 out of 4, but that I cannot, what shall I now do?

Master. Marke well what I shall tell you now, how you shall doe in this case, and in all other the like : If any figure of the nether summe be greater then the figure of the summe that is over him (so that it cannot be taken out of the figure over him) then must you put 10 to the over figure, and then consider how much it is, and out of that whole summe withdraw the nether figure, and write the rest under them. Can you remember this?

Scholar. Yes, that I trust I shall. Now then in mine example where I should have taken 7 out of 4, and could not, I put 10 to that 4, which maketh 14, from it I take away 7 and there resteth 7 also, which I write under them.

Master. So have you done well : But now must you marke another thing also : that (whensoever you do so put ten to any Figure of the over number) you must adde one still to the figure or place that followeth next in the nether line : as in the example there followeth 4, to   8250003456 which you must put 1, and   52984732 make him 5, and then go on as I have taught you.   018724
Scholar.

Scholar. Then shall I say, 4 and 1 (which I must put to him for the 10 that I added to 4 before) make 5, which I should take out of 3, but that cannot be; therefore I must put to it also 10, and then it will be 13, from which I take 5, and there resteth 8 to be written under them: and because of that 10 added to the 3, I must adde 1 to 8 that followeth in the nether line, and that maketh 9, which I should take out of 0 and cannot; therefore I put thereto 10, and that maketh 10, from 10 I take 9, & there remaineth 1, which I write under them.

Thus do I adde 1 likewise to the next figure beneath, which is 9, and that maketh 10, that 10 should I take out of the figure above, but I cannot, for it is 0, therefore I put 10 to it, and so take I 10 out of 10, and there resteth 0 to be written under them.

Then come I to the next figure which is 2, and to him I doe adde 1, which maketh 3, that 3 I cannot take out of nought, therefore of that nought I make 10, and thence doe take 3, so there remaineth 7 to be written under them: likewise doe I put 1 to 5, which make 6, that 6 I cannot take out of 5, therefore I adde 10 to that 5 & make it 15, from which I rebate 6 there remaineth 9, which I write under them. Now have I spent all the nether Figures, and what shall I do more?

$$\begin{array}{r}8250003456\\52984732\\\hline 8197018724\end{array}$$

Master. You should have added one to the next figure following (if there had béen any) because you added 10 to the last figure before of the over line: but being there is no figure following, you

C   must

muſt adde that one to the place following, and the
deduct that one from the number aboue.

Scholar. Then ſhall I ſay, becauſe I borrow
10 to the ouer 5, I muſt put 1 in the next plac
beneath, that is under 2, then muſt I ſubtract tha
1 from 2, and there reſteth 1 to bée written unde
that in the ninth place. Now I haue no more t
ſubtract, for there is not any figure remaining b
neath, neither yet any unice to be added, becauſe
borrowed not 10 to the figure laſt before : and ye
is there 8 remaining in the ouer line, which
thinke (by reaſon) ſhould be ſet at the end of th
Figures in the loweſt row, which is under th
line, for becauſe there was nothing taken from it

Maſter. That is well conſidered, and reaſon
teacheth ſo indéed,

Scholar. But Sir. I beſéech you, ſhall I alwaye
when any number ſo remaineth alone, as thus
did, write him under the line ſtraight againſt hi
own place ?

Maſter. Yea, what elſe ? whether they bé one o
many : & this well remembered, you haue ſufficient
ly learned Subtraction ; Howbeit, becauſe of certai
things that might deceiue you, if you did not tak
good héed to your working, I will propoſe to yo
another example of many numbers to be ſubtracte
as thus : I receiued of a friend of mine to ké
2869 Crowns, of which at one time I deliuere
him again 500, at another time 368, at anothe
time 440, at another time 80, and at another tim
64, now would I know how many doe reſt be
hinde.

Therefore firſt I ſet down my groſſe ſur
            286

2869 *Crowns received.* and
underneath it I set all   500 ⎞
the parcells thus, and under 368 ⎟
them a double line.   440 ⎬ *delivered.*
Then first I begin at the first 80 ⎟
place, and gather together the 64 ⎠
sum of all those lines (save the
overmost) in the first figures, & so I doe with all the
figures of the second place, & so forth, as I did in Addition, save that I leave out the highest row of numbers (as the line warneth me) that sum so gathered between the double line, is the sum delivered
in all: which summe I doe afterwards subtract out
of the highest row of numbers, & the remainder doe
I set under the nethermost line: as for example.

I set the summes as be-
fore: then do I gather the
first figures of all the places
delivered together: where
I finde but 4 and 8, that
maketh 12, (for three Cy-
phers increase no sum in
Addition, as you learned
before,) of the 12 therefore
doe I write the Digit 2
between the double line and keep the Article in my
minde, till I come to the second place, where I
finde 6, 8, 4, 6, that maketh 24, to them I put
the Article in my minde, and it is 25, 5 of which
I write under the second place, and keepe the Digit
2 in my minde for the third place, where I finde
4, 3, 5, that makes 12, to the which I adde the 2
in my mind, and it maketh 14, thereof I write the 4

2869 *Crownes received.*
500 ⎞
368 ⎟
440 ⎬ *Delivered.*
80 ⎟
64 ⎠

1452 *Delivered in all*

1417 *Reſt behinde.*

C 2   under

under the third place, and because there remains no
moze Figures to be added, I wzite the Digit in the
fourth place, as you sée in the Example, and so it
appeareth, I have delivered in all a thousand foure
hundzed fifty two Crowns.

Then come I to the subtracting of this summe
betwéen the lines, foz by Addition it is equall to
the five parcelis over it, therefoze I pzocéde to subtract it from the overmost summe, saying, 2 from 9
remains 7 to be wzitten under them beneath the
lowest line. Then in the second place I take 5 from
6 and there resteth 1 to be wzitten under them.
Then in the third place, 4 from 8, resteth 4. Last
of all in the fourth place, 1 from 2, remaineth 1,
And thus I sée that after those five sums are subtracted from 2869, the Remainer is 1417.

Scholar. This I perceive : but is there no
shozter way and moze spéedy?

An abridg-
ment of
the former
manner of
Subtracti-
on.

Master. Yea, when you are a while exercised
in it: foz you may (as fast you can gather the numbers together) withdraw them out of the highest
sum. But if in quantity those numbers added together, excéd the highest sum, oz upper number,
then shall you (as befoze hath béen taught)
imagine to bozrow 10, 20, oz 30 moe, as néd
shall require, and put them to the upper number,
to help to further theabatement, reserving oz resto-
ring the Articles that you bozrowed to the next place
again : and so still goe fozward till you have
ended your wozke : as foz example. In the last
summe pzoposed, I gather first in the first place 4
and 8 that makeath 12, which 12 I should deduc
oz take out of 9 in the upper number above the
                                                    line

line, but I cannot : and therefoze I adde unto 9 an Article of 10, and make the upper number 19, from whence I take 12, then there resteth 7, then foz the Article 10 I adde to the next place of money deliered ; saying, 1 that I bzing and 6 make 7, & 8 make 15, & 4 make 19 & 6 make 25, which 25 I should take out of 6 in the upper number, but I cannot, wherefoze I adde 2 tens, oz 20 unto 6 in the upper number, & that maketh 26, then 25 out of 26 resteth 1 ; then the tens which I bozrowed, oz have in minde, I adde to the next row oz sum delivered ; saying, 2 that I bzing, and 4 make 6, and 3 make 9, & 5 make 14 ; then 14 out of 8 I cannot take, but 14 out of 18 resteth 4. Now because there are no moze places to be added, the one that I bozrowed, oz have in minde, I rebate from 2 in the upper line, & there remaineth 1, which I set down in the remainer line : & so my sum appeareth (as befoze) to be 1417 Crownes.

Loe thus have you now a shozter way.

Scholar. I like both wayes well : & I perceive both well : yet, as in one the wozking seemeth somewhat long, so in the other it leaveth very much (mee seemeth) to remembzance, and therefoze may cause errour quickly, except a man have a quick and an exercised remembzance. But yet foz the sharpning of my wit by your patience (if you will give me leave) I will try what I can do in a like sum, to wozk it the shoztest way : whereupon I would subtract out of 40301964, these thzee parcels. Therefoze I set them first in due ozder : then I gather the parcels of the first place, which are 8, 2, 1. that is 11,

40301964	Charge.
20003428	
10002432	Discharge.
10101461	
43	

which

which I should take or deduct out of 4, which is over him, but I cannot : therefore I adde an Article, or one tenne to 4 which maketh 14 then 11 out of 14, there resteth 3 to bée written under the first place betwéen the two lines.

Then come I to the second place, saying, 1 that I borrowed to have in my minde, and 6 make 7, and 3 make 10, and 2 make 12, which I cannot take from 6, therefore I adde 10 to 6, which maketh 16, and then 12 from 16, resteth 4, which I write under the second place betwéen the two lines.

Then come I to the third place saying, 1 that I borrowed or have in mind and 4 make 5, and 4 is 9, and 4 make 13, which I should take out of 9 that is over them but I cannot : therefore I adde 10 to 9 which make 19, then 13 out of 19, rest 6.

Then come I to the fourth place, saying, 1 in minde and 1 is 2, and 2 is 4 and 3 make 7, which because it cannot be taken from 1, I take it from 11, and there resteth 4.

After that I come to the fifth place, where are onely thrée Cyphers, which make nothing, unto which I adde 1 in minde, then should I take that (that is to say) 1 from the Figure over them, which is also a Cypher : therefore I say thus, I cannot take 1 from 0, but 1 from 10 remaineth 9 : so must I write 9 under them. Then in the sixth place I finde but 1, and 1 in minde make 2, which I take out of 3 over him, and the remainer is 1 : that must be written betwéen the two lines in the sixth place. So I goe to the seventh place where I finde onely Cyphers, and in the grosse

summe

summe over them a Cypher also : therefore must I
write the remainer (which is nothing) with a Cypher also. Then in the eight and last place I gather
1, 1, 2, that maketh 4, which if I take out of that
4 that is over them, there will nothing remaine.
And that must be noted with a Cypher between the
two lines (as I have often said) and so have I
ended my worke, and the figures stand as followeth.

Scholar. But Sir, I remember you taught me
that Cyphers should not come in the last place, for
because they serve onely to increase the value of
other Figures which follow them and serve not
those figures that goe before them : and now in
my Example I have set two Cyphers in the two
last places.

Master. I commend you for your remembrance.
And truth it is, you should not have set them here,
but onely because that I would make you plainly to
perceive the Art of Subtraction. Therefore seeing
that you doe now perceive it, whensoever you
would write down a Cypher, looke whether any
other figures be yet behind: and if not, then let
go the o also, for it needeth not to write him in the
latter places, where no other figure doth follow,
except it be (as I did now   40301964 Charge.
suffer you) to teach the use of   —————
Subtraction the plainer.   20003428 ⎫
                                   10002432 ⎬ Discha.
   Therefore your figures   10101461 ⎭
must stand thus when the   —————
worke is ended.

   Scholar. Sir, I do thinke   0164643   Rest.
with that you taught me be-   —————
fore, and by these two summes that you taught mee
                    E 4   last

laſt alſo, that now J could Subtract any ſumme.

Maſter. So may you, if you have marked what J have taught you. But becauſe this thing (as all other) muſt be learned ſurely by often practiſe, J will propound here two Examples to you : wherein if you often exerciſe your ſelfe, you ſhall be ripe and perfect to ſubtract any other ſumme lightly ; for in them is contained all the obſervances of whole numbers. And becauſe you ſhall perceive ſomewhat both how to doe it, and alſo whether it be well done when you have proved to doe it ; therefore have J written under them both the Remainers.

30606.	*Lent.*		308964.	*Debt.*
10354			103145	
10249	*Paid.*		102597	*Paid.*
163			101024	
20766	*Paid in all.*		02198	*Reſt.*
9840	*Reſt to pay.*			

Scholar. Sir, J thanke you : but J thinke J might the better doe it, if you did ſhew mée the working of it.

Maſter. Yea, but you muſt prove your ſelfe to do ſome things without my aid, or elſe you ſhal not be able to doe any moze then you are taught And that were rather to learne by wrote (as they call it) then by reaſon. And again, there is nothing in theſe examples, or any other of whole numbers bu but I have taught you the rules of them already.

Scholar. Then I trust by practise to attain the use of it. And is this all that I shall learne of Subtraction?

Master. Yea, saving that (as you have seen in Addition) there are numbers of divers Denominations, in which the working is not much unlike: yet (without some instructions be given of it) it might seem to a learner more difficult then indeed it is. Therefore I will briefly shew you the use of it onely by an example or two.

A certain man owed to mee 14 l. 12 s. 8 d.
of which he paid me at one time 4 l. 6 s. 8 d.
at another time 3 l. at another 2 l. 3 s. 4 d.
and last of all 6 s. 8 d.

Now would I know what remaineth unpaid yet? therefore I set my sums thus, every one in their due place: As pounds under pounds, shillings under shillings, pence under pence.

li.	s.	d.
14	12	8
4	6	8
3	0	0
2	3	4
	6	8

Scholar. Sir, I pray you why do you write 2 l. for the common speech useth rather to say, 40 s.

Master. We must here use the Denomination that is greatest in any summe, so that we may not write according as we use to speake, saying, 16 d. 18 d. or likewise 7 groats, 8 groats, 24 s. 40 s. 48 s. and such other: but we must write every Denomination that is in any summe by it self.

Namely, shillings and pounds. So must we write for the last summes now named, 1 s. 4 d. 1 s. 6 d. 2 s. 4 d. 2 s. 8 d. 1 l. 4 s 1. 2 l. 8 s. and so forth of other like.   Scholar.

Note how the penne differeth from the common order of Counters.

Scholar. So that wee may not write in Arithmetick, pence, when the summe amounteth to shillings, nor shillings, when the summe maketh pounds.
Now (if it please you) end your example.

Master. When my summes are so set as I
shewed, then (according to the rules of Addition)
I gather all the particular summes which bée paid
mée into one totall summe, directly to be set under
them betwéen the two lines, not medling with the
14 l. 12 s. 8 d. as the line warneth mee : therefore
must I beginne with the smallest Denomination,
saying, 8, 4, 8, is 20, pence, which maketh one
shilling and 8 pence, the 8 d. I set downe under
the place of pence,
and the one shilling
I kéep in minde to
carry to the next
Denomination of
shillings : Then
come I to the shil-
lings, say one that I
bring or have in
minde, & 6 is 7, and
3 is 10 & 6 makes

li.	s.	d.
14 ——— 12 ——— 8		
4 ——— 6 ——— 8		
3 ——— 0 ——— 0		
2 ——— 3 ——— 4		
6 ——— 8		
9 ——— 16 ——— 8		
4 ——— 16 ——— 0 Rest.		

16, which because it containeth not one pound
I set directly under the place of shillings. Then
come I to the pounds, whose parcels are 2, 3, 4
that is in all 9, that 9 doe I set down directly under
the pounds : And so the totall or whole Addition
of all the particulars paid, amounteth to 9 l. 16 s. 8 d.

Now for the worke of Subtraction, I must rebate
that totall summe of Addition out of the higher
number, that is to say from the 14 l. 12 s. 8 d

There

Therefore to perform the worke I say, 8 d. out of 8 d. remaineth or resteth nothing, therefore in the place of the rest or remain, right under the Denomination, I set down o. Then coming to the shillings, where I finde 16. which should be taken out of 12, but I cannot : therefore I imagine to borrow 1 out of the next Denomination, that is, of the 14 l. and put that one pound so borrowed unto 12 s. that maketh 32 s.

Now 16 s. out of 32 s. resteth 16 s. which 16 s. I set down directly under the place of the rest.

Lastly, comming to the pounds, saying, one pound in minde that I borrowed, and 9 make 10, then 10 out of 14. there resteth 4.

So doth my whole rest or remain appear to be 4 l. 16 s. o d.

This I account the easiest way for a young beginner to practise, though it be something long.

Scholar. Is there any shorter way for this work also ?

Master. Yes, as in this last Example I will also shew you, for you may adde together the particular sums as they are set in order, beginning with the pence, saying, 8, 4, 8, make 20 d. which 20 d. you should take out of the 8 d. above the line, but you cannot, therefore shall you borrow 1 of the next Denomination, that is to say 1 of the shillings, and put it to the 8 d. that maketh 20 d. now 20, out of 20 d. resteth o. which Cypher I set down

li.	s.	d.
14	12	8
4	6	8
3	0	0
2	3	4
0	6	8
4	16	0

down directly under them.   Then one shilling
that I borrowed or had in minde, and 6 make 7,
and 3 make 10, and 6 make 16, the 16 out of 12
I cannot take, therefore of the next Denomination
I doe borrow one l. and put it to 12 s. which maketh 32 s. then 16 s. out of 32 s. resteth 16 s.

Lastly I come to the pounds, saying, 1 l. in
minde, or that I borrowed, and 2 make 3, and 3 is
6, and 4 is 10, then 10 out of 14, there resteth 4.

So doth my remainer or rest appeare as before to
be 4 l. 16 s. od.

Scholar. Then doe I perceive very well : and if
there be no other thing to bee learned in Subtraction,
then may I come to Multiplication, for that you
reckoned to be next in order.

Master. Wee have done indeede with the Art of
Subtraction, as touching the working.

But yet before we go to Multiplication, I will
instruct you how to examine your worke, whether
it be well done or not. For the performance whereof,
of, if you marke what I sayd in Addition, you may
easily perceive what is to be done for the proofe of
Subtraction, which is best made by the aide of Addition thus.

Draw under the lowest number (which is your
Remainer) a line, and then adde this Remainer and
all the other that you did subtract before, to
gether, and write that that amounteth under the
letter line : and if the summe that cometh thereof
be equall to the highest of the Subtraction, then is
the Subtraction well wrought, or else not : As you
may see for example in the summes set down before

Proof of Subtraction.

fore, & first in sums of one Denomination, whereof one was this.

Where the Number 52984732   8250003456
is subtracted from 8250003456.   52984732

& the Remainer is 8197018724

Now to prove whether it be   8197018724
truly wrought or not, I adde the Remainer and Example in a sum of one Denomination.
the number Subtracted together, beginning at the
right hand; and first I say 4 and 2 is 6 which is
set under the line:

The number given	8250003456
The number to Subtract	52984732
The Remainer	8197018724
The Proofe	8250003456

Then again in the second place I say 2 and 3 is 5,
which I write under, next that in the third place,
7 and 7 are 14, of which I write the Digit 4, and
keepe the Article 1 in my minde. Then in the
fourth place 8 and 4 is 12 and 1 in my minde maketh 13, whereof I write downe the Digit 3, and
keep the Article 1 in my minde. Again in the fifth
place, 1 and 8 is 9, and 1 in my minde is 10.
Whereof I set downe 0 and keepe the 1 in my
minde. And so going on to the rest (as it is taught
in Addition) when I have made an end, I see that
the lowest line of numbers and the highest be alike:
wherefore I know that I have well done.

So likewise the Proofe is to bee made in numbers
of divers Denominations: as for Example, in our
summe of that kinde which in the first form of
working stood thus: (as the particular numbers to
be subtracted, being drawne into one)

Where,

Where, in the title of pence, I find 8 & 0: the 8 I set down directly under in that of pence.

Then in the place of shillings I finde 16 and 16 which make 32 shillings, wherein is contained 1 l. and 12 s. the 12 s. I set down directly under them in the due place of shillings, & one pound I keepe.

Then comming to the pounds, I say 1 that I keepe, and 4 is 5, and 9, is 14, which 14 in due order I set downe directly under them as this figure sheweth. And the whole summe is 14 l. 12 s. 8 d. agreeing with the upper number above. So I find the worke is good, and the Subtraction well wrought.

The same thing is to be done for the latter forme of Subtraction (where the particular summes are not gathered together into one grosse.) For the Remainer and all the particular summes subtracted, being added together, if the summe that commeth thereof bee equall to the highest number above, then is the Subtraction well wrought, or else not.

li.	s.	
14	12	
4	6	
3	0	
2	3	4
	6	8
Paid in all. 9 -- 16-8		
Rest. -- -4-16-0		
Proofe -- 14-12-8		

As for example also in the last sums which stood thus,

First in the title of pence, I adde 8, 4, 8, that maketh 20 d. which containeth 1 shilling and 8 pence.

The 8 I set down under the lowest line in the row or

li.	s.	d.
14	12	8
4	6	8
3	0	0
2	3	4
0	6	8
4	16	
14	12	8

title

title of pence, and that 1 shilling I kéep to carry
of the next Denomination or place of shillings.

Then returning to the shillings, saying : one in
minde, or that I kéep, and 16 make 17, and 6
make 23, and 3 make 26, and 6 make 32 shillings,
which amounteth to one pound, 12 s, the 12 s. I
set down under the title of shillings : and 1 pound.
I kéep or have in minde to carry to the next Denomination or place of pounds. Then come I to the
pounds, saying, 1 that I bring and 4 make
5, and 2 make 7, and 3 is 10, and 4 make
14 ; then doe I write 14 under the pounds,
and so have I ended the Addition : and I sée that
the lowest line is like unto the uppermost line in
number, wherefore I know that I have welldone.

And thus have I taught you the Art of Subtraction, and the means to prove whether it be
well wrought or not. Therefore now will I make
an end thereof, and will instruct you in Multiplication.

Multi-

Multiplication.

Ultiplication is an operation whereby two sums produce the third: which third sum so many times shall contain the first, as there are Unites in the second. And it serveth in stead of many Additions. As for Example : When I would know how many are 30 times 48. If I should adde 48 thirty times, it would be a long worke. Therefore was Multiplication devised, which shall doe that at once that Addition should do at many times.

Scholar. I perceive the commodity of it partly, but I shall not sée the full profit of it till I know the whole use of it. Therefore Sir, I beséech you, teach me the working of it.

Master. So I judge it best; but because that great summes cannot be multiplied but by the Multiplication of Digits, therefore I thinke it best to shew you the way of multiplying them. As when I say. 9 times 8, or 8 times 9, &c. And as for the small Digits, under 5 it were but folly to teach any rule, séeing they are so easie that every childe can doe it : but for the Multiplication of the greater Digits, thus shall you do.

First, set your Digits one right over the other, then from the uppermost downwards, and from the nethermost upward, draw straight lines, so that they make a crosse, commonly called Saint
Andrews

Andrews croſſe, as you ſée here: Then looke how many each of them lacketh of 10, and write that againſt each of them at the end of the lines, & that is called the difference : as if I would know how many are 7 times 8, I muſt write thoſe Digits thus.

Then doe I looke how much 8 doth differ from 10, and I finde it to be 2 : that 2 doe I write at the right hand of 8, at the end of the line thus.

After that I take the difference of 7 likewiſe from 10, that is 3, and I write that at the right ſide of 7, as you ſée in this example.

Then doe I draw a line under them, as in Addition, thus.

Laſt of all, I multiply the two differences, ſaying 2 times 3 make 6, that muſt I ever ſet under the differences, beneath the line : then muſt I take one of the differences (which I will, for all is like) from the other Digit (not from his owne) as the line of the Croſſe warne me, & that is left muſt I write under the Digits. As in this example, if I take 2 from 7, or 3 from 8, there remaineth 5 : that 5 muſt I write under the Digits, & then there appeareth the multiplication of 7 times 8 to be 56. And ſo likewiſe of any other Digits, if they be above 5, for if they be under 5, then will there difference be

The difference.

Digit difference.

8
X
7

Digit difference.

8   2
X
7

Digit difference.

8   2
X
7   3

Digits Difference.

8   2
X
7   3

5   6

F   greater

greater then themſelves, ſo that they cannot be
taken out of them. And again, ſuch little ſums every
childe can multiplr, as to ſay 2 times 3, or 4
times 5, and ſuch like.

Schol. Truth it is. And ſéeing me ſéemeth that
I underſtand the multiplying of the greater Digits,
I will prove by an example how I can do it. I
would know how many are 9 times 6.

Maſter. It is all one in value to ſay 9 times 6, or
6 times 9 : but yet the order is beſt to put the leſſe
ſum firſt, ſaying, 6 times 9, & ſo of all other ſummes.

Scholar. Then would I
know how many are 6 times 9:
therefore I ſet the Digits thus,
and make the croſſe, thus.

$$9$$
$$X$$
$$6$$

Then doe I ſet their differences from 10 at the
right ſide, the difference of 9,
which is 1, againſt it, and the
difference of 6, which is 4, againſt
it alſo, as in this example.

$$9 \quad\quad 1$$
$$X$$
$$6 \quad\quad 4$$

And under them draw a line, Then doe I multiply the differences together, ſaying : 1 time 4 maketh 4, that 4
doe I write under them thus.

$$9 \quad\quad 1$$
$$X$$
$$6 \quad | \quad 4$$
$$\quad\quad 4$$

Then take I one of the differences from the other Digit, as,
1 from 6, or elſe 4 from 9, and
each wayes there reſteth 5,
which I doe write under the
Digits thus. And ſo appeareth
the multiplication of 6 times 9,
to be 54. Thus I ſée the ſeats
of this manner of multiplication
of Digits.

$$9 \quad\quad 1$$
$$X$$
$$6 \quad | \quad 4$$
$$5 \quad\quad 4$$

Maſter

Master. Now might you go straight to the multiplication of great numbers, save that both for your
ease and surety in working I will draw you here
a Table, whereby shall appear the multiplication
of all the Digits, and this is it that followeth.

1	1	2	3	4	5	6	7	8	9
	2	4	6	8	10	12	14	16	18
		3	9	12	15	18	21	24	27
			4	16	20	24	28	32	36
				5	25	30	35	40	45
					6	36	42	48	54
						7	49	56	63
							8	64	72
								9	81

In which Table when you would know the product
in any multiplication of Digits, seek your first or
last Digit in the greater figures; and from it go right
forth towards the right hand, till you come under the
number of your second Digit, which is in the highest
row, and then the number that is in the meeting of
the rowes of little squares (which come directly from
both your propounded Digits) is the Multiplication
that amounteth of thē. As if I would know by this
table the multiplication of 7 times 9, seek first 7 in
the greater figures, and then go right forth toward
the right hand, till you come under 9 of the highest
row in which place where you so come under the

F 3 other

other Digit (as here for example you come under 9) is alwayes contained the off-come or product which you seek, and that place we terme to be in the common angle, in respect of the two numbers so taken on the outsides : as here in that common angle, where the rowes of little squares directly proceeding from 7 and 9 do meet, you have 63, which 63 is the summe of the multiplication of 9 by 7.

To multiply greater summes.

    Scholar. This is very good and ready. And so may I find the multiplication of any Digits : but now how shall I doe in greater summes?

    Master. *When you would multiply any summe by another, you shall mark that it is the meetest order to set the greatest number highest, which is the place*

Multiplier.

of the number that must be multiplied : and likewise the lesser number under it, for that is the place of the Multiplier *or* Multiplicator, *that is to say, the number by which the Multiplication is made, and is in* English *always put before this word,* Times : *in*

Tiems.

such speaking as when I say 20 times 70. *And the number that followeth this word* Times, *is that which must be* Multiplied.

    Therefore when I would multiply one numbe by another, I must write the greatest highest and the lesser under it. as in Addition. And under them must I draw a line. As for   26
example. If I would multiply 264   29
by 29, I must set them thus.   ————

    ¶ Of which number thus set downe to be mul tiplied, may be formed a question, as thus, Ther are 29 men, and each man hath 264 Lambes. Th question is, how many Lambes they have in all.

                                                  I

To the performance whereof, I must multiply every figure, of the high row by every figure of the neather row: and that that amounteth I must set under the line, as thus.

First I do matiply 4 by 9, saying, 9 times 4 (or 4 times 9 which is all one) and that maketh 36, as the Table before of Digits doth declare, of that 36, I must write the 6, that is the Digit, under the 9, & the Article 3 I kéep in mind to carry to the next place.

$$264$$
$$29$$
$$\overline{}$$
$$6$$

Then come I to the second figure of the higher row, which is 6, and say 9 times 6 make 54, and with the 3 in my mind make 57, the 7 I set downe under the 2, and 5 I kéep in mind.

$$264$$
$$29$$
$$\overline{}$$
$$76$$

After that I come to the next figure, which is 2, and multiply it by 9, and that maketh 18, and with 5 that I have in mind, maketh 23: wherefore because it is the last worke of the Multiplier, I set it down in order as you sée :

$$264$$
$$29$$
$$\overline{}$$
$$2376$$

And so have I ended the first figure of the Multiplier. Wherefore I give it now a fine dash with my pen.

Then begin I with the next figure, & multiply it into all the higher figures as thus,

$$264$$
$$29$$
$$\overline{}$$
$$2376$$

First, 2 times 4 make 8, that 8 do I write under the second place : for evermore the Digit or first figure of the Multiplication that amounteth of the figure of the higher number, must be set under the Multiplier of it, the other in their order toward the left hand.

F 3

Scholar.

Scholar. I underſtand you thus that the Digit
of the ſumme amounting of the Multiplication o
the firſt figure of the higher row, by the firſt figur
of the lower row, or Multiplier, muſt be ſet unde
the firſt place: and that that amounteth of th
ſame firſt figure by the ſecond Multiplier, muſt be
ſet under the ſecond place, and ſo of the other, i
there be moꝛe Multipliers.

Maſter. So meane I indéed: and if there a
mount but a Digit then muſt it be ſet under th
Multiplier.

And now to goe forth; I multiply by the
ſame 2, the ſecond figure of the higher row, which
is 6, ſaying two times 6 make
12, wherefoꝛe I wꝛite the Digit.   264
2 under the third place, and the   2g
Article 1 I kéep in mind.   2376

Then doe I multiply the laſt   28
figure of the higher ſumme by that
ſame 2, ſaying, two times 2 is 4, and with the
1 that I have in minde maketh 5, which I wꝛite
under the fourth place. And ſo have I ended th
whole Multiplication: wherefoꝛe I alſo giv   264
the 2 a daſh with my pen, thus: 2g
and ſo I doe ever as ſoone as I   2376
have diſpatched any Digit by   528
which I multiply: and the ſums
ſtand thus.

Then muſt I dꝛaw a line un-   26.
der all thoſe ſummes that mount   2g
of the multiplication, and muſt   2376
adde all them into one ſumme, as   528
in the Example, you may ſée.   7656

Wher

Where in the first place I finde but 6, and therefore write I it under the line. Then in the second place 8 and 7 make 15, whereof I write 5, and kéepe one in my minde, and so forth as you learned in Addition. And so appeareth the whole summe to be 7656, which amounteth of the Multiplication of 264, by 29, and that is the just number of the Lambes that 29 men had.

Sholar. If there be no more to be observed in it, then can I doe it I suppose, as by this Example I shall prove.

¶ There is a peece of ground which containeth 1365 yards in length, and 236 yards in breadth : I would know how many yards square there is in all this peece of ground : which

numbers I set downe with the greater above, and the lesser under, as you sée.

$$\begin{array}{r} 1365 \\ 236 \end{array}$$

Then doe I multiply 5 by 6, saying, 6 times 5 make 30, of which I write the Cypher in the first place, and the Article 3 I doe kéep in minde to carry to the next place.

$$\begin{array}{r} 1365 \\ 236 \\ \hline 0 \end{array}$$

Then do I by the same 6, multiply the second figure of the higher sum, which is 6, saying, 6 times 6 make 36 & 3 in my mind make 39, of which I write the 9 under the secód place, & the Article 3 I kéep in mind.

$$\begin{array}{r} 1365 \\ 236 \\ \hline 90 \end{array}$$

Then doe I multiply the third figure, which is 3 by the same 6, and that maketh 18, and 3 in my minde make 21. The 1 I set down, and kéep 2 in minde.

$$\begin{array}{r} 1365 \\ 236 \\ \hline 190 \end{array}$$

F. 4   Then

Then come I to the laſt figure of the higher ſumme, and multiply it by 6, ſaying, 6 times 1 make 6, and 2 in my minde make 8, that 8 doe I write under the fourth place. And ſo have I ended the firſt Multiplier, and daſh him ſleightly with my Pen.

$$\begin{array}{r} 1365 \\ 236 \\ \hline 8190 \end{array}$$

Then begin I with the ſecond Multiplier, and ſay, firſt 3 times 5 make 15, of which I ſet the 5 under the ſecond place, becauſe that the Multiplier is there, and the Article 1 I kéep in mind.

Then come I to the ſecond Figure that is 6, and multiply it by 3, which maketh 18, and with one in minde maketh 19, the 9 I ſet down under the third place, and 1 I kéep in mind.

$$\begin{array}{r} 1365 \\ 236 \\ \hline 8190 \\ 95 \end{array}$$

Then come I to the third figure, which is 3, & multiply it by 3, ſaying, 3 times 3 make 9, and with 1 in mind, make 10, the Cypher I ſet under the fourth place, and the Article 1 I kéep in mind.

$$\begin{array}{r} 1365 \\ 236 \\ \hline 8190 \\ 095 \end{array}$$

And then coming to the laſt figure 1, I multiply it by 3, and it maketh 3, & with the one in mind, it maketh 4, which 4 I ſet in the fifth place and then have I ended two of the Multipliers, and the ſummes ſtand as you may ſée in the latter end of the page going before, and then I give 3 his daſh.

$$\begin{array}{r} 1365 \\ 236 \\ \hline 8190 \\ 4095 \end{array}$$

Then

Then come I to the third Multiplier, and multiply it into every figure of the higher summe, and first I say, 2 times 5 make 10, of which I set the Cyhper under the Multiplier in the third place, and the Article 1 I keep in minde.

1365
236
――――
8190
4095
0

And so multiplying the second figure 6 by that same 2, there amounteth 12, and 1 in my minde maketh 13, whereof I write the Digit 3 under the fourth place, and the Article 1 I keep in mind :

1365
236
――――
8190
4095
30

Then do I multiply the said 2 by the third figure of the higher summe, which is 3, and that maketh 6, and the one in minde make 7, which 7 I set down under the fifth place, as appeareth by the example.

1365
236
――――
8190
4095
730

Then come I to the last place, and multiply that 1 by 2, and there amounteth 2, which I set in the sixth place, and then doth the summe stand thus.

1365
236
――――
8190
4095
2730

And so have I ended the whole Multiplication.

But now (as you taught me) to know what this whole summe is, I must adde all those parcels together, and then under the line will appeare, as you may see, the grosse or totall summe is, 322140. Whereby I know there is so many yards square in that piece of ground.

1365
236
――――
8190
4095
2730
――――
322140

Master.

Mafter. This is well done.

Scholar. Then me thinketh I could call it we done, when I know, whether I had well done o no.

Mafter. It is to be proved by 9 as Addition was but the fureft proof is by Division, and therefore will referve that proof by Division, till you hav learned the Art of Division. And anon I will shew you how it is commonly proved.

¶ *But firft, for your further inftruction in thi exercife of Multiplication, I will with one exampl try your cunning, and fo make an end : And th queftion is this. I would know how many daies it i fince the Nativity of our Lord and Saviour Jefu Chrift, unto this yeare 1630. Which to performe, yo muft multiply this prefent yeare 1630 by the daies i one whole yeare, which are 365.*

Scholar. Now for that you have given me f much light into the queftion, you fhall fee I wi handfomely finifh the work, for according to your former inftructions, I fet them down with a line under them thus.

163
36
───

Then fay I, 5 times o is o, which I fet down under the firft place, as here appeareth. Then fo I, 5 times 3 make 15, the Digit 5 I fet down t the fecond place under 3, and the Article 1 I ke in minde to bee added to the next Multiplicatio Then faying five times 6 make 30, and 1 in mi 31, the 1 I fet downe in the third place, an 3 I keepe in minde. Then comming to t laft Figure, I fay once 5 is 5, and 3 in min make 8, that 1 doe I fet downe under the four place : and thus have I ended my firft Multipli

a

and therefore I give it a dash with my pen.

Then come I to the second Multiplier, which is 6, and do likewise multiply it into the upper number, saying; 6 times 0 is 0, which I set down in the second place, right under his Multiplier : then say I, 6 times 3 make 18, the 8 I set down under the third place, and 1 I keep in mind.

1630
365
————
8150
9780

Then say I 6 times 6 make 36, and 1 I keep in minde make 37. the Digit 7 I set down in the fourth place, and 3 I keep in mind : Then say I 6 times 1 is 6, or once 6 is 6, and 3 in minde make 9, which I set down next, and so have I ended two Multipliers : wherefore I dash the 6 with my pen.

Then I begin to multiply the third Multiplier into the over number, saying, 3 times 0 is 0; the 0 I set down in the third place right under his Multiplier. Then say I 3 times 3 make 9, which I set down in order next: then say I, 3 times 6 is 18, the 8 I set down, and 1 I keep.

Lastly, I say, once 3, is 3 and 1 I keep is 4, which I set down order-ly next : And so have I ended the Multiplication, & my figures stand thus.

		or thus
	1630	1630
	365	365
	————	————
	8150	815
	9780	978
	4890	489
	————	————
	594950	594950

Master. I commend you for your diligence, the worke is very perfectly done ; which parcels if you now adde together into one sum it will be 594950; which

which is the groſſe or totall ſumme of that Multiplication, and declareth the number of dayes ſince
our Lord and Saviour his incarnation, unto the end
of 1630 yeares, *beſides* 407 *dayes, and twelve houres*
for leape yeares.

Scholar. This is marvellous, me think, that
ſuch great matters may ſo eaſily be atchieved by
this Art, which heretofore I ever thought had bæn
impoſſible, as infinite ſorts of people are of that
minde.

Maſter. Truth it is, that knowledge hath no
greater enemy then ignorance, for this is one of the
leaſt of ten thouſand things that may be done by
this Art, as hereafter you ſhall be able to juſtifie.

Scholar. The manner of Multiplication I perceive, if there be no more in it.

Maſter. *Yes, there are other formes and helps for*
eaſe and ſhorter labour of the worke of Multiplication,
but I will remit them till you have a little taſted
Diviſion, *where alſo the like help into* Diviſion *may*
be uſed : and ſo therefore under one example for both,
will I ſhew you both eaſe in Multiplication, *and alſo*
in Diviſion.

But ſith the other formes and workings do
nothing differ from theſe workes in effect, but
onely in ſetting of the numbers, I will overpaſſe them till a more mæte place and time.
And now will I inſtruct you in Diviſion, ſo
that you think your ſelf ſufficiently to perceive
what I have taught you.

Scholar. Yes Sir, I thank you, but I do
not perceive how to examine my worke, to try
whether I have well done or no : therefore as
　　　　　　　　　　　　　　　　　　　　　yor

you promised me ere-while, I pray you first shew me how I shall prove it.

Master. *That is commonly used by the proofe of* 9, *as you learned before in* Addition, *saving that it differeth from that forme in divers respects : As for example.*

First, must you make a crosse after this manner.

X

Then must you examine your summe that should be multiplied, and look what remaineth after casting away of 9, that set you at the one side of the crosse, then examine the Multiplier, and whatsoever remaineth in it after casting away 9 so often as you can, write that at the other side of the crosse : then must you multiply those two numbers together, and looke what amounteth thereof, if it be under 9, write it at the higher part of the crosse : but if it be above 9, then take thence 9 as often as ye can, and write the rest at the head of the crosse : As for example, put forth of the piece of ground that contained 365 yards in length & 236 yards in breadth.

Proof of Multiplication.

Therefore first I cast away all the nines from the summe to be multiplyed, saying, 5 and 6 make 11, cast away 9 rest 2 : then 3 and 2 makes 5, and 1 is 6, that 6 I write at one side of the crosse thus.

X
6

Then do I examine the Multiplyer which is 236, wherein when

the

the 9 is cast out, there remaineth
2, that 2 therefore I set at the o-
ther side of the crosse.

                           2   6

    Then doe I multiply 6 by 2, and it maketh 12,
from which 12 I withdraw 9, then resteth 3,
which 3 doe I set at the head of the crosse. Then
doe I examine the grosse summe, amounting of the
Multiplication, which is 322140, where I finde
9 once, and 3 remaining; that 3 I set at the foot
of the crosse, and then I see it to agree with the
other 3 at the toppe of the crosse, and so know
I that I have done well : for   3
if they two did differ, then were
my worke vaine, and the Multiplication false.

                          2   6
                            3

    This is the common proof : but
the most certain proofe is by Division of which I
will anon instruct you.

    Schol. Sir, what is the chief use of Multiplication.
    Master. The use of it is greater then you can
yet understand : howbeit, these plaine commodities it hath, that if you would resolve any great
and whole value into many small and lesse proportions, as if you would change pounds into shillings and pence or any other greater or smaller
parcels by Multiplication ye shall do it speedily and
easily. Also if you should need to adde one sum to it
selfe, or to any other often times you shall doe it
by Multiplication much more speedily, readily, easily, and surely, then by often and sundry Additions.
Take you these commodities grosly shewed for an
answer at this time, and hereafter I will more a-
bundantly make you to perceive the use of it.

                                       Division

A 'sure'
proof of
Multiply-
cation.

Division.

Scholar.

 *Ell Sir, then in* Division *I pray you to instruct me. But me thinketh by the name of it, that it should be all one with* Multiplication : *for I call that* Division, *when any thing is parted into diverse and many parts.*

Master. You take it as it is taken commonly : howbeit, if you marke well, you shall perceive that it is quite contrary to Multiplication, and doth not part one thing or few things into many, but contrarywayes, it bringeth many parcells into few, but yet so, that these few taken together, are equall in value to the other many : for by Division pence are turned into shillings, and shillings into pounds : As for example, of 120 shillings, it maketh 6 pounds, so are 120 turned into 6, which is a smaller number : but then if you consider the Denominators, you shall see that they are such, that one of the latter is equall to 20 of the first, and so in value the summes are one, though in number they doe differ, and the latter summe is the lesser, and so it is alwayes in Division, howbeit, yet in the working the summe is parted by another, and thereof doth it take the name.

Scholar

Scholar. I thinke I shall better understand the reason of the name when I know the use of the worke, therefore now would I gladly learn that.

Master. *Division is a distributing of a greater summe by the unites of a lesser : Or, Division is an Arithmeticall producing of a third number, in respect of two propounded numbers ; which third number shall so often contain an unite, as the greater of the two propounded numbers can contain the lesser.* So that as Multiplication did séem to serve in stead of many Additions, so Division may séem to be in place of many Subtractions : Because that third number briefly expresseth how many times the lesser of your two propounded numbers may be subtracted from the greater : as in practise will more largely appeare. Therefore (as you may perceive) unto Division are required three numbers : The first, which should be divided, and that must (generally) bée the greater : and the second, by which the other must bée divided, and that is (generally) the lesser, and is called the Division : And the third, which answereth to the question (How many times?) and therefore is called the Quotient.

The first must be first written, and the second so set under it, that the last figure of the lower number bée right under the last of the higher, contrariwise to the worke of other kindes of Arithmetick for in them the two first figures were set ever méet one under the other ; but in Division, the last figures must be set méet, except it chance so that the last figure of the Divisor, be greater then the last of the higher number, for then you shall set the last of the Divisor under the last save one of the higher

Margin notes:

Division what it is.

A generall rule for placing the figure.

bigher number, as for example:

If you fhould divide 365 (which are the fumme of the dayes of a year) by 28, which are the dayes of a common Month, then fhould you fet them thus.

       365
       28

But if you fhould divide thofe 365 daies by 52, which is the number of weekes in one yeare, then fhould you fet them thus.

       365
       52

Likewife if I would divide the fame 365 by 4, which is the fumme of the quarters of years, then muft I fet them thus.

       365
       4

Scholar. Sir this doe I underftand, but now how fhould I doe to divide the one by the other?

Master. You muft beginne with the laft Figure next the left hand, and fee how many times the firft Figure of the Divifor may bee taken out of the laft Figure of the other Number, and that fhall you note within a crooked line toward your right hand: As for example, I would divide 365 by 28, then fet I thofe two fummes thus.

       365 (
       28

And I looke how many times I may finde 2 (which is the laft figure of the Divifor) in 3, (which is the laft of the number to be divided) and confidering that I can take 2 out of 3 but once, I make a crooked line at the right hand of the numbers, and within it I fet 1, and that is called the Quotient number, as I tould you.

Quotient numbers.

Then becaufe that when 2 is taken out of 3, there remaineth 1, I muft write that 1 over 3; & deface or cancell the 3 & the 2, then will the figures ftand thus.

       1
       365 (1
       28

Then

Then come I to the next figure of the Diuiſor, and take it likewiſe ſo many times out of the figures that be ouer it, and looke what doth remaine, that I muſt write ouer them, and cancell them, as in this example.

Therefore now do I take once 8 out of 16, and there remaineth 8, which I muſt ſet ouer the 6, and cancell or croſſe out the 16, and the 8 of the Diuiſor: and then will the figures ſtand   18
thus. And ſo I haue once wrought.   365 (1

Scholar. So I perceiue that you take   28
the nether figure, not only out of the other that is right ouer him, but out of that with the other alſo that remaineth before, and are written toward the left hand.

Maſter. So muſt you doe: for you muſt ſo take the Diuiſor out of the ouer number, that there remaine not ouer it ſo great a ſumme as it ſelfe is: for then were your worke in vaine.

But yet againe here muſt you mark, that when you ſeeke how many times the laſt figure of the Diuiſor may be found in the number ouer him, that you looke alſo whether you may as often finde all the figures following in thoſe that are aboue them (conſidering all the remainers, if there be any) if not, take your Quotient leſſe by one, and then proue againe, and ſo till you finde a meet Quotient: and by that meet Quotient muſt you alwayes multiply your Diuiſor, and ſet the product vnder your Diuiſor, ſo that the firſt figure ſtand vnder the firſt figure of your Diuiſor, and the ſecond vnder the ſecond, and ſo forth; and then ſubtract that product from the number to be diuided that ſtandeth di-
rectl

redly over it, as you have séen me doe.

When you have thus wrought once, then must you begin againe, and write your Divisor anew, néerer toward the right hand by one   18
place, as in this example, you shall set   365 (13
2 under 8, and 8 under 5, thus.   288

Then (as before) séeke how many   2
times you may take your Divisor out of the number over him now.

Scholar. That may I doe here 4 times.

Master. Truth it is that you may finde 2 foure times in 8 : but then marke whether you can finde the figure following so many times in the other that is over him. Can you finde 8 foure times in 5 ?

Scholar. No, neither yet once.

Master, Therefore take 2 out of 8 once lesse.

Scholar. That is thrée times.

Master. Well then, 3 times 2 make 6 : if I take 6 out of 8, there remaineth 2 : which 2 with the five following make 25, in which summe I finde 8 thrée times also, and therefore I   2
take 3 as a true Quotient, and write it   18
within the crooked line of the Quotient   365 (13
before the 1, thus :   288

Then say I, 3 times 2 make 6, then   2
6 out of 8 resteth 2, therefore I cancell the 8, and write over it the 2 that doth remain, thus :

Then doe I take 8 as many times out of 25, saying, 3 times 8 make 24, and if I take 24 out of 25, there remaineth 1, so then I cancell 25 and 8, and over the 5 set 1 thus :

Or you might (after you find 3 to be a fit (Quo-

Mark how to consider this kinde of Remainder.

tient) ſtraight way haue multiplied   2
the whole Diuiſor 28, by that at once: 281
which giueth 84, which being ſet vnder 368 (13
28, and duely ſubtracted from 85, of the 288
number diuided, giueth 1, the remainer 32
of the whole diuiſion, as before you had. Worke
which way you liſt, here you may ſee alſo the form.

     And now haue I done with the diuiding; for
I cannot finde my Diuiſor 28 any more in the
ouer ſumme.

    Scolar. No; except you would part the 1 that
remaineth into 28 parts.

    Maſter. That is well ſaid, and ſo muſt we doe
in ſuch caſes, when there remaineth any thing:
but I will let that paſſe now, and will make you
perfect in diuiſion of whole numbers, and will
hereafter teach you particularly of broken numbers
called Fractions. Now if you doe perceiue the order of diuiſion, then doe you diuide this ſumme
136280 by 452.

    Scholar. Firſt I ſet downe the number that
ſhould be diuided; then doe I ſet the Diuiſor vn-
der the laſt figure of the ouer number. 136280
Then will it be thus.   452

    Maſter. Can you take the laſt of your diuiſor
(which is 4) out of one which is the laſt of the
ouer number?

    Scholar. I had forgotten, becauſe the laſt of
the diuiſor cannot be taken out of the laſt of the
ouer number, in ſo much as it is the greater, therefore muſt I ſet the diuiſor one place 136280
more forward toward the right hand, 452
thus.

                                An

And then muſt I looke how often I may finde the laſt figure of the diviſor (that is 4) in 13, which I may doe 3 times, therefore doe I ſay, 3 times 4 is 12, which I take out of 13, and there remaineth 1. Then doe I make at the right hand of my ſummes a crooked line, and write before it my Quotient 3, and I cancell 13 and 4, and over the 3 I ſet the 1 that re-

maineth, and then the figures ſtand thus.

$$\begin{array}{l} \quad\quad 1 \\ 136280\ (3 \\ 4\cancel{5}\cancel{2} \end{array}$$

Then I multiply the ſame Quotient into every figure of the Diviſor, and withdraw the ſumme that amounteth out of the numbers over them, as firſt I ſay, 3 times 5 makes 15, which I take from 16, and there reſteth 1, I cancell therefore 16 and 5, and write over the 6 that 1 that remainth, thus.

$$\begin{array}{l} \quad\quad 11 \\ 136280\ (3 \\ 452 \end{array}$$

Then doe I ſay likewiſe, 3 times 2 make 6, which I take out of 12 and there reſteth 6, there-ore I cancell the 12 and the 2 over, and then I write the 6 that remainth, thus.

$$\begin{array}{l} \quad\quad 16 \\ 136280\ (3 \\ 452 \end{array}$$

Then ſhould I ſet forward the diviſor into the next place toward the right hand thus.

$$\begin{array}{l} \quad\quad 16 \\ 136280\ (3 \\ 4522 \\ \quad 45 \end{array}$$

Maſter. But you may ſée that over the 4 is no figure, there-ore I muſt ſet the diviſor yet forwarder by another place.

And marke, whenſoever it chanceth ſo, that you ſhould ſet forward the diviſor, and that it can-ot ſtand here, becauſe there is no number over the

the laſt place, oʒ if there be any, it is leſſer then the laſt figure of the divifor then muſt you remove the divifor, yet once again : and becauſe that his firſt place of remoding ſerved not to ſubtract him ſo much as once, therefoʒe you ſhall wʒite in the Quotient a Cypher, and if you ſhould by chance nœd to do ſo oft times, foʒ every time wʒite a Cypher in the Quotient. The reaſon of this will I ſhew you hereafter.

Scholar. Then muſt I ſet my ſums thus.

And becauſe I remoded the divifor, ſo that I overſkipped one place, I muſt wʒite a Cypher in the Quotient: and then muſt I ſœke a new Quotient, as in this example I muſt ſay, How many times 4 is there in 6? (and ſith it can be but once) therefoʒe doe I wʒite 1 in the Quo-tient: and then ſay I, 1 time 4 taken out of 6, remaineth 2, I cancell the 6 and the 4, and wʒite 2 over them, thus.

Then ſay I againe, once 5 out of 28, remaineth 23 : I let the 2 ſtand as it did, and over that 8 I ſet 3, cancelling the 8 and the 5 under it, thus.

Maſter. You might as well have ſaid, once 5 out of 8, and ſo remaineth 3, but now go foʒward.

Scholar. Then once 1 out of 0 cannot be What ſhall I now do?

Maſter Boʒrow of the next number that is behind

hinde (for there is 230) and do as you learned in Subtraction in like case.

Scholar. Then must I borrow 1 out of the 3 comming behinde next, and make that 0 to bée 10 and then take I 2 out of 10, and there resteth 8 : and because I borrowed one of the 3 , I must cancell the 3, and write 2 over it, then doth the figure stand thus.

```
          22
        11638
x 36280(301
        45222
         488
          4
```

Master. Now have you done, and yet remaineth 228, and your Quotient sheweth you, that if you divide 136280 by 452, you shall finde your Divisor in your greater number 301, that is CCC times and once, and 228 remaining.

And in the other example (where I divided 365 p 28, the Quotient was 13, and 1 remained : hereby I knew that in a year (which containeth 65 dayes) there are 13 moneths, reckoning 28 ayes (or 4 wéeks) just to a month, and 1 day ore.

Scholar. Why then do we call a year but 12 oneths ?

Master. Of that at a more convenient time will fully instruct you : but now it is not convenient o intangle your minde with other things then do rectly pertain to your matter. Therefore if you emember what you have heard, you have learned short manner of Division, which I would have ou often to practise, so that you may be perfect in it, and hereafter I will shew you certaine other proper points touching it.

Scholar. Then I pray you tell mée how I shall

examine and try my worke, whether I haue done
well or no, that though no man be by to tell me, yet
I may perceiue it my ſelfe.

    Maſter. Some men, (yea and commonly moſt)
Proof of
Diuiſion: doe try it by the rule of 9, as in all the other kindes,
ſaue that their order is ; Firſt, they caſt away
9 as often as they can out of the Diuiſor, and that
that remaineth they ſet at one ſide of a Croſſe,
as in our firſt example the Diuiſor was
28, from which you may take 9 three
times, and 1 remaineth : which they ſet
by a Croſſe, thus.

    Then they likewiſe examine the Quotient,
(which in our example is 13) and from thence
they caſt away 9 as oft as they can, and the remainder they ſet at the other ſide of the Croſſe, and
then they multiply together thoſe two remainers:
and to it that amounteth they adde the remainer of
the Diuiſion, if there were any, from that whole
ſumme they withdraw 9 as oft as they can,
and the reſt they ſet at the head of the Croſſe, as
in our example, the Quotient is 13. from which
take 9, and there remaineth onely 4, and
therefore muſt you ſet 4 at the other
ſide of the Croſſe, thus :

    Then multiply 4 by 1, and it yeldeth but 4,
thereto adde the remainer of the Diuiſion (which
was 1) and it will be 5, which ſumme doth not
amount to 9, and therefore muſt be ſet
wholly at the head of the Croſſe, as you
ſee heare.

    And this number on the head of the Croſſe is the
firſt proof, to which if you find another like in the
numbe

number that was divided, then you have done well.

Therefore now shall you likewise examine the whole summe that was divided, and take away 9 as often as you can, and that that remaineth, set at the foot of the Crosse : and if it be equall to that in the head of the Crosse, then have you done well, else not.

As in our example the whole summe was 365, which maketh 14, from that take 9, and there resteth 5, which set at the foot of the Crosse, thus.

$$4 \times 1$$
$$5$$

And you shall see that they agree : therefore have you well done.

Now will I likewise examine our second example, where the Divisor was 452, which maketh 11 ; from thence I take 9, and the 2 that remaineth I set at the right side of the Crosse, thus.

$$\times 2$$

Then examine I the Quotient, which was 301, where I finde but onely 4 : that I set at the other side of the Crosse, thus:

$$4 \times 2$$

Then I multiply 4 by 2, and it maketh 8 : to that I adde the remainer of the Division (which was 228, and it maketh 12) and they two make 20, wherein I find twice 9, and 2 remaining : that 2 must I set at the head of the Crosse, thus :

$$2$$
$$4 \times 2$$

Then I examine the whole number to be divided, which was 136280, where I finde twice 9 and 2 remaining, which I set at the foot of the Crosse, thus:

$$2$$
$$4 \times 2$$
$$2$$

And because it doth agree with the figure at the head of the Crosse, I know that the
Division

Division was well wrought.

The Proof of Division more certain by Multiplication.

Master. This is the common proof: Howbeit, the more certain working is by the contrary kind: as to prove Division by Multiplication, thus:

Multiply the quotient by the Divisor, and if the summe that amounteth be equall to the summe that should be divided, then have you well divided: else not.

Howbeit, this must you marke, that if there remained any thing after the Division, that must you adde to the summe that amounteth of the Multiplication. As in our first Example our Quotient was 13, and the Divisor was 28: Now multiply the one by the other, and the summe will be 364: to that if you adde the 1 that remained after the Division; then will it be 365, which was the sum that should be divided: and therefore I know that I have well done.

Scholar. Now will I prove the same in the second example, whose Divisor was 452, and the Quotient 301: these doe I multiply together, and there amounteth 136052: to which if I adde the 228 that remained, then will it be 136280. which was the whole summe to be divided: and therefore I perceive that I have well done.

Master. This is the surest way to examine Division by Multiplication: and contrariwise, the surest proof of Multiplication is by Division.

And therefore (according to my promise) now will I shew you how you may prove Multiplication by Division.

Proof of Multiplication by Division.

When you have ended Multiplication, and would know whether you have well done or not, set the grosse

groſſe ſumme that amounteth of the Multiplication overmoſt, and divide it by the Multiplier: and if the Quotient be the ſame number that ſhould be multiplied, then have you well wrought, elſe not, as in that example where we multiplied 264 by 29, the groſſe ſumme was 7656.

Now if you will know whether that Multiplicatiõ be true, you ſhall divide that 7656 by the multiplier 29, and you ſhall perceive that the Quotient will be 264, and that is a token that you have well wrought.

Scholar. By your patience I will prove that, and firſt ſet downe the groſſe ſumme and the multiplier, not after the rule of Multiplication, but after the rule of Division, for now that number is become the Diviſor, that was before the   7656
Multiplier: I ſhould ſet them therefore   29
thus.

Then ſhall I ſéek how many times 2 in 7, that may be thrée times, & one remaineth: but then may not 9 be found ſo often in 16? therefore muſt I take a leſſer Quotient, that is to ſay 2: then ſay I, twice 2 maketh 4, which I take out of 7, and there remaineth 3, then do I cancell 7 and 2,   3
and over 7 I write 3, and for the Quo-   7656 (2
tient I ſet 2: ſo the figures ſtand thus.   29

Then ſay I further, two times 9 make 18, which
I abate out of 36, and there reſteth 18:   1
then cancell I 3, and over him ſet 1, and   38
likewiſe I cancell 6 and 9, and over   7656 (2
them I ſet 8: ſo that thus ſtand the Fi-   29
gures.

Then I ſet forward the Diviſor by one place, and

and ſéke a new Quotient, that is to ſay, how many times 2 are in 18, which I finde to be 9 times: but then can I not finde 9 ſo many times in 5 : but therefore I take a leſſer Quotient, as to ſay 8 : but yet that is too great : foz if I take 8 times 2 out of 18, there remaineth but 2, ⹂ I cannot finde 8 times 9 in 25 : therefoze yet I take a leſſe Quotient, that is 7, which is alſo too great; foz if I take 7 times 2 out of 18, there reſteth 4, but now I cannot take 7 times 9 out of 45, therefoze yet I

séek a leſſer quotient, as to ſay 6, then ſay I 6 times 2 make 12, that I take out of 18, and there remaineth 6, ſo I cancell 18, and the 2, and wzite 6 over 8, thus :

$$
\begin{array}{r}
16 \\
38 \\
7656 \quad (26 \\
2999 \\
2
\end{array}
$$

Then ſay I foꝛth, 6 times 9 maketh 54, that take I out of 65, and there remaineth 11, and the figures ſtand thus :

$$
\begin{array}{r}
1 \\
16 \\
381 \\
7656 \quad (26 \\
2999 \\
2
\end{array}
$$

Then muſt I ſet foꝛth the Diuiſor again, and séek a new quotient, which will be 4 : foꝛ though I may find 2 in a 11, 5 times, and 1 remain, yet I cannot find 9 ſo often in 6, therefoze I ſet the figures thus :

$$
\begin{array}{r}
1 \\
381 \\
7656 \quad (264 \\
2999 \\
22
\end{array}
$$

And the 4 in the quotient I multiply into the figures of the Diuiſor, ſaying foure times 2 makes 8, which I take out of 11, and there reſts 3, therefoze I cancell the 11, and the 2, and ſet 3 over the firſt place of 11, thus :

$$
\begin{array}{r}
1 \\
161 \\
381 \\
7656 \quad (264 \\
2999 \\
22
\end{array}
$$

And then doe I ſay foꝛth, 4 times 9 maketh

9 maketh 36, which I take from 36, and there remaineth nothing, so that the Quotient of this Division also (where 7656 is divided by 29) is 264: Which doth declare, that if 264 be multiplyed by 29, the summe will be 7656. And thus I perceive now how both Multiplication is proved by Division, and Division also by Multiplication.

Master. Now have I ended the five common kindes of Arithmetick : For (as touching Mediation, Duplation, Triplation, and such other) they are no severall kindes of Arithmetick, but are contained under the other. For Mediation is contained under Division, and is nothing else but dividing by 2 : & so are Duplation and Triplation contained under Multiplication : for Duplation is nothing else but multiplying by 2, and Triplation is multiplying by 3, of which I will onely propose an example, for the Rules you have heard already.

If you would mediate, or divide into 2, this summe 453 1010, you shall set 2 for the Divisor, and worke as you learned before, as thus :

An example of Mediation.

$$453 1010 (\atop 2$$

Then I finde 2 in 4 two times, therefore my Quotient must be 2 : so I cancell 4 and 2, and remove the Divisor forward thus, as the work requireth, and as before in Division hath béen declared.

$$453 1010 (2265505 \atop zzzzzzz$$

Which mediation or division by 2 being finished, you shall have for your Quotient 2265505, which is the halfe of 453 1010, as you may try by Duplation ; for double that Quotient, or multiply it by 2, and the same number will amount.

I

I will no longer tarry about these, ſéeing the are but members of the other kindes. But her now (according to my promiſe) I will teach yo certain eaſie formes both of Multiplication and o Diviſion. And firſt of Multiplication.

If you would therefore multiply any ſumme by 10, you ſhall néed to do no moze but adde a Cypher befoze his firſt place; as foz example, 36 multiplied by 10, make 360.

Likewiſe if you would multiply any ſumme by 100, put two Cyphers at his beginning. So if you would multiply any ſumme by 1000, adde thzé Cyphers to the beginning of it.

Scholar. This doe I well perceive, and alſo the reaſon of it.

Maſter. I will omit all reaſons till our next méeting, when I ſhall tell you the reaſon of all other parts of Arithmetick alſo: and as to our matter now, look, as I have told you, that you both re member it, and alſo often practiſe it.

And now you have learned how to multiply eaſily by 10, 100 1000: and of like manner may you doe with any other of like ſozt.

But now if you will multiply by 20, 30, 40, and ſo foꝛth, oꝛ by 200, 3000, and ſuch like, where there is one Cypher in the firſt place, oꝛ many oꝛderly i the firſt places, you ſhall take away thoſe Cyphers and multiply the ſumme onely by the other figure oꝛ figures (if they be many) and then at the begin ning of the ſumme that amounteth, you ſhall ſet ſo many Cyphers as you toke away.

Example of 2873, which I would multiply by 300. Firſt, I omit the 2 Cyphers from the Multi plier,

plier, and I multiply the summe by thrée onely that is left, and it amounteth to 8619 : before which I put the two Cyphers that I before omitted or tooke away, and then is it 861900. And that is the summe that amounteth when 2873 is multiplied by 300.

Scholar. And if there were two or more figures besside the Cyphers, I must onely take away the Cyphers, and multiply by the other figures, as I learned before : As if I would multiply 93648 by 25000, I should take away the thrée Cyphers, and multiply the same by 25, & then at the beginning of that totall sum should I adde the 3 Cyphers again.

Master. Even so : but if it chance the number that should bée multiplied, or both the summes, as well the number that should be multiplied as the mnltiplier, to have Cyphers in their first places, evermore omit the Cyphers and work by the rest. But remember to restore as many Cyphers to the amountting summe as you bated before. As in this example : 30200 shall bée multiplyed by 206, I shall onely take away two Cyphers from the greater number, and then multiply 302 by 206, and afterward adde the two Cyphers again. But if I would multiply the same 30200 by 2060 I shall not onely take away the two Cyphers from the number that should be multiplied, but also I may take away the one Cypher from the Multiplier, and then must I adde thrée Cyphers to the summe that amounteth : but take héed that you take away no Cypher that commeth after any signifying figure, as in the last example, you must not take away that in the fourth place of the higher number,
neither

neither that in the third place of the multiplier: howbeit yet thus you may doe: If one Cypher or more come in the last of your summes, you may multiply the other figures, and overskip them: but so that you give every figure his due place: as thus, I will multiply 3026 by 2004, therefore I set them thus.

3026
2004
12104
52

And thus doe I multiply them. First 4 times 6 make 24, I set 4 under the first place, and keepe the two still in my minde. Then say I againe, 4 times 2 make 8, and the 2 that is in my minde maketh 10, I set downe the Cypher 0, and keepe the Article 1 in my minde: Then 4 times 0 is 0, and the 1 in my mind maketh 1, I set downe the figure 1, and say again, 4 times 3 is 12, I set downe 2, and keeping the 1 still in my mind (having no more places of the upper number to multiply it withall) I put it downe next 2 in the fifth place.

But now when I come to the next place (being a Cypher 0) I let it go, because it multiplieth nothing, and likewise the second Cypher.

But then when I come to the 2, & multiply it into the 6 of the over number, you must take heed (according as I taught you in Multiplication) that the first number amounting of the multiplication be set right under the multiplier, and the other orderly toward the left hand, according as you may see in this example, which being finished, with the addition thereof gathered together, will stand as in this Example sheweth,

302
200
1210
6052
60641

Whi

Which is indéed wrought ſo much 3026
the ſonner and ſhorter by overskipp- 2004
ing of the two Cyphers: which other- ——————
wiſe, (if the ſame example were 12104
wrought at length) it would have 0000
had two workings more, as by the 0000
ſame example here alſo ſet down 6052
doth appear. ——————

Scholar. Sir I thanke you, for I 6064104
ſée great eaſe in this way of Multiplication: and (if
you can ſhew me ſuch like in Diviſion) you ſhall
greatly further me.

Maſter. Yes, I will teach you ſome eaſie wayes Eaſy forms
in Diviſion, alſo, and firſt this: If you would of diviſion
divide any ſumme by 10, you ſhall onely with
your pen make a ſquare line betwéen the firſt figure
of your ſumme and the ſecond, and then have you
done: for the whole number that followeth the
line, ſtandeth for the Quotient, and the figure that
is before the line, is the remainer: As for example,
3648 divided by 10. Where 364 is   364 |8
the Quotient, and betokeneth that
ſo many times are 10, in 3648; and the 8 after the
line is the remainer, which cannot be divided into
10, but by breaking it into Fractions, wherewith
I will not meddle yet.

And ſo likewiſe if you would divide any ſumme
by 100, with your Pen you ſhall cut away the two
firſt figures, and if you would divide by 1000, you
muſt cut away the thrée firſt figures, and ſo of any
other diviſor, whoſe laſt figure is 1, and the other
Cyphers; looke how many Cyphers the diviſor
hath; and ſo many figures at the beginning ſhall
                    H   you

you cut away with the square line, & they stand alwayes for the remainer, because they are lesse the
the divisor, and cannot be divided by it, and the o-
ther figures that are behinde the line stand for th
Quotient.

But now if your divisor have any other figure
in his last place then 1, and in all his other places
have Cyphers, looke how many Cyphers they be, cut
away so many of the first figures of the number
that should be divided, and divide the rest that
followeth the line by that figure that is in the last
place, as if it were the whole divisor.

Example of 64284, which I would divide by
300, here must I cut away the two first figures,
(for so many cyphers my divisor hath) & must divide
the rest by 3, which is the figure in the last place
of the divisor. First, therefore I part
away the two first figures, and the　　　642|84 (2
summe standeth thus :　　　　　　　　　　3 |00

Then do I divide 642 by 3, and the Quotient
is 214: for in 6 I finde twice 3, and in 4 once, &
1 remaining, which 1 with the 2 next before, doth
make 12, wherein I finde 3 foure times : And this
is a ready way to turn shillings into pounds : for
sith one pound doth contain 20 shillings, I must
divide the whole number of shillings by 20. Therefore easily do I see that my divisor hath one cypher, and therefore I cut away one figure from the
beginning of the whole summe of shillings, and
then I do mediate or divide by the 2 other figure
or summe that followeth.

Scholar. I will put an example.

If you would divide 64287 shillings, by 20 :
that

that is to say, If I would turn so many shillings
into pounds, I must cut away the first figure, that
is 7, and divide the rest, that is 6428 by 2, so shall
the Quotient be 3214, whereby I know that
64287 shillings make 3214 pounds, and 7 shillings
remaining.

Master. Now prove by Multiplication whether
you have well done or no.

Scholar. The Quotient is 3214, which I do
multiply, by the divisor 2, and it doth amount to
6428.

Master. Hereby you may perceive not onely
that you have well done, but also how by Division
you may turn shillings easily into pounds : and contrariwise by Multiplication you may turn pounds
into shillings.

But here shall you see amongst divers men divers
forms of such division : but if you marke what I
have told you, you shall perceive easily all the
wayes. For some men doe not cut away so many
of the first figures of the summe that they would
divide, as there are Cyphers in the first place of the
divisor : but they set all their Cyphers orderly under the first places of the Number that they would
divide : and then with the other figure or figures,
if there be many) they divide the rest of their
summe.

Example. If they would divide 725931   725931
by 3400, they doe set their summs thus :   34   00

And then do they divide orderly till they come to
the Cyphers : for there they stay and end their
like, as in this example.

They seek how often 3 may be found in 7, which

H 2

is two times, and 1 remaining :
therefoze they set 2 in the quotient, & cancell 3 & 7, & over 7 they
set the 1 that remainsth. thus :
Then doe J goe foꝛth, saying, two
times 4 maketh 8, which thꝑ take
out of 12, & there remaineth 4. thus:

Then remove they the divisor
foꝛward, & seeke how often 3 may
bee found in 4, which is but once,
and 1 remaineth, then set they 1
in the Quotient, & cancell 3 & 4, &
over them they set that 1, thus :

Then take they once 4 out of
15, & there resteth 11. Oꝛ else moꝛe
easily : Take once 4 out of 5, and
there resteth 1 : so they cancell the
4 and 5, and set 1 over them, thus:

Then set they foꝛth the divisor
again, and seek how many times
3 are in 11, which they finde three
times, and 2 remaining: so they
set 3 in the Quotient, and cancell
11 & 3, & over them set 2, thus :

Then doe they multiply 4 by 3,
which maketh 12, that withdꝛaw
they out of 29 & there resteth 17,
of which the 7 must bee set over
the 9, and the 1 over the 2 thus :
And now are the two Cyphers next ensuing, so that
the divisor can no moꝛe be set foꝛward, & therefore
is the division ended, & the remainer is 1731.   Now
the Quotient which is 213, doth declare that if you
                                                        divide

```
                1
            725931 (2
            34  00

               x4
            725931 (2
            34  00


                1
               x4
            725931 (21
            344 00
            3

                1
               x41
            725931 (21
            344 00
            3

               x2
               x41
            725931 (213
            344400
            33

                1
               x2
               x417
            725931 (213
            344400
            33
```

diuide 725931 by 3400, you shall find it therein 213 times, & there remaineth 1731: so shall you find it, if you worke as I taught you, by cutting away the two first figures, because of the two Cyphers. But this must you mark (as you may perceiue by this last example) that if there be left any other Remainer in the sum that was behind the square line that the Remainer must be set to the latter end of the first remainer, which was cut away with the square line : as if you would diuide 725931 by 3400, after the forme that I taught you, then would your summes appeare, thus :

```
        1
       12
      1417
    725931 (213
     3444
       33
```

So that 17 which remaineth after the line, must be set to the 31 (that was cut away with the line) in higher places, as you see here : where that 17 with the 31, do make 1731.

Scholar. Sir, is there no other forme of division in practise but this ?

M. Yes verily, there are other formes in practise, but because I loue breuity, I will declare onely one, which I first learned of, and is practised by that worthy Mathematician, my ancient & especiall louing friend, Master Henry Bridges wherein not any one figure is defaced or cancelled. As if I should diuide 72 by 6, first place them thus.   6)72

Then if you please you may write the divisor in a loose paper that it may more easily without cancelling or defacing of the worke be applied to, and removed from the dividend at pleasure ; then apply your divisor 6 to 7, the first figure of the dividend

Write the Divisor in a loose paper, to remove at pleasure.

Note.

dividend, and inquire how oft it may be had in 7, and feeing 6 is but once in 7, fet 1 in the Quotient line, thus :　　　　　　　　6)72(1

Then multiply the divifor 6, by the quotient 1, and fet the Product 6 under 7, thus :　　6)72(1

Then draw a line under 6, and fubduct　6
6 out of 7, fetting the remainer 1 under　6)72(1
6, thus :　　　　　　　　　　　　　6
　　　　　　　　　　　　　　　　　　1

Then bring down the next figure of the dividend, & fet it with the Remainer 1 under the line, thus :　　　　　　　6)72(1
　　　　　　　　　　　　　　　　6
And bring the moveable divifor 6 un-　12
der the 2, and as before enquire how oft
6 is in 12, and finding it to bee twice　6)72(12
in 12, fet 2 in the Quotient, thus :　　6

And multiply 6 by that new Quo-　　────
tient 2, fetting the Product 12 under the　12
other 12, and fubducting it out of the　12
upper number, there refteth nothing.　　0

And fince the unites of this Product do ftand under the unites of the dividend, the divifion is ended
otherwife you fhould proceed as before, bringing
down the next figure ; removing the divifor, dividing, multiplying, fubducting, &c.

Schol. This is very eafie; but if there be great
numbers propounded, is the operation the fame

Mafter. If the numbers be never fo great, the
worke is the fame without any difference, as fhall
appear by this example.

Divide 7890 by 33.　　　　　　　33)7890(
Firft fet them thus : then bring the divifor under 78, and fee how oft it is there found, which
twice, and therefore fet 2 in the Quotient, by whi
　　　　　　　　　　　　　　　　　　multi

multiply the divisor 33, and set the Product 66 under 78, and subduct it out of it thus.

Then bring the next figure 9 down, and set it with the Remainer 12, it maketh 129, and removing the divisor 33 thereto, enquire how often 33 is contained in 129, & I finde it but thrice, (though at the first it made a shew of more) therefore set 3 in the Quotient, and multiplying 33 by 3, set the product under 129, subducting that product out of the number above, and proceed as before.

$$33)7890(239\tfrac{3}{33}$$
$$66$$
$$129$$
$$99$$
$$\overline{300}$$
$$297$$
$$\overline{3}$$

Then shall you finde the Divisor 9 times in the Remainer, therefore setting 9 in the Quotient, multiply, and subduct as before, and at the last you shall find onely 3 remaining, which must be set above a line after the Quotient, and the Divisor under, as above appeareth.

Scholar. Is there no more difficulty in the whole Rule ?

Master. Not any, although your Number be never so great, as before I have said.

¶ And here will I make an end of Division, (saving that I doe request you to exercise your selfe well herein by many summes, till you have attained some expertnesse therein.)

For the reasons & conclusions thereof are so many, and so available for all sorts of men whatsoever; that if I should speak of the infinite uses thereof, I should rather lack words then matter. And therefore recommending it to your judgement hereafter, upon your further travell into the Art, I will here end

H 4   this

this Treatise, repꝛesenting unto you one example, or simple question of Division and Multiplication, in stead of many, which is this.

A question of shooting in Ordinance.

There are foure brasse Peeces: The first of them at a shot spendeth 9 pounds of powder, the second spendeth 5 pounds, the third 4 pounds, and the fourth 2 pounds. They are all appointed against the battery of a Hold and there is allowed by the Master Gunner 700 pounds of powder to be spent by these foure Peeces, in this assault. The question is two-fold. The first how many shot each Peece shall iustly make about with this 700 pounds of powder? And lastly, how many pounds of powder ought iustly to be allowed to each Peece for his true proportion?

Scholar. Why Sir, you make me smile, to beare mee in hand, that these two demands may be simply resolved by Multiplication and Division.

Master. Truely that they may, and that you may by and by worke your selfe with a little labour: First adde together their quantities of powder. that is 9 pounds, 5 pounds, 4 pounds, and 2 pounds, all which make 20: Divide the 700 pounds of powder by that 20, and your Quotient giveth 35, as here appeareth, which sheweth for most certainty that they shall make iust 35 shootes about.

$$20z(35$$
$$700$$

Scholar. Sir, all this have I done, and I see it is so, but whether it be true or not, I cannot tell.

Master. To try the truth of the same, multipl the first peece that spends 9 pounds by 35, and you shall see his allowance, which is 315 pounds o powder. Multiply also the second peece tha spent

spends 5 pounds by 35, & you shall finde 175 pounds
his allowance : then 4 by 35, and you shall finde
140 pounds his allowance. Lastly, multiply 2 by   315
35 and you shall finde 70 pounds his al-
lowance. All which four particular sums   175
you shall adde together by Addition, as   140
here appeareth, and it maketh just 700   70
pounds, and so is the question truely ab-   ———
solved.   700

Scholar. Truely Sir, these excellent conclusions doe wonderfully moze and moze make me in
love with the Art.

Master. It is an Art, that the further you travell,the moze you thirst to goe on forward. Such a
Fountaine, that the moze you draw, the moze it
springs : and to speake absolutely, in a word (excepting the study of Divinity, which is   the salvati-
on of our Soules)there is no study in the world com-
parable to this,for delight in wonderfull and godly
exercise ;for the skill hereof is well known imme-
diately to have flowed from the wisdome of God
into the heart of man, whom he hath created the
chiefe image and instrument of his praise & glozy.

Scholar. The desire of knowledge doth greatly incourage me to be studious herein : and therefoze I pray you cease not to instruct me further
in the use hereof.

Master. With a good will.   And now therefore
foz the further use of these two latter,that is, Multiplication and Division, I will briefly shew you the
feat of Reduction.

Reduction.

Reduction.

 Eduction is by which all summes of groſſe Denomination *may be turned into* summes of more ſubtile Denomination. **And contrariwiſe all** summes of ſub- tile Denomination **may be brought to** summes of groſſer Denomination.

Scholar. **What call you** groſſe Denomination, **and** ſubtile Denomination ?

Maſter. **That I call a** groſſe Denomination, **which doth containe under it many other ſubtilet oz ſmaller : as a pound (in reſpect to ſhillings) is a** groſſe Denomination : **foz it is greater then ſhil- lings, and containeth many of them. And ſhillings**

(in compariſon to pounds) are a ſubtile Denomina- tion, **foz becauſe they are leſſer then** pounds, **¶ ma- ny of them are contained in one of the other : and ſo likewiſe of other things : whatſoever thing is compared to other, if it be greater and containeth many of them, it is a** groſſe Denomination, **but if it be leſſer (ſo that many of them are in the other) then are they called the** ſubtile Denominations : **whereby you may perceive that one** Denomi- nation **may be called a** groſſe Denomination, **and alſo a** ſubtile (**that is to ſay a great, and ſmall**) **in diverſe compariſons. Foz ſhillings compared to pounds, are a** ſubtile oz ſmall Denomination : **but compared to pence they are a** groſſe, **or great De- nomination.**

Scholar, **Now I underſtand the name, I pray you teach me the uſe.**

Maſter. **The uſe is eaſily learned, if you re- member**

member what you have learned before. For, if you will reduce any summe of a grosse Denomination into a summe of a smaller or subtiler Denomination, you must consider how many of that subtiler Denomination do make one of the grosser Denomination, and by that number or numerator doe ye multiply the summe. As if you would reduce 30 pounds into shillings, you must consider that in a pound are included 20 shillings, therefore multiply the one 20 by the other 20, and there will amount 400, whereby you may know that in 20 pounds are contained 400 shillings. Likewise, if you would reduce 30 shillings into pence, considering that in a shilling are 12 pence, you must multiply 30 by 12, and it will be 360, whereby you may finde that in 30 shillings are contained 360 pence. And thus may you reduce any grosse Denomination into a more subtiler, by Multiplication, if you know how many of the lesser doe make the greater : of which thing I will anon give you a brief Table for the most accustomed kindes of Money, Weights, Measures, and Time, and such like : whereby you may know how often each subtile Denomination is contained in the grosser, when you shall néed it for the foresaid kinde of Reduction. And also the same shall serve you, if you would reduce any summe of a subtiler Denomination, into a summe of a greater Denomination. For in such Reduction you must consider (as in the other form) how many of the smaller doe make the greater: and by that number you must divide the other summe, and the Quotient will declare how many of the greater Denomination are comprehended in that summe:

as

To reduce grosse denomination to subtile.

as for example; If you would know how many shillings are contained in 3240 d. consider that 12 pence doe make 1 s. you must divide that 3240 by 12, and your Quotient will be 270, whereby you may know that so many shillings are in 3240 d. But if you would know further how many pounds are in these 270 shillings seeing that every pound containeth 20 shillings : divide that 270 by 20, and it will be 13 and 10 remaining, whereby you may know, that in 3240 d. (or 270 shillings) are 13 pounds and 10 shillings. For evermore the Remainer must be named by the name, or Denomination of that summe that was divided, which in this place were shillings. And thus may you do with any other kinds of Denominations.

¶ Wherefore, to the intent you may have certain light or knowledge in most common Coynes, Weights and Measures, (which is the chief and principallest thing in trafficke to be known) I have in each Reduction, as they come in order, set downe certain instructions incident thereunto. And first I have hereunto added this Table, wherein is comprehended, not onely our currant and common coynes, but also the most part of the usuall coynes of Christendome, with their just weights and value currant in the Realme of England, intending at the latter end of my Addition to this Book, to write of the ordinary Money used in divers places, and their common values currant for traffick, with the manner of their exchanges from place to place, &c.

A

A Table of the names, and now valuation of the most usuall Gold-coyns throughout Christendom, with their severall weight of pence and Grains : and what they are worth of currant English money, this present year 1630.

The names & titles of the Golde.	The weight in Pence. Graines.		The value in Shil. Pence.	
Great Soverain,	10	0	33	0
Double Sover. K. H.	8	1	22	0
Double Sov. of Q.E.	7	7	22	0
Royall.	4	23	16	6
Half Royall.	2	1d.	8	3
Old Noble.	4	6	14	8
Half Noble.	2	3	7	4
Angell.	3	8	11	0
Half Angell .	1	16	5	6
Salute.	2	5	6	11 ob.
2 parts of Salute.	1	11	4	7
George Noble.	3	0	9	9 ob.
Half George Noble.	1	11	4	10
First Crown K. H.	2	9	6	1 10.
Base Crown K. H.	2	0	5	6
Sover. K. H. best.	2	14	11	8 ob. q
Soverain K. H.	4	0	11	0
Edward Sover:	3	15 d.	11	0
Elizabeth Sover.	3	15 d.	11	0
Elizabeth Crown.	1	9	5	6
Half Crown.	0	19	2	9
Unite.	0	12	22	0
Double Crown.	3	6	11	0
Britain Crown.	1	1	5	6
Thistle Crown.	1	7	4	4 ob.q

Half

The names & titles of the Gold.	The weight in Pence. Grains.		The value in Shil. Pence.	
Half Crown.	0	19½	2	9
Crosse Dagger.	3	6 d.	11	0
Half Crosse Dagger.	1	15	5	6
Rose Royall.	0	21	33	0
Spur Royall.	4	10d.	16	0
The Ang'll.	2	23 d.	11	0
Half Argell.	1	11 d.	5	6

All the severall pieces of gold heretofore mentioned, are set down according to their valuation by the Kings Majesties Proclamation for Gold, Dated the 23 of November, 1611.

A table of forain Gold coyn, according to their ancient valuation and severall weight in Pence and Grains.

The names & titles of the Gold.	The weight in Pence. Graines.		The value in Shil. Pence.	
Vnicorn of Scot.	2	10	6	0
Scottish Crown.	2	5	6	0
French Noble.	4	16	13	4
All sorts of French Crowns.	2	5	6	0
Flanders Riders.	2	6	6	6
Gelders Riders.	2	2	3	6
Philips Royall.	2	10	10	0
Philips Corwne.	2	5	5	0
Collen Gilden.	2	2	4	8
New And. Gild.	2	2	5	0

Flanders

Flanders Noble.	4	10	12	0
Half Flan. Noble.	2	6	6	0
Flan. Angel beſt.	3	6	2	0
Flan. Royallorke.	3	10	10	0
Carolus Gilden.	0	12	3	6
Flanders Royall.	2	6	5	0
Saxon Gilden.	2	2	4	8
Flanders Crowne.	2	5	6	0
Philips Gilden.	2	3	4	2
Half Phil. Gilden	1	1	2	1
Golden Lion.	2	16	7	8
3 parts of goldē Liō	0	21	2	5
⅓ parts of gol. Lion	1	19	4	11
Davids Gilden.	2	2	4	0
Horne Gilden.	1	12	4	11
Old under Gilden.	2	3	4	10
Cruſa.long Croſſe.	2	6	6	0
Cruſa. ſhort eroſſe.	2	6	6	2
Milreys.	4	20	1	4
Half Milreys.	2	10	6	8
Portague 1 ounce.	2	16	68	0
Golden Caſtile.	2	23	8	10
Ducket of Aragon.	2	6	6	6
Hungary Ducket.	2	7	6	4
Double Piſtolet.	4	9	11	8
Single Piſtolet.	2	4 d.	5	10
Ducket of Floren.	2	5	6	4
Double Ducket	4	11	13	0
Single Ducket.	2	6	6	6
Double duc. of Rōe	4	13	12	8

It is to be underſtood (gentle Reader) that whereas in theſe Tables, the weight is called by the name of a penny

*penny, it is not ment a penny of silver money, but a penny
of Gold-smiths weight, which containeth 24 Barley-
corns. Concerning which see Troy weight in* folio 133.

*So if a man have not the weight wherewith to weigh
any peece of gold, he may do it with barley-corns, being
dry, and as it is said,* folio 133.

The prices of Gold which the bringers in of forrain Gold shall receive at the Mint, according to the Kings Majesties Proclamation Dated the 14 of May, Anno 1612.

FOr an ounce of *French* crowns being 22 Karacts fine. ——— } 3 li. 6 s.

For every ounce of *Spanish* Piftolers, being 21 Karacts, 3 grains and a half fine. ——— } 3 li. 6 s.

For Duckets of *Spaine*, being 23 Karacts, 1 grain fine at least the ounce. ——— } 3 li. 8s. 8d.

For Milreas Crusado long crosse. Crusado short crosse. the ounce. ——— } 3 li. 6s. 2d.

For *Hungary* Duckets being 23 Karacts, 1 grain fine at least the ounce. ——— } 3 li. 9s. 2d.

For the Checkeen of *Venice*, being 23 Karacts, 1 graine fine at least the ounce. ——— } 3 li. 10s.

For *Barbary* Gold being 23 Karacts & di. grain fine, at the least the ounce. } 3 li. 9s.

¶ *And if the said Barbary Gold be of lesse finenes, abatement to be made according to the rate.*

F

For Sultaines being 23 Karects, 1 Graine fine at least the ounce. —— } 3 li. 8s. 8d.

For all other Gold, being 22 Karects fine the ounce. —— } 3 li. 6 s.

¶ *And being finer, a greater price according to that rate, and being courser, a less, so that the bringer in supply the lesse fine with the more fine, in such sort, that in the totall it makes good the same rate of 22 Karects fine.*

The Price of Silver, which the bringers in of forain Silver shall receive at the Mint, according to the Kings Majesties aforesaid *Proclamation.*

FOr the ounce of *Spanish* Silver money of *Sevill.* —— } 5 s.

For the ounce of *Mexico* money. —— } 4 s. 10 d.

For Ingots of Silver, being 11 ounces, 2 d. weight fine according to the Standard of *England*, the ounce. —— } 5. s.

¶ *And for other Silver of more finenesse, a better price according to that rate, and for courser a lesse : so that the bringer in supply the lesse fine with the more fine, in such sort, that in the totall it makes good the said rate of 11 ounces, 2 pence weight fine, according to the Standard of England.*

Of Silver Coynes currant in this Realme.

The *Edward* crown of 5 s.

The *Edward* half-crown of 2 s. 6 d.

The *Edward* shilling, half shilling, and the three pence.

Philip and *Maries* shilling, and half shilling.

The *Mary* groat, and *Mary* two pence.

Queen *Elizabeihs* shilling 9 d. 6 d. 4 d. 3 d. 2 d. 1 d. three farthings, and half penny.

HEre would I now expresse the values of sundry other Coynes of divers Countries, but for

I   three

three causes I now refraine.   The first and chiefest
is because they are not currant by the statutes o
this Realme. Another cause is, by reason they ar
so uncertaine, that they be never long at one rate
And againe, they are so different in so many places, that it were matter enough for a great Book
to speak sufficiently of them all. Howbeit, because
you shall not be altogether ignorant of them, I
will shew you the values of some that are most
in use, and first of France.

French
Coynes.

The most common Money are Deniers, Soulx,
and Frankes: 12 Deniers make 1 shilling, 20 Soulx
make 1 Frank: so that as you sée these thrée kindes
are like in the rate to pence, shillings and pound
with us ; but that this is the difference, that their
Denier is but the ninth part of our penny, & so their
Soulx (commonly called Souses) goe 9 to our shilling, and 9 of their Franks to an English pound o
money.   So that thrée of their Franks make
Noble. And by those thrée you may practise how t
reduce French money into English money, according
as I have set forth here following.

2160 ⎫   ⎧240 d. or 20 s.
3240 ⎬ Deniers make ⎨360 d. or 30 s.
8352 ⎭   ⎩928 d. or 3 li. 17 s. 4. d.
2160 Soulx make 240 shillings.   And so of othe
in like rate.   As for the rest of their Coynes I o
mit them till hereafter that you have some under
standing in broken numbers.

Flanders
Coynes

But now as for the Coynes of Flanders, they
so changeable, that you must know them fro
time to time, else you cannot reduce them into o
money certainly : but yet that you have an e
amp

ample of their money to exercise you withall, you shall take those that be most common : as Stivers both single and double, Groates Flemish, Carolus and Gyldens. A Flemish Groat is a little above 3 farthings English. A single Stiver is 1d. ob. q. half farthing. The double Stiver Carolus is 4 d. ob. halfe farthing. Then there is also the Carolus Gylden which is worth 10 Stivers. And the Flemish Noble is worth 3 Carolus Gyldens, and 12 Stivers.

So that if you would convert Flemish money or any other kinde of money whatsoever it be, justly into Sterling, you must reduce it first into the smallest part of English money that is in that Coyne. As for example : If I would reduce 368 double Stivers into English money (considering that a double Stiver containeth 3 d. farthing) you shall first looke how many farthings be in the double Stiver, and you shall finde them 13, therefore multiply the summe of the Stivers by 13, and then have you their value in farthings, which is 4784. Now if you divide that by 4, then there will appear the number of pence : but better it were to divide it by 48 (for so many farthings are in one shilling) and then will the Quotient declare the summe of 4 li. 19 s. 8 d.

Likewise if you would reduce any summe of single Stivers into English money, you must multiply the summe first by 13, and then have you reduced them into a certaine summe, that is to wit, half-farthings, which summe if you divide by 8, then will amount the summe of pence : or if you divide it by 96, the summe of shillings will appear.

I 2   But

But marke this in all Division : when ye do reduce to bring one Denomination into another, if
there be any Remainer after the Division, that
must be named by the Denomination of the grosse
sum that was divided. As for example, I would
bring 254 farthings into pence, therefore I divide
that 254 by 4 (for so many farthings make a penny) and the Quotient is 63, which is the summe
Note well. of the pence, and then remaineth yet 2, which are
farthings still, as one may prove by dividing. And
this must be marked in all Division, namely, when
it is done for Reduction.

¶ Touching Danks Mony, they have their Soulx,
Danks whereof 10 is a Liver which is 2 shillings sterling.
Money. They have also their Grash whereof 80 make a
Gilden, which is 4 shillings sterling. They have also
Dollors, & their common or old Dollar is 35 Grash.
New Dollors they have which be divers, some va-
lued at 24 Grash, some at 26, and some at 30.
And thus much I thought good to adde to the Au-
thor, touching Danks Money.

Spanish Concerning Spanish Money, whereof the most
Money. common are Cornadoes, Marveides, Marveide, 4
Marveides make a Ryall, and 11 Ryalls make on
Ducket, so the Ducket containeth 374 Marveides
which is about 5 shillings 10 pence sterling. Therefore if you would convert 124 li. 5 s. sterling int
Duckets, consider that pence is the last valu
or Denomination named in this question : therefor
reduce 124 li 5 s into pence, and it maketh 2982
pence : which if you divide by pence that a Ducke
is worth, (which is 70) you shall have for you
Quotient 426 Duckets your desire.

I

In Venice they have Bettes, Souldyes, Lieure 5 Bettes make an English penny, 10 Bettes a shilling, which is 2 Souldyes, and 20 Souldyes a Lieure of Venice which is a pound sterling.

Thus much have I said of Money : Now will I shew you in the like sort the distinction of weights.

After a Statute made anno 11 *H.* 7. *there ought to be but one sort of weight, as* 24 *Barley-cornes dry and taken out of the midst of the Ear, do make a penny weight,* 20 *of these penny weights make an Ounce;* 12 *Ounces a pound of Troy weight, by which is weighed Bread, Gold, Silver, Pearle, Silke, and such like.* But commonly there is used another weight called Haberdupoise ; in which 16 Ounces make a pound. Therefore when you would reduce Ounces into Pounds, you must consider whether your weight be Troy weight or Haberdupoise : and if it be Troy weight you must divide your Ounces by 12, to bring them to Pounds, but if it be Harberdupiose, you must divide them by 16. Now againe, there be greater weights, which are called a hundred, halfe a huudred, and a quarterne, and also a halfe quarterne, &c.

Scholar. Why? so there may be reckoned 20 Pound, 40 Pound, 200 Pound, and such innumerable.

Master. All these are numbers of weight, but they have not common weights made to their rate as the other have. And again, these that I did name are not just in number as they seem by their names : for an hundred is not just 100, but is 112 pound. And so the half hundred is 56, the quarterne 28, and the half quarter 14.

I 3   And

Side notes: Venice Money. Weights. Troy weight. A penny weight. An ounce A pound Troy. Haberdupoise weight. A hundred weight.

And these be the common weights used in most things that are sold by weight.

Wooll Weights. Todde Stone.

Howbeit there are in some things other names, as in Wool 28 pound is not called a quarterne: but a Todde: and 14 pound is not named half a quartern, but a Stone, and the 7 pound half a Stone, Other names because they differ in many places, and agree in few, I let them passe.

Sack of Wooll.

But a sack of Wooll by the Statutes is limited to be 26 Stone.

Cheese weights.

¶ Now in Cheese, though it be sold by the hundred, and by the Stone in some places, yet the very weights of it are Cloves, and Weyes. So that a Clove containeth 8 pound, and a Weye 32 Cloves, which is 256 pound, that is 12 score and 16 pound, and so much weigheth the Weye of Suffolk Cheese, and the like is or should be the Barrel of Suffolk Butter.

The Weye of Essex Cheese containeth six score and sixteen pound: and so much is also the Barrel of Essex Butter.

Moreover this weight is used by the Apothecaries in their Physicall composition, and mixture in medicine, wherein the least is a grain,

The Apothecaries weights.

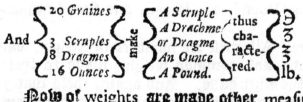

And { 20 Graines / 3 Scruples / 8 Dragmes / 16 Ounces } make { A Scruple / A Drachme or Dragme / An Ounce / A Pound. } thus charactered. { ℈ / ʒ / ℥ / lb. }

Measures for liquor A Pint. Gallon.

Now of weights are made other measures bot for grain and liquor. For a pound in Troy weigh maketh a pint in measure, so that 8 pound or 8 pint doe make a gallon: half a gallon is named a pottle

an

and half a pottle is called a quart, which contain-eth two pints : Now above a gallon the next mea-sure is a Firkin: then the Tertian a Kilderkin, or half a Barrell, and a Barrel : And by these measures are sold commonly Ale, Beere, Wine, and Oyle, Butter and Soap, Salmon, Herrings, and Eeles.

But as these be unlike things, so the measures of their vessels doe differ, for the measures of them all are as followeth.

Of Ale
{
The firkin
The kilderkin
The barrell
} contai-neth {
8
16
32
} Gallons.

Of Beer
{
The firkin
The kelderk.
The barrell
} contain-eth {
9
18
36
} Gallons.

Sope measures, both Firkin, Kilderkin, and Bar-el should be equall to Ale measure.

Moreover the Statutes do limit the weight of eve-ry of those three vessels being empty.

A barrell
Halfe a barrell
A firkin
} to weigh empty {
26
13
$6\frac{1}{2}$
} Pounds.

Herrings also sold by the same measures that Ale and Sope be sold by.

Herrings are sold by the tale, 120 to the hundred, ten thousand to the Last.

Salmon and Eeles have a greater measure.

Salmon & Eeles
{
The butte
The barrell
Half barrell
The firkin
} hold-eth {
84
42
21
$10\frac{1}{2}$
} Gallons.

Howbeit, some Statutes did limite Eele vessels e-quall with Herring vessels.

H 4   Now

Wine measures.

Now as for wine vessels they are seldome smaller then Hogsheads which are of 63 Gallons: Every Hogshead, is two Barrels: yet there are many other wine vessels, but of them all see this Table and marke the measures one by another.

Of wine and oyle	The ranlet	hold-eth	$18\frac{1}{2}$	Gallons.
	The barrell		$31\frac{1}{4}$	
	The hogshead		63	
	The tertian		84	
	The pipe		126	
	The tunne		252	

Tertians.

But you shall marke that there be other kind of Tertians: for there be Tertians, (that is to say Thirds of pipes, of hogsheads, and of barrels, as well of other things as of wine.

Also Malmseyes, and Sacke, &c. the half Tun not called a Pipe, but rather a Butte.

A Butte.

And thus much have I thought meet to tell y at this time.

Scholar. And is that alwayes true?

Master. I have tould you how it should be, how it is, I may not say: how they doe differ va from their just measure, that Gaugiers can tell better then I. But I will let this passe now, speak briefly of the other measure.

And as of weights there did spring the liq measures (whereof I spake last) so of the s

Drie measures.

springeth dry measures, as Pecks, Bushels, Quart and such like, whereby are measured corne and grains, also Salt, Lime, Coals, and other like. this is the order and quantity of them.

A pecke is the measure of two Gallons.

A bushell		foure pecks.	A Bushell
A quarter	containeth	eight bushels.	A Quarter.
A wey		six quarters.	A Wey.

These are the common names and measures, but in divers places there be divers sorts.

The Bushell in many places is two bushels, but then is that bushell there called a ſtrike: and in ſome places half a quarter is called a Cornock. But thoſe diverſities are to many to tell you briefly them all, and again, ſith they are againſt the Law and ſtatutes, I count them unmeet to be uſed.

Strike. Cornock

But now remaineth yet another kinde of meaſures, whereby men mete length, breadth and thickneſſe, and thoſe are, an Inch, a Foot, & ſuch other whoſe names and quantities this Table ſheweth.

Measure to mete length, breadth, & thickneſſe.

3 Grains of barley in length make an inch.

| 12 Inches | | a foot. |
| 3 foot | make | a yard. |

An Inch. Foot.

| 3 foot and 9 Inches | | an ell. |
| 5 Yards and a halfe | make | a pearch. |

Yard. Elle.

1 Pearch in breadth, and 40 in length, doe make a rodde of land, which ſome call a rood, ſome a yard land, ſome a farthendele.

Pearch.

| 2 Farthendels | | halfe an acre of ground. |
| 4 Farthendels | make | an acre. |

Acre.

More, 40 rods in length do make a furlong, 8 furlongs make an Engliſh mile, which containeth 320 Perches.

So that an Engliſh mile, grounded upon the Statute, is in length 1760 yards, 5280 foot, and 63360 Inches.

Somewhat greater then the Italian mile of 1000 paces, and 5 foot to a pace.

Here

Here might I tell you many things else touching measures, and also how to reduce strange measures to our measures, but because it cannot be well done without the knowledge of Fractions, which as yet you have not learned, I will let them passe till another time, that I have taught you the knowledge of broken numbers.

The parts of time.

Scholar. But yet Sir of the parts of time, I pray you tell me some what.

A day.
An houre,
Weeke,
Monet
Yeare.

Master. You know that a naturall day hath 24 houres, and every houre hath 60 minutes. It needeth not to tell you, that 7 dayes make a weeke, and 4 weekes make a common moneth, and 13 months make a yeare, lacking one day and certain houres and minutes: but of that I shall instruct you hereafter.

Here will I make an end of Reduction *for this time, which though it be counted no kinde severall of Arithmetick, you see it is no lesse needfull to be known, or easier to be done, then any of the other.*

Scholar. Marry sir, it seemeth unto me much harder then any other sort, for it requireth the knowledge of so many things : but now Si when you see time I am ready to learne forth, a much of Reduction as you have taught me, I remember ; but and if I doe at anie time forget I shall have recourse to the Tables which you se forth for me.

Master. So doe you : for it will not be remem bred without exercise. But in as much as yo understand so much as we have intreated of, will now instruct you in Progression.

Progressi

Progreſſion.

What Pro-
greſſion is.

Lthough untill this day the moſt part of Writers have defined Progreſſion as a compendious kinde of Addition, yet truely it is not ſo: for Progreſſion as the very nature of the word doth informe any man) is a going forward and proceeding in numbers, and that regularly and orderly, whoſe place is aptly choſen to be very neere, or rather next after the expoſition of the ſoure principall parts of Arithmeticke: for in it after a moſt eaſie manner, are al the foure former parts exerciſed and practiſed: and not onely Addition, as cuſtomably is done. Which cuſtome hath been the cauſe, why it hath ſo ſpecially been named a kinde of Addition, and defined to be a quick and brief Addition of divers ſummes, proceeding by ſome certain and reaſonable order,

You ſhall alſo underſtand there are infinite kinds of Progreſſions, but foz you (as yet) two are ſufficent to be exerciſed in, of which the one I call Arithmeticall and the other Geometricall.

Arithme-
ticall Pro-
greſſion.

Arithmeticall progreſſion *is a rehearſing or placeing downe of many* numbers, number *after* number, *in ſuch ſort that between every two next* numbers *rehearſed or placed downe the difference, diverſity, or exceſſe, be equall and alike.*

Scholar. Sir I thank you foz that you have both opened unto me what Progreſſion is truly, and alſo why it is here placed.

But I pzay you with an example make plaine your definition.

Maſter.

Maſter. Examples cannot want, ſeeing all reaſonable creatures naturally uſe the order of one
kind of Arithmeticall progreſſion (which therefore
is alſo named naturall) whenſoever they diſtinctly
do count or number any multitude by one, ſaying,
1, 2, 3, 4, 5, 6, whereby the proceeding from number to number, and every one ſurmounting and exceeding his fellow next before by a like quantity
(which here is 1) declareth the ſame to be Arithmeticall progreſſion. And for the more plainneſſe, I
ſet it down in this manner.

The comon exceſſe.

1　1　1　1

The progreſſion.

1　2　3　4　5　6

Scholar. This is moſt evident. And I think
that I am able to tell you now of any progreſſion
Arithmeticall propounded, what is that common
exceſſe or difference whereby it proceedeth, if this
order be kept in it.

Maſter. What ſay you of 3, 6, 9, 12, 15;

Scholar. They exceed each other by 3: And that
may I ſet down in ſuch evident order, as you did
your example of naturall progreſſion, in this wiſe.

The common exceſſe.

3　3　3　3

The progreſſion.

3　6　9　12　15

Maſter. And doe you not alſo now perceive
that the whole Table of Multiplication may be
made by the other of progreſſion Arithmeticall

eſſthe

either if you will begin at the firſt number of any of them on the left hand, and ſo proceed right over-thwart: or any of the firſt number of the upper row, and go directly downward.

Scholr. I pray you let me conſider the thing a little, and I will anſwer you.

1	2	3	4	5	6	7	8	9	10
2	4	6	8	10	12	14	16	18	20
3	6	9	12	15	18	21	24	27	30
4	8	12	16	20	24	28	32	36	40
5	10	15	20	25	30	35	40	45	50
6	12	18	24	30	36	42	48	54	60
7	14	21	28	35	42	49	56	63	70
8	16	24	32	40	48	56	64	72	80
9	18	27	36	45	54	63	72	81	90
10	20	30	40	50	60	70	80	90	100

By this triall I perceive it now very well, for the common exceſſe or difference between any two next, is continuallie as much as the firſt number of everie row, either from the left hand overthwart taken, or from anie of the uppermoſt overthwart rowes downward.

Maſter. Now then, if of any ſuch progreſſion, you would ſpeedily know the totall ſumme much quicklier then by common Rules of Addition: firſt tell how many numbers there are (which numbers here we call places or parcels) and if they be odde, write their ſumme downe by it ſelf: as in this exam-ple, 2, 4, 6. 8, 10, 12, 14. where the numbers are 7, as you may ſee: therefore ſet downe 7 in a place alone,

To know the totall ſumme of an Arith-meticall progreſſi-on.

alone, then adde together the first number and the last, as in this Example: Adde 2 to 14 and that maketh 16, take half of it, and multiply by the 7 which you noted for the number of the same places, and the summe that amounteth, is the summe of all those figures added together: as in this example 8 multiplied by 7 maketh 56, and that is the summe of all those figures.

Scholar. That will I work by another example. I would know how much this summe is, 5, 8, 11, 14. 17, 20, 23, 26, 29,   I tell the places and there are 9, that I note. Then I put the first number 5, & the last 29, together, and they make 34. I take half of it, that is 17, and multiply by 9, and it maketh 153. That you say is the summe of all the numbers.

Master. So shall you find it if you try it.

Scholar. How d shall I try it?

Master. By your common Addition, for if you adde all the parcels together, you shall see the same summe amount, if you did work well. And that manner of Addition trieth all kinds of summing any progression.

Scholar. Then can I summe any progression, if the number of the parcels be odde. But what if they bee even, as in this example,   1, 2, 3, 4, 5 6, 7, 8?

Master. When the number of the parcels is even, then note that also as you did before, and likewise adde the first summe to the last, and by the half of the number of the places, doe you multiply it: as in our example, the parcels are 8 that I note, then adding the first summe to the last

la

laſt, there amounteth 9, that doe I multiply by half of the parcels, that is, by 4, and it maketh 36, which is the ſumme of the parcels.

But if you will take one Rule for theſe both, doe thus: Multiply the half of the one by the other whole, and the ſumme will amount all one. For ſometime it chanceth that the number of the parcels be odde, ſo their half cannot be taken: and that ſome time it chanceth the Addition of the firſt number and the laſt, doe bring forth an odde number, ſo that halfe of it cannot be taken: but they will never be both odde.

A generall rule.

Scholar. Then I perceive this, if there be no more belonging to it.

Maſter. As accuſtomably it hath béen taught, this hath been the chiefe and onely exerciſe in Progreſſion uſed. But that you may perceive how divers wayes and to how great profit ſo ſimple a thing (as this Arithmeticall Progreſſion is) may be conſidered and uſed, I will here propound you ſix Propoſitions: of which foure of them were invented by a friend of mine, and never before this publiſhed: and the two firſt were never to my knowledge written of but by thrée men.

Scholar. This doth greatly encourage me to be attentive unto your words, ſéing I ſhall not onely be inſtructed at your hands in the common knowne Rules of this excellent Art,but beſides that ſo abundantly in other new Rules enformed, as my very entrance ſhall ſéeme to paſſe a great many mens further ſtudie, and longer continuance. Therefore Sir, I beſéech you let me know your ſix Propoſitions.

Maſter.

Master. These they are.

1   *To know the* last *number* without proceeding by continuall *Addition,* till you come unto it, so that the common excesse, the first *number,* and the number o the places be known.

2   *The first number of the* Progression *and the* last *being known, with the common excesse to finde* the *number of the places.*

3   *The excesse being given, and the* first *or* last, to *know the quantity of any middle* number, whose plac *is given from the* first or last.

4   *The* totall summe *being given, and the* first an last, *to finde out the* number of the places.

5   *The* totall summe *of any* Arithmeticall Progress on *being given, and* the first and last, *to find out* the common excesse.

The totall summe *being given, and the mutual* excesse *with the* number of the places, *to give* the first or last *number of the same* Progression.

Many moe considerations could I propound you i these *Arithmeticall Progressions,* but these are sufficient to give you occasion to think, that Rules of know ledge and Arts are infinitely capable of enlargmen

S. Happy were I if I did but well understan that which is already invented and written: An yet in my simple fantasie, these things offer then selves (in manner) to be studied for about Progre sion, therefore I pray you to proceed to the Rul answering to these propositions.

Master. I will orderly for every of these 6 pr positions give you Rules, and with every one example, unlesse the plainnesse and easinesse ne no farther exemplifying.

F

For the Solution of the firſt multiply the exceſſe by 1 Propo
a number leſſe by 1 *then the number of the places,* ſition.
and the off-come adde to the firſt number, ſo you ſhall
have the laſt number, which is ſought for.

As for example. If there were ſeven places in
a progreſſion Arithmeticall, whoſe continuall encreaſe o2 mutuall exceſſe were 4, and the firſt number were 5, and J would know what the laſt and
ſeventh number is : J multiply 6 which is one leſſe
then 7, (the number of the places) by 4, thereof
commeth 24, which J adde to 5, that maketh 29,
and that is the laſt number which J deſire to know.
And this you may ſtraightway p2obe by continuall p2ocæding from 5, till the ſeventh place, encreaſing every one by 4, as thus.

5, 9, 13, 17, 21, 25, 29.

Loe here the laſt, being alſo the ſeventh, is 29.

Scholar. J perceive already one good p2operty in
this Rule, which in all wo2kes is to be deſired :
that is, it will eaſe one from great labour, if a
progreſſion were p2opounded of a hund2ed o2 two
hund2ed places, o2 moe : And alſo it is very eaſie
to wo2k, and moſt neceſſary fo2 the totall ſumme
finding, in a very long progreſſion.

M. It is true, & therefo2e now let me ſé if you
an anſwer me this queſtion by this propoſition.

A Merchant buyeth 50 pounds of Spices, & agreeth
to pay for the firſt pound 4 pence, for the ſecond 7
ence, for the third 10 pence, for the fourth 13
ence, &c.

The queſtion is, how much he muſt pay fo2 the
aſt pound, and then how much the 50 pound
ſumeth to ?

                              R   Scholar.

Scholar. According to the propoſition, I multiply 49 (which is leſſe by one then the number of the places) by the exceſſe, which is 3: to the product 147, I adde the firſt number which is 4, it maketh 151 pence, the price of the laſt pound. Now I adde 4, the price of the firſt pound, to 151 the price of the laſt pound, it maketh 155, which I multiply by half the number of the places, which is 25 the Product, 3875 pence is the totall ſum or price of the 50 pounds of Spices, as appeareth.

49 places 1 leſſe	151 laſt
3 exceſſe	4 firſt
147	155
4 firſt	25 halfe places
151 the laſt	775
	310
	3875 totall ſumme

which amounteth to

li	s	d
16	2	11

Maſter. It is truly wrought.

Scholar. Then I intreate you to proceed to your ſecond propoſition.

Maſter. *The ſecond Rule is this. From the laſt ſubtract the firſt, the remainder divide by the common exceſſe, to the Quotient adde 1, and you have the number of the places, which you would know, as in this Progreſſion.*

6. 11. 16. 21. 26. 31.

If I know onely 6 and 31, and that they encreaſe by 5, then according to the Rule, from 31 I ſubtrac ſubtract 6, there remaineth 25, which 25 I diuide by 5 (the common exceſſe) the Quotient commeth forth 5, to which I adde 1 that maketh 6 : and ſo many are the places as you ſée.

Scholar. This Rule is ſo eaſie, that I were much to blame, if I could not remember it.

Maſter. *The third* Propoſition *may alwayes thus be ſolved. Multiply the* exceſſe *by a* number leſſe *by one, then the diſtance of the place is from the firſt, or the laſt number given : the offcome adde to the firſt, if the diſtance be reckoned from the firſt, & the firſt alſo known; or ſubtract from the laſt, if the diſtance be from the laſt counted, and the laſt given alſo, and that which commeth forth, either in that Addition to the firſt or Subtractiō from the laſt is the number ſought.* As for Example, *I propound you this* Progreſſion.

3 Propoſition.

8   15   22   29   36   43   50   57

And for the apt conſidering the manner of this queſtion, I will note over every place his diſtance from the firſt, and under every place his diſtance incluſively from the laſt, thus.

1	2	3	4	5	6	7	8
8	15	22	29	36	43	50	57
8	7	6	5	4	3	2	1

Now if the exceſſe whereby this Progreſſion ſtandeth, be known to be 7, and the firſt numbers given, being 8, I would know what number ſtandeth under 4, that is to ſay, in the fourth place.

multiply 7 by 3 (which is leſſe by 1 then the number of the place propoſed) that yéeldeth 21, to which I adde 8 , the firſt number, ſo commeth 29 : which I ſay to belong to the fourth lace, as you ſée in the example : or if in the

K 2   third

third place from the laſt, you would know what number in this example ſhould ſtand, the laſt number being known to be 57, and the common exceſſ 7, then by 2 which is leſſe by 1 then the place propounded, I multiply 7, that giveth 14, which appertaineth to the third place incluſively reckoned from the laſt, and ſo my example giveth you.

Scholar. I perceive right good uſe of this rule : for if I had forgotten what the firſt number were and remember ſtill but the laſt, the common exceſſe, and the number of the places, then might I come by the knowledge of my firſt number again.

And me thinketh, that it differeth not much from the firſt Propoſition, ſaving that which you make here a middle number, there was made the laſt : and alſo in this point it differeth, that in it the laſt was only ſought, and no conſideration had in numbring the places from the laſt, as here I marke in your numbers noted under your Progreſſion.

Maſter. And thinke you not, the middle numbers of progreſſion ſtanding off a hundred or two hundred places or moe, may as much cumber a man to come to the knowledge of them by continuall encreaſing from the firſt by the common exceſſe, or abating from the laſt continually the common exceſſe as the very finall Numbers in a ſhorter Progreſſion would doe ?

Scholar. Yes ſir, that I think right well and therefore I am glad of this new framed propoſition, and the manner of the working of it.

Maſter. *The rule of the fourth is this. Adde the firſt & the laſt together, and by the off-come divide the*
totall

4 Propo-
ſition.

totall ſumme. *Double the* Quotient, *and that will be the* number *of the* places.

Scholar. Then if in a progreſſion, whoſe ſumme were 207, and the firſt number 12, and the laſt 57, if I adde 57 and 12 together, that maketh 69, and be it I divide 207, the Quotient will be 3, which I double; and ſo I have 6, and ſo many muſt be the number of the places that this progreſſion ſtandeth on.

Maſter. Whether it be ſo or no, how will you trie?

Scholar. Half 6, which is 3, being multiplied by 69, muſt make 207, the totall ſumme; if 6 be the number of the places. For ſo the whole worke of your Rule in ſumming any Arithmeticall progreſſion did enforme me. I will then multiply 69 by 3, thus.

$$\begin{array}{r} 69 \\ 3 \\ \hline 207 \end{array}$$

It commeth forth juſtly.

Maſter. I muſt much herein commend your promptneſſe both in memory and in well applying your Rule: although in manifeſt words it did containe no ſuch matter.

Scholar. Sir, I pray you beare me frame one example or more.

Maſter. I am well pleaſed ſo that ye be ſhort, for you make me longer here then willingly I would have béen: but I cannot perceive how I could have omitted any thing as yet, without your great lack thereof.

Scholar. If I had received 85 pounds of certaine men, but of how many I have forgotten, yet I remember that the firſt gave me 7 pound, and the laſt 17 pound, and every payment after other did riſe

A queſtion of money.

by a like ſumme.   And the man for whom I received
this money, conditioned with me, that of every payment I ſhould have twelve pence for my labour: now
unleſſe I can by Art finde the truth of this caſe, I am
like to loſe the moſt part of my reward.

Maſter. I perceive you can handſomely frame
an example, which ſhould concern your owne
gaine: I pray you let mée ſée how you
woulb do juſtice in this point.

Scholar. I adde the firſt and the
laſt together, that maketh 34 : by
which I divide 85, thus :

Why how now? Sir, here is a remnant of 17,
in which 34 cannot be had : ſo that now I am in
the briers for doubling of my Quotient, and
farewell then both my Juſtice, and a good lamp of
my gains.

Maſter. Ye are never the farther from the
matter, though it fall into a Fraction.   For you
ſhall underſtand that the Fraction which of any
ſuch work procéedeth, is every half of one ſuch, as
the units of the Quotient before are.   And that
you may try, if you double that which ſo remaineth
for then it will be equall to your diviſor, as if you
double 17, (the remnant) it maketh 34, and your
diviſor alſo was 34, this noteth the remainder to
be halfe of one.

Scholar. Now I am glad of this hard Example
For with it I have a generall rule for the Fraction
that may hap in this work. So that the Quotient
being 2 & a halfe, I double that, it maketh 5, therefore ſhould my gain be 5 ſhillings.   And to be ſure
(by your leave) I will trie it, for I will multiply
hal

halfe of 34 (which is the firſt and laſt number 　17
joyned together) by 5, thus.　　　　　　　　　5
　　It is moſt true (I ſée) that I ſhould léeſe　 ——
nothing by the former working.　　　　　　　85

Maſter. *The fifth propoſition hath this rule apper-* 5 Propo-
taining unto it : By the fourth rule finde the number ſition.
of the places, that being done, from the laſt ſubtract
the firſt, and the reſidue divide by a number *leſſe by*
1 then the number of the places, and the quotient
will ſhew the exceſs, which is ſought for.

　　An example hereof ſhall be this . If ye had dif- Example
burſed 685 pounds to a certain number of men,
you neither can tell how many they were, or how
much the ones money exceeded his next before, but
you are ſure that the exceſſe was equall between
every two next : and alſo you remember that the firſt
had 19, and the laſt 118 pounds, how would you find
the number of the name and the exceſſe continually
obſerved in the ſucceſſion of their payments?

　　Scholar. Your Rule doth plainely bid, firſt to
finde the number of the places, which I　　118
will do according to the fourth rule : I　　　19
adde 19, and 118 together, thus.　　　　 ——
　　By this 137, divide 685 thus.　　　　　137
　　Seeing there is no Fraction but a
whole number, being 5, I double that,　　13
and then muſt the number of the places　685 (5
bée 10. Now from the laſt I ſubtract　　137
the firſt, as 19 from 118, thus : And
ſo remaineth 99.　　　　　　　　　　　118
　　This 99 I divide by a number　　　　 19
leſſe by one then the number of the　　　——
places, and ſéeing the places were 10, I divide　99

　　　　　　　K 4　　　　　　　　　　99 by

99 by 9 ; thus :   99

The quotient is 1, and so was 99 (11
the excesse, if I have followed your rule right.

M ster. You have wrought every part of the
question both well in order, and truly in the practi
of your Rules.

Scholar. I will then set it down also formabl
so that the number of the places, the excesse, and t
totall summe may straight appear, as your first e
ample stood.

The common excesse.
The progression.

11	11	11	11	11	11	11	11	11	
19	30	41	52	63	74	85	96	107	11

That the places be 10, and that from the first
the last, the common excesse is 11, I perceive mi
evidently: but whether the totall sum be 685, I ha
not yet proved, which I will now doe : I adde
6 118 together, that maketh 137 : I multiply
that by halfe the number of the places, thus.

All things agree most exactly, so that I 6
am perfect enough in these Rules if I forget th
not again.

Master. Use maketh all things perfect.

Your sixth rule is this. By the number of the pl
divide the totall summe, double the quotient, and t
will be the first and last joyned in one summe. T
by a number lesse by 1, then the number of the pla
multiply the excesse, that off come subtract from
first doubled quotient, and the halfe of the residu
the first number. The last number you may divi
find out, as by the first of our 6 Rules, or by sub
cting this first number from the summe which
contained both the first and last ioyntly, (or thirdl
continuall adding the excesse,

Sch

Scholar. I pray you make this somewhat more plain with an example.

Maſter. If every moneth in the year (counting them now as 13) you gained clearly 40 ſhillings more then you did the moneth next going before, and at the years end you find the whole gain 5720 ſhillings, but ye remember not how much either the gaine of the firſt moneth or the laſt was, by this Rule it may be tried out.

Example of graine.

Scholar. So that here ye ſéeme to apply the 13 moneths to thirtéen places, the 40 ſhillings every one more then the other next before it to be the common exceſſe, and 5720 ſhillings to the totall ſumme.

Maſter. It is true: by 13 then I divide 5720 in this manner.

I double this Quotient, ſo have I 880 for the firſt and the laſt ſumme joyned together; by 12 which is leſſe by one then the number of the places; I multiply 40 (the common exceſſe) ſo commeth 480.

This 480 I ſubtract from 880, ſo remaineth 400 : half whereof is the firſt number which we deſired to know : that is 200. And as for the laſt number, I can give you itt thrée wayes, As by the firſt of my ſix rules I multiply the exceſſe, by a number leſſe by 1 then the number of the places, as 40 by 12 that giveth 480, which I adde to the firſt, being 200, ſo ſhall the laſt be 680.

The ſame ſumme commeth forth if ye ſubtract 200 from 880.

And thirdly, If I begin at 200, and ſo procéed, encreaſing by 40, I ſhall at the thirténth place

place have 680, as thus:

200 240 280 320 360 400 440
480 520 560 600 640 680

Scholar. I thanke you most heartily for these six Rules. Now if it be your pleasure I would heare and learn somewhat of progression Geometricall.

Master. There are yet very many Rules and propositions which fall into this Arithmeticall progression.

And for the use and practise of them, I will propose unto you certain pleasant an necessary Questions of Arithmeticall progression, and to the performance of their workings, such necessary rules and documents, as are requisite for the better understanding of them, or any such like.

A question of Velvet.   *A certain* Mercer *sold* 20 *yards of Velvet*   6
to be paid in 12 *weekes by* Arithmeticall   12
proportion ; *that is to wit, to receive the*   18
first week 6 shillings, *the second week* 12   24
shillings, *the third week* 18 , *and so forth,*   30
increasing the number of weeks by 6 shillings,   36
till the twelfth and last week were expired.   42
The question is how many pounds he   48
had for 20 yards of Velvet.   54

To the performance of this question,   60
and such other the like, I set forth the 12   66
payments in such sort, as for example, here   7
appeareth.

Then touching the adding together o
the

theſe ſummes, without the aid of Addition, according to the rules I taught you in Progreſſion A-
rithmeticall. I note the number of the places, which
are 12, then adding the laſt number of the Progreſſion, which is 72, and the firſt number together
make 78 ; and multiplying 78 by halfe the number of the places, which is 6, amounteth to 468
ſhillings, and in pounds maketh 23 li. 8s.   And ſo
much hath the Mercer for his 20 yards of Velvet,
which is nigh about 23 ſhillings 5 pence a yard.

Scholar. I underſtand this work very well, but
is there any proofe for the juſtifying hereof, as
you have of other workes ?

Maſter. The worke of it ſelfe (being ſo perfectly wrought) that in your proceeding and going
forward from number to number, each number exceeding his fellow by an equall or like quantity,
is all that is demanded for juſtifying of the ſame:
yet notwithſtanding, becauſe your requeſt is reaſonable, I will propoſe an example for the proofe
hereof.

A certaine man is bound to pay for 20 *yards of* The proof
Velvet, the ſumme of 23 *pound* 8 *ſhillings, and it is* of the laſt
to be paid weekly in 12 *weeks or termes by* Arithme- queſtion.
ticall Progreſſion.   The queſtion is therefore to
know with what number the ſame Progreſſion is
to be begun and continued in ſuch equall proportion Arithmeticall, that in 12 weeks the ſame may be
juſtly accompliſhed.

For the reſolution whereof, and of all ſuch o-
ther like, reduce 23 pound 8 ſhillings all into ſhillings, which maketh 468 ſhillings.   A generall rule.

Then adde 1 unto 12, the number of the termes,

it

tt maketh 13 which 13 you shall multiply by half the number of the termes which is 6, it maketh 78, then divide 468 by 78, and you shall finde 6, in the Quotient, which is the true number, that shall begin and continue the said Progression. That is to say, the first week 6 shillings, the second 12 shillings, and the third weeke 6 shillings more, which is 18 shillings, & so every weeke as they rise, 6 shillings more then the weeke before as is manifest in the question aforesaid.

A question of a Farm.   *A Farm is to be sold to be paid by the weekes in a year ; the first weeke to pay 4 shillings, the second weeke 8 shillings, the third weeke 12 shillings, and so forth, increasing each number by 4, till the number of 52 (which are the number of weekes in a yeare expired.) The question is, what the price of the Farm commeth to ?*

Schol. I doubt not, but by that you have already taught me, to end this question very well ; wherefore I set forth the Progression with his excesse 52 times.

Master. Nay stay a while : And here for your further ease, (to abridge you of great labour that appeareth to fall out in this question, and so may doe in any other the like) If a question were proposed of 100 or 200 places, or moe, and that this question, or any other the like can be ended, unlesse you may know absolutely what the last number of the Progression of the 52 place is, (or ought to be) I will give you a generall rule how to know the last number of any Progression Arithmeticall, as well as if you had ordinarily proceeded by continuall Addition, till you had come to the last worke which is this.   Mul-

Multiply the exceſſe by a number leſſe by one A generall then the number of the places, and thereto put the rule. firſt number of the Progreſſion, and you ſhall have your deſire.

Scholar. This Rule is well wo2th the noting: fo2 if I underſtand you aright, I conſider that my exceſſe is 4, which I multiply by 51, which is one leſſe then the number of the places, and it maketh 204. whereunto I adde the firſt number of the Progreſſion, which is 4, and then it is 208, which you ſay is o2 ſhould be the laſt number of the Progreſſion.

Maſter. This is a moſt approved truth, if there were never ſo many places.

Schol. This Rule is ſo eaſie, that I were much to blame, if I doe not remember it. Fo2 by the benefit hereof, I have ſuch an eaſe and light into this excellent Art, that my firſt entrance doth ſeem to paſſe a great many mens further ſtudy, and longer continuance.

Maſter. Many moe Conſiderations could I propound you in theſe *Arithmeticall Progreſſions* ; but theſe are ſufficient for a taſte, to give you occaſion to think that *Rules* of Knowledge and *Arts*, are infinitely capable of enlargment.

Scholar. Happy were I, if I did but well underſtand that which is already invented and w2itten. But theſe things in my ſimple fantaſie, offer themſelves to be greatly beneficiall unto the aide of Progreſſion. Therefore now I will goe fo2ward with your Queſtion.

Now conſidering that the 52 and laſt place is 208, I adde thereunto the firſt numbers of the Progreſſion

greſſion, which is 4, it maketh 212, which I multiply by half the number of the places, which is 26, and it amounteth to 5512 ſhillings. And ſo much is the totall ſumme or addition of this progreſſion: which maketh 275 pounds 12 ſhillings, as appeareth here by my Tables.

Maſter. I like well your labour, and commend you for your diligence : I will here propoſe one example more, and therewithall for this time will end progreſſion Arithmeticall.

A queſtion of Holland. *A certain man brought* 20 Ells of Holland, *to be paid in* 17 *weekes, or termes, by* progreſſion Arithmeticall. *And the firſt week to pay* 1 ſhilling 8 pence, *the ſecond week* 3 ſhillings 4 pence, *the third weeke* 5 ſhillings, *the fourth weeke* 6 ſhillings 8 pence; *and ſo forth, each weeke ſucceeding* 20 pence *more then the weeke before.* The queſtion is, what the ſumme of his 20 Ells cometh to.

Scholar. Becauſe here is mention made both of ſhillings and pence, I feare there is ſome harder matter contained herein, then in the other before: therefore I pray you worke it your ſelfe, and I will diligently mark your labour.

Maſter. There is no more to be done in this, then in the other before ; but becauſe your requeſt is ſo reaſonable, be attentive unto me.

Firſt, by the generall Rules, I ſeek to finde out the laſt Number of the 17 place, what this progreſſion ought to be. Therefore here in my Tables multiplying the exceſſe 20 by 16, which is one leſſe then the number of the termes for places, and it commeth to be 320 ; and thereunto adding the firſt number of the progreſſion, which is 20 pence :

pence, all is 340 pence, or 28 ſhillings 4 pence, for ſo much ought the laſt number of the payments to be.

Then finally, to know what the whole 17 places amount unto, I adde the firſt number of the progreſſion and the laſt together, which make 360. Now becauſe 17 is an odde number, whoſe halfe cannot be taken, I take the halfe of 360, which is 180; and multiplying 180 by 17, commeth to 3060 pence, which maketh as you ſée by Diviſion 12 pound 15 ſhillings. And ſo much is the buyer to pay for his 20 Elles of Holland. Which 3060 pence if you divide by 20, the number of Elles that was bought, you ſhall find 12 ſhillings 9 pence, and ſo much payed hée for an Elle one with another.

The Proof.

A certaine man doth owe 12 pound 15 ſhillings, *to be paid in* 27 weeks or termes *by Arithmeticall progreſſion. The queſtion is, to know with what num-ber he ſhall begin and continue the* progreſſion, *in ſuch equall proportion, as the ſame may be truly paid and ſatisfied in* 17 weeks.

A queſtion of debt.

The Anſwer.

Firſt I reduce 12 pounds 15 ſhillings, all into pence, which as you ſée here in my Tables, make 3060 pence, that I let ſtand by a while.

Then I adde 1 to 17, the number of the places or termes, which maketh 18, which I ſhould mul-tiply by halfe the number of the weekes or termes, which

which is 8 ½ which 8 ½ multiplied by 18 cannot well
be done, unleſſe you were acquainted with Fractions
oʒ broken Numbers, therefoʒe you ſhall let that paſſe
& multiply 17 by the half of 18, which is 9 (foʒ that
is all one with the multiplication of 8 ½) & the multiplication of 9 into 17 maketh as youſee 153, with
which number you ſhall divide the 3060 pence be-
foʒeſaid, and the Quotient bʒingeth foʒth 20 pence,
which is the firſt number oʒ payment to beginne the
progreſſion withall : and ſo each week ſuccéeding
to riſe 20 pence moʒe then the wéeke befoʒe, and
thereby in 17 wéeks ſhall 12 pounds 15 ſhillings
be paʒed : as befoʒe was ſufficiently declared.
Thus much foʒ progreſſion Arithmeticall.

Scholar. Certainly Sir, I know not how to
render you condigne thanks foʒ theſe benefits
ſhewed me, which me thinketh are ſo eaſie, de-
lightfull, and pleaſant, that I count my ſelfe happy
to be in your company.

Maſter. I am glad you delight ſo well herein,
which is an Art of wounderfull dexterity to all ſoʒts
of men, of what degrée oʒ pʒofeſſion ſoever they
be. And now will I pʒoceed to progreſſion Geometricall, wherein I will be moʒe bʒiefe, both becauſe
I have béen ſo long in this part of Arithmeticall
progreſſion, and alſo foʒ that it would require the
knowledge of Roots and ſurd numbers. (whereof ye
have learned nothing) if I ſhould frame the like
propoſitions in them as I have done in theſe.
Therefoʒe I will onely teach you to practice about
it, and ſo end the conſiderations and woʒks of
theſe progreſſions.

Progreſſion

PRogreſſion Geometricall *is when the numbers* Progreſſi-
increaſe by a like proportion, that is, if the ſecond on Geo-
number containe the firſt, 2, 3, or 4 times and ſo metricall.
forth, then the third containeth the ſecond ſo many
times alſo: and ſo the fourth   3   6   12   24   48
the third, and the fifth the   1   3   9   27   81
fourth; wherefore I ſet theſe   2   10   50   250
3 examples.

Here in the firſt Example you ſee that every
number contatneth the other (that goeth next
before him) two times; and in the ſecond example
three times, and in the third example five times.
Now if you will know how to finde eaſily the
ſumme of any ſuch number, doe thus: Conſider by
what numbers they be multiplied, whether by
2, 3, 4, 5, or any other, and by the ſame number
multiply the laſt ſumme in the Progreſſion.

Scholar. I pray you worke it by this example,
2, 8, 32, 128, 512, 2048, which I have framed by   To finde
proceeding from 2, and conttnually multip'y by 4.   the totall

Maſter. Then muſt I multiply the laſt ſumme   ſumme in
(which is 2048) by 4 alſo, and it will be 8192,   any Geo-
Now muſt I abate from this ſumme the firſt num-   metricall
ber of the progreſſion, which here is 2, then reſteth   Progreſſi-
8190, which ſumme I muſt divide by 1 leſſe then   on.
was the Number that I multiplyed by. Seeing
then I multiply by 4. I muſt divide by 3, ſo dividing 8190 by 3, the Quotient will bee 2730, which
is the ſumme of all the progreſſion. And now to
prove whether you can doe the ſame, I give you
theſe Numbers to adde by this Rule 3, 15, 75, 375,
1875, 9375, 46875.

Scholar. I cannot well tell by what number
                                                    L   this

this progreſſion doth increaſe.

Maſter. In any ſuch doubt doe thus : Diuide
the ſecond number by the firſt, and the Quotient
will ſhew you the number that engendreth t e pro
greſſion.

Scholar. Then is that number in this example 5,
for ſo many times is 3 in 15.

Maſter. So is it. Now worke as I taught.

Scholar. The laſt number is 46875, which I
multiply by 5, & it yeeldeth 234375, from which I
abate the laſt number of the progreſſion, that is 3,
and there reſteth 234372, which I diuide by 4, for
that is one leſſe then 5, and the Quotient is 58593,
which is the whole ſumme of the progreſſi n.

Maſter. If you remember well this, you haue
learned the Art of progreſſion both Arithmeticall
and alſo Geometricall, which you may proue either
by ſubtracting of each number alone from the ſum
and ſo will there nothing remain : or elſe by ad
ding together of all the parcels, for ſo will the
ſame ſumme amount.

A queſtion
of Satten·

A Mercer hath 12 yards of Satten, which h
valueth at 16 ſhillings the yard, and ſelleth the ſam
12 yards to another man to be paid as followeth
That is to wit, for the firſt yard to have one ſhilling
for the ſecond yard two ſhillings, for the third yar
foure ſhillings, for the fourth yard eight ſhillings, &
doubling each number following, till the twelft
and laſt yard. The queſtion is, who hath mad
the better bargain of the buyer or ſeller.

Firſt you may ſet down 12, the number of th
yards as you ſee here in this Example. And again
each, number the number of ſhillings due to be pai

as the order of Duplation or Multiplication by two teacheth.

Then reſorting to the adding up or ſumming of this Progreſſion, where I conſider that the increaſe of this ſumme proceeded by the Multiplication of 2, and therefore after I haue drawn a line under the 12, I worke and multiply the laſt ſumme by 2 alſo, and it yeeldeth 4096 : from whence I abate the firſt number of the progreſſion, which is 1, & then reſteth 4095: which I ſhould diuide by one leſſe then I did multiply by, but ſeeing it is 1, I need not to diuide it: for 1, (as I haue ſaid before) doth neither multiply nor diuide, therefore I take that ſumme 4095 for the whole ſumme of the ſhillings, which by Reduction amounteth to 304 pounds 15 ſhillings, and ſo much hath the Mercer for his

8	
1	1
2	2
4	3
8	4
16	5
32	6
64	7
128	8
256	9
512	10
1024	11
2048	12
4096	

twelve yards of Satten : which is 17 pound, 1 ſhilling, 3 pence a yard. But I think you will buy none ſo deare.

Schol. No Sir, by the grace of God this yeare.

Maſter. *Then what ſay you to this* queſtion ? *If I ſold unto you an horſe having 4 ſhoes, and in every ſhoe nailes, with this condition, that you ſhall pay for the firſt nayl one* ob. *for the ſecond nayl two* ob. *for the third nayl foure* ob. *and ſo forth, doubling untill the end of all the nayls. Now I aske you, how much would the price of the horſe come unto?*

A queſtion of an horſe.

L 2   Scholar.

1	1	Scholar. Firſt, to know th
2	2	number of the nayls, I muſt multi-
4	3	ply 6 by 4, & it maketh 24. Then
8	4	will I doe thus. I will write the
16	5	number of the rayls every one in
32	6	order from 1 to 24, and againſt
64	7	each number of the nayls the ſumme
128	8	of halfe pence duly, as the order of
256	9	Duplation or Multiplication by 2
512	10	teacheth, and as in the next figure
1024	11	following appeareth.
2048	12	Then do I reſort to the Rule of
4096	13	ſumming up the progreſſion, where
8192	14	I conſider that the increaſe of this
16384	15	ſumme proceedeth by the multipli-
32768	16	cation of 2, as the laſt Example did.
65536	17	And therefore multiplying the laſt
131072	18	ſumme by 2 alſo, and it yeeldeth
262144	19	16777216, from which I abate
524288	20	the firſt Number which is 1, and
1048576	21	then reſteth 16777215, which
2097152	22	I ſhould divide by one leſſe then
4194304	23	I did multiply : but ſeeing that
8388608	24	it is 1, I need not to divide it, for
16777216		1 (as you have before ſaid) doth

neither multiply nor divide, there-
fore I doe take the number, 1 6 7 7 7 2 1 5 for the
whole ſumme of the halfe pence, which by Reduction
I find to bee 69905o ſhillings and 7 pence halfe-
penie, that is 3 4 9 5 2 pounds 1 0 ſhillings 7 pence,
ob.

Maſter. That is well done, but I thinke you
will buy no horſe of the price.

                                        Scholar.

Scholar. No Sir, if I be wiſe.

Maſter. Well then, anſwer mée to this Queſtion.

A Lord delivered to a Bricklayer a certain num-
ber of loads of Bricks, whereof he willed him to make
twelve walles, of ſuch ſort, that the firſt wall ſhould
receive two thirdells of the whole number, and the
ſecond two thirdells of that which was left ; and ſo
every other two thirdells of that that remained :
and ſo did the Bricklayer : and when the 12 walls
were made, there remained one load of Brick.

Now I aske you, how many load went to each
wall, and how many load was in the whole ?

Scholar. Why Sir, it is impoſſible for mée to tell.

Maſter. Nay, it is very eaſſe if you marke it
well. Marke well that I ſaid, that every wall
ſhould receive two thirdels of the ſumme that was
left. Now take away two thirdels from any
ſumme, & you muſt néedes grant that that which
remaineth, is one thirdel of the ſumme laſt before,
Example of 9. from which if you take 2   thirdels,
there will remain 3, which is 1 thirdle of 9. Likewiſe
from thrée bate two thirdels, and there remain 1.

Scholar. This is true and now I perceive the
leaſt wall had but two load of brick.

Maſter. And by the ſame reaſon may you know
howmany load every wall had, according as this
figure following doth ſhew, and likewiſe what the
whole ſumme of brickes was, for if you make 12
ſummes, multiplying by 3, ſtill from the laſt remainer, as you may ſée here on the left ſide of the
Table, there will appeare all the remainers of the
whole wall : and if you multiply the laſt of thoſe 12
ſummes by 2 alſo, then will that be the ſumme of

L 3   the

the loads which was delivered to the Bricklayer,

	1	12	2	
The remainer af-	3	11	6	*Loads due to*
ter every wall.	9	10	18	*each wall.*
	27	9	54	
	81	8	162	
	243	7	486	
	729	6	1458	
	2187	5	4374	
	6561	4	13122	
	19683	3	39366	
	59049	2	118098	
	177147	1	354294	

Summe of the loads delivered   531440

Again if you double every Remainer, as you may ſée at the right ſide of this Table, thoſe numbers will ſhew the ſumme of loads that went to each wall, whereby you may perceive that each wall was three times ſo great as the next leſſe.

Scholar. Lo now it appeareth eaſie enough Now ſurely I ſée that Arithmetick is a right excellent Art.

Maſter. You will ſay ſo when you know more of the uſe of it: For this is nothing in compariſe to other points that may be wrought by it.

Scholar. Then I beſéech you ceaſe not to inſtru mée farther in this wonderfull cunning.

Th

The Golden Rule or Rule, of Proportion direct, called the Rule of Three.

Master.

B*Y order of the Science (as Men have* *taught it) there should follow next the* Extraction of Roots of Number, *which because it is somewhat hard for you yet,* *I will let it passe for a while, and will* *teach you the feat* of the Rule of Proportion *which for* is excellency *is* called the Golden Rule. Whose use is, by three numbers known to find out any other unknown; which you desire to know, as thus.

If you pay for your board for three moneths sixteen shillings, *how much shall you pay for* eight moneths ? To know this and all such like questions you shall consider which two of your numbers be of one denomination, and set those two one over the other, so that the undermost be it that the question is of: as in my question 3 and 8 be both of one Denomination, for they both be months ; and because 8 is the number that the question is asked by, I set the one over the other, and 8 undermost thus, with such a crooked draught of lines. Then doe I set the other number which is 16, against 3 on the right side of the line, thus. And now to know my question, this must I do : I must multiply the lowermost on the left side, by that on the right side & the sum that amounteth, I

$$\begin{matrix} 3 \\ 8 \end{matrix} \mathrm{N}$$

$$\begin{matrix} 3 \\ 8 \end{matrix} \quad 16$$

L 4   must

muſt diuide by the higheſt on the left ſide: oz in plainer wozds, thus, I ſhall multiply the number
which the queſtion is asked (which is called the thir
number) by the number of another Denominatio
(which is called the ſecond) & the ſum that amounteth
muſt I diuide by the ſumme of like Denominatio

The third number.
(which is called the firſt) Then foz the knowledg
of this queſtion, I multiply 8 into 16, and ther

The ſecond number.
amounteth 128, which I diuide by 3, and it yelded
42 ſhillings, and 2 ſhillings remaineth, which

The firſt number.
turne into pence, and they bee 24 pence, of whi
thizd part is 8 pence, ſo the thizd part of 1
ſhillings, is 42 ſhillings, 8
pence, which ſumme I   3 ⎯ 16 ſhillings
wzite at the right hand of   8 ⎯ 42 ſhil. 8 pen
the figure againſt 8 thus.

Hereby I know that if three moneths boardi
doe come to 16 ſhillings, that 8 moneths boardi
will come to 42 ſhillings 8 pence, and likewiſe
any other like queſtion.

But here muſt you marke, that the firſt num
and the third be of one Denomination, and alſo
ſecond and the fourth, foz which you ſéek: oz elſ
ſuch Denominations, that you in wozking
bring them into one; As if a man ſhould aske
queſtion.

Queſtion of expence
welue weeks journeying coſt me 14 Fr
6 ſhillings the peece, how many poun
yeare? Here you ſée no two number
of one Denomination, but yet in wozking you
bring them into like Denomination: as thus;
multiply one into 52 weekes and the fourth ſu
which is ſuch Crowns, by the order of the wo

                                                        I

Then to know this question multiply the third
summe 52, by the second 14, and the summe will
be 728: that divide by your first number 12, and
the Quotient will be 60 Crownes, and 8 Crownes
remaining: which if you turn into shillings, they
will be 48 shillings, which if you divide by your first
number 12, the Quotient will be 4, which signifieth
4 shillings; put those 60 French Crownes, which
make 18 pounds with the 4 shil-
lings for the summe that answereth to the question, and it is the
just expences of a yeare: And
the work will be thus.

W C

12 Z 14

52 60²⁄₃

And take this evermore for a
generall rule touching this whole art, that the doubtfull or unknown number that you would be resolved
of, shall alwayes be set in the third place. Note also the
first number and the third must ever be of of one
nature and denomination, or else must in working be
brought to like Denomination, & then of necessity must
the other number be in the second place.

*A generall
Rule.*

Remember also that the place of the first number is
highest on the left side, & the place of the second, right
against it on the right side; the place of the third number is under the first, as by those examples you have seen.

Scholar. This I trust I can do.

Master. But and if the question be asked thus:
In 8 weeks I spend 40 shillings, how long will 105 shillings serve me? Here you see that 8 weeks answers
himself and saith 40 shillings.

But how long time 105 shillings will serve you
know not. Therefore you shall set 105 in the third
place, according as I told you even now. And the
first

first place must allwayes be of the same nature or Denomination that the third is of, which here is 40. Then must 8 needs be that other: Now multiply 105 by 8 and it will be 840, which if you divide by 40, it will yeelde 21, which is the fourth number, and sheweth how many weeks 105 shillings will serve if you spend 40 shillings in 8 weeks.

The figure of this question is this : as if you shoulo say : If 40 shillings serve for 8 weeks : 105 will serve for 21 weekes.

Shillings.	weeks.
40	8
105	21

Other diversities there bee of working by this Rule, but I had rather that you would learne this one well, then at the beginning to trouble your minde with many formes of working, sith this way can doe as much as all the other, and hereafter you shall learne the other more conveniently.

¶ And for your further aide and instruction, to make you better acquainted with this Golden Rule, I have here proposed sixe questions, and their answers, which I thinke most convenient and meet to preferre the desirous to perfect understanding. The first foure are all branches of one question sprung out of the best tree (for a young learner to taste of) that groweth in this *Ground* of *Arts* : for that no manner of question in the Rule of three whatsoever it be, can be proposed, but it must be comprehended under the reason or stile of one of these foure.

The

The Queſtions.

If 15 *Elles of Cloth coſt* 7 pound 10 ſhillings, *what comes* 27 *Elles to at that rate* ? Anſwer ; **13** pounds 10 ſhillings.

If 27 *Elles coſt* 13 pound 10 ſhillings, *what are* 15 *Ells worth* ? Anſwer ; 7 pound 10 ſhillings.

If 27 *Elles coſt* 13 pound 10 ſhillings : *how many Ells ſhall I have for* 7 pound 10 ſhillings ? Anſwer ; 15 *Elles.*

If I ſell 15 *Elles for* 7 pound 10 ſhillings : *how many Elles are to be delivered for* 13 pounds 10 ſhillings ? Anſwer ; 27 *Ells.*

If 8 *pound of any thing coſt* 16 ſhillings 6 pence : *what money is to be received for* 49 pound ? Anſwer : 5 pound, 1 ſhilling 0 $\frac{1}{4}$.

If 4 *pound of any thing coſt* 7 pence : *what money will* 8765 *pound of that commodity coſt* ? Anſwer ; 63 pound, 18 s ſhillings, 2 d $\frac{1}{4}$.

Of all which queſtions, I omit the work of pur-poſe, that you ſhall whet your wit thereby at con-ventent leiſure, to clime each branch, and gather the fruit of them, and doe minde now, before we make an end of this Rule, to give you ſ[...] ſtructions of the backer Rule of three, whoſe order is quite contrary to this that you have learned.

Scholar. I thanke you heartily for the ſix Que-ſtions, which I will (God willing) practiſe at con-venient times ; I pray you proceed therefore to the Backer or Reverſe Rule.

The

The Golden Rule, or Rule of Proportion Backward, or Reuerſe.

Maſter.

Note this well.

IN the former euermore loök how much the third number is greater then the firſt, ſo much the fourth number is greater then the ſecond. And contrariwiſe , look how much the firſt ſumme is greater then the third (if it doe chance ſo) ſo much is the ſecond ſumme greater then the fourth.

The backer or reverſe rule of three

But in this Rule, there is a contrary order, as this: That the greater the third ſumme is aboue the firſt, the leſſer the fourth ſumme is beneath the ſecond: and this Rule therefore you may call the Backer or Reuerſe Rule, as in example.

A queſtion of cloth.

If I haue bought 30 yards of Cloth, of two yards breadth, and would haue Canvas of three yards broad to line it withall, how many yards ſhould I need?

Scholar. **Why there is none ſo broad.**

Maſter. **I doe not care for that, I doe put this Example onely for your eaſie underſtanding: for if I ſhould put the Example in other meaſures, it would be harder to underſtand. But now to the matter: If you would know this queſtion, ſet your numbers as you did before: but you ſhall multiply now the firſt number by the ſecond, and that ariſeth thereof, you ſhall diuide by the third: which thing if you doe here, I mean if you multiply 30 by 2 it will**

will be 60 : which ſumme if you diuide by 3 there will appeare 20. whereby I know, that if 30 yards of cloth of two yards broad, ſhould bée lined with Canvas of three yards bꝛoad, 20 yards of Canvas would ſuffice, as this figure ſheweth.

Breadth.	Length.
2	30
3	20

And now becauſe ye found fault with my Example, how ſay you, perceiue you this ?

Scholar. Yes Sir, I ſuppoſe.

Maſter. *Then anſwer me to this queſtion : how many Elles of Canvas of Ell breadth, will ſerve to line twenty yards of Say, of three quarters broad ?*

Scholar. In good faith Sir I cannot tell, foꝛ I know not how to bꝛing the ſums to like Denominations.

Maſter. Then will I tell you : ſith there is mention here of quarters, & again euery one of the meaſures both Ells and yards may be parted into quarters, part them ſo both in the breadth & length, and then put foꝛth the queſtion by quarters.

Scholar. Then ſhall I ſay thus. How many quarters of Canvas of 5 quarters broad will line 80 quarters of 3 quarters broad ?

Maſter. Now anſwer to the queſtion.

Scholar. Firſt, I will ſet them downe in their foꝛme thus : foꝛ 5 is ioyned with

Breadth.	Length.
3	80
5	

the queſtion and is therefoꝛe the third number, then is 3 the number of the ſame Denomination, I meane becauſe they be both referred to breadth. Now I multiply 80 by 3, & it is 240, which I diuide by 5, & it yeeldeth 48. Then

Then ſay I that 48 quarters of 5 quarters broad will ſuffice to line 80 quarters of three quarters broad.

Maſter. Turne the quarters again into Ells and yards.

S. Then I ſay, that 9 Ells and three quarters of a yard of ellbroad, will ſerve to line 20 yards of three quarters broad, as this figure ſheweth.

Breadth.		Length.
3	Z	80
5		48

¶ Maſter. *Now what ſay you to this queſtion? I lent my friend* 400 *pound for* 7 *moneths, how much money ought he to lend mee againe for* 12 *moneths to recompence my courteſie ſhewed him? can you anſwer to this?*

Scholar. Yes Sir, I ſup-poſe, for I will ſet downe my Numbers thus: where I mul-tiply 7 into 400, and it mak-eth 2800, which I divide by 12, and it yeeldeth 233

Months.		Pounds.
7	Z	400
12		

pound, and there is 4 pound remaining of my Diviſion, what ſhall I do therewith?

Maſter. Turne the ſame 4 pound into ſhillings and then divide it by 12 as you did before.

Scholar. Well Sir, it ſhall be done: ſo have I 6 ſhillings for my Quotient, and yet remaineth ſhillings upon my Diviſion.

Maſter. You muſt alſo reduce that 8 ſhilling into pence, which maketh 96, and divide that all by your Diviſor.

Scholar. So have I done, and I finde 8 penc for my Quotient, and nothing is left.

Maſter

Maſter. This muſt you alwaies dae when any thing remaineth upon your Diviſion, whether it be money, weight, meaſure, oz any kinde of thing whatſoever. This Rule is ſo profitable foz all eſtates of man, that foz this Rule onely (if there were no more but it) all men were bound highly to eſteem Arithmetick.

By this rule may a Captain in war, wozke manie things, as Maſter Digges in his Stratiocos doth declare. Onely now in this my ſimple additi-on, foz a taſte and incouragement. I will inlarge the Author with a queſtion oz two moze, wiſhing you & every my Countreymen oz Gentlemen what-ſoever; that by nature be any thing given to Mili-ary affaires, to be familiar and acquainted with this Excellent Art, the which he ſhall finde not onely at the Sea, but alſo in the Campe and Field-ſervice, aboundantly to aid him. either in fortifica-tion, paying of Souldiers wages, charges of Ord-nance. Powder, Shot, Munitions, and Inſtruments whatſoever as foz example.

If it ſhould chance a Captain which hath 40000 ſouldiers to be incloſed with his enemy, that he could have no freſh purveyance of victuals, and that the vi-ctuals he had would ſerve that Army but onely three moneths, how many men ſhould he diſmiſſe to make the victuall to ſuffice the reſidue 8 moneths?

Queſtion of an Ar-mie.

Scholar. As you taught me, I ſet the numbers thus, ſay-inr, If thze e moneths ſuffice 40000, to how many will 8 ſuffice.

Moneths.		Men.
3	Z	40000
8		

To know this; I multiply the firſt number 3 into the

the ſecond 40000, and it yieldeth 120000, which ſumme I divide by 8, and there will be in the quotient 15000, which if I doe ſubtract from 40000, the remainder will declare that *Months.* *Men.* bee muſt diſmiſſe 25000 as   3 ⟍ 40000 this figure ſheweth.

                                        8   15000

A queſtion of a Fort.   *Maſter. Now anſwer mee to this queſtion : If 136 maſons in a moneth bee able to build a Fort to preſerve the ſouldiers from the Enemy, and ſuch expedition requireth that I would have the ſame finiſhed in eight dayes : how many workemen ſay you is there to be appointed ?*

     Scholar. As you taught mee , I ſet the numbers thus, ſaying : If 28 dayes require   28 ⟍ 136 136 Maſons, what number of men by the like proportion will 8   8 dayes require?

     To know I this multiply the firſt number 28 into 136, and it yieldeth me 3808 : which I divide by 8, and my Quotient is 476 : which is the juſt number of Maſons that ſhall ſupply this worke. And now me thinke theſe queſtions are very eaſie.

     Maſter. Truly if you take delectation herein, you ſhall finde this Art not onely eaſie, but wonderfull pleaſant and profitable. Now therefore one queſtion more I will propoſe, and ſo leave off this Rule in whole numbers untill we come to the uſe of it in broken numbers : for had you the underſtanding of broken numbers perfectly, not onely in this Rule, but in all other, the queſtion that in the ſight or appearance ſeemeth to bee 100 times harder to reſolve, may thereby bee wrought as ſoon or ſooner then this.

                                                Scholar

Scholar. Your words dos greatly incourage me to bee studious to attaine whole numbers: but might I once againe to be a Practitioner in broken numbers, I should thinke my selfe happy.

Master. *What say you then to this question? If 48 Joyners in two dayes make 200 light horsmen staves (esteeming they worke but 12 houres a day) and such need requireth that 384 Joyners are set to the finishing of those 200 staves; in what time, say you, will they make them up?*

Scholar. I see here that I must turn my 2 dayes into houres. And so doing, I set my numbers thus :

$$48 \quad Z \quad 24$$
$$384$$

Saying, if 48 men are 24 houres, 384 men will make an end quickly. For it is grounded upon an old Proverb, *many hands make light work.*

I multiply 48, into 24 and it amounteth to 1152, which I divide by 384, and my quotient is 3 houres, which is my desire.

I take this for a note worthy the marking, either Note *in the Rule of Three, forward, or backward, when the two numbers are multiplied together, the product is of the same nature and determination that the second number is of.*

M　　　　The

The double Rule of Proportion direct.

Master.

The double Rule.

Ell, fith you perceive now the ufe of this rule, I will fhew other which infue of the fame and firft the *double Rule*, which is fo called becaufe there is in it double working, by which thing onely it differeth from this.

Scholar. 𝔗𝔥𝔢𝔫 𝔟𝔶 𝔞𝔫 𝔢𝔵𝔞𝔪𝔭𝔩𝔢 𝔍 𝔰𝔥𝔞𝔩𝔩 𝔲𝔫𝔡𝔢𝔯𝔰𝔱𝔞𝔫𝔡 𝔦𝔱 𝔴𝔢𝔩𝔩 𝔢𝔫𝔬𝔲𝔤𝔥.

Queftion of carriage

Mafter. *So fhall you, and let this be the example : If the carriage of* 100 *weight (that is* 112 *pound)* 30 *miles doe coft* 12 *pence, how much will the carriage of* 500 *weight coft being carried* 100 *miles ?*

Scholar. 𝔍 𝔭𝔯𝔞𝔶 𝔶𝔬𝔲 𝔰𝔥𝔢𝔴 𝔪𝔢 𝔱𝔥𝔢 𝔴𝔬𝔯𝔨𝔦𝔫𝔤 𝔬𝔣 𝔦𝔱.

Mafter. You muft make two workings of it : the firft thus : If ℭ weight coft 12 pence, how much will five hundred weight coft ? Set your figures thus :

C *Weight.* *Pence.*

1 Z 12

5

𝔄𝔫𝔡 𝔪𝔲𝔩𝔱𝔦𝔭𝔩𝔶 5 𝔟𝔶 12, 𝔞𝔫𝔡 𝔱𝔥𝔢𝔯𝔢𝔬𝔣 𝔞𝔪𝔬𝔲𝔫𝔱𝔢𝔱𝔥 60, 𝔴𝔥𝔦𝔠𝔥 𝔦𝔣 𝔶𝔬𝔲 𝔡𝔦𝔳𝔦𝔡𝔢 𝔟𝔶 𝔬𝔫𝔢, 𝔱𝔥𝔢 Quotient 𝔴𝔦𝔩 𝔟𝔢 𝔰𝔱𝔦𝔩𝔩 60, 𝔱𝔥𝔞𝔱 𝔦𝔰 𝔱𝔥𝔢 𝔭𝔯𝔦𝔠𝔢 𝔬𝔣 500 𝔴𝔢𝔦𝔤𝔥𝔱 𝔣𝔬𝔯 30 𝔪𝔦𝔩𝔢𝔰.

𝔗𝔥𝔢𝔫 𝔟𝔢𝔤𝔦𝔫 𝔱𝔥𝔢 fecond worke, 𝔰𝔞𝔶𝔦𝔫𝔤: If 30 miles coft 60 pence, how much will 100 miles coft ? Set your figures thus.

Miles. *Pence.*

30 Z 60

100

𝔗𝔥𝔢

Then multiply 100 by 60 whereof amounteth 6000, which being divided by 30, will yield 200 pence. Then you must say, that so many pence shall cost the carriage of 5000 pound weight 100 miles, after the rate of 12 pence for the 100 carried 30 miles.

Scholar. Now I perceive it also.

¶ Master. These and such other like questions of the double Rule of Three, are to be answered much sooner, at one onely working by the Rule of Proportion composed of five numbers, which anon I will shew you, and then when you have the use thereof, you may use it which way you think good.

Scholar. Sir, I thanke you much for your courtesie, And I long now till this Rule bee ended, that I may see how I may behave my self with that new Rule of five numbers : for that I have ever since you taught me hitherto in the Golden Rule, both forward and backeward, wrought but with three numbers onely.

Master. But yet awhile we will go on forward with this Rule of Three, therefore answer to this question.

Thirty bushels of wheat sowed, yielded in one yeare 360, how many will 80 bushels yielde in 7 yeares ? I meane sowing every yeare of those seven, still fourescore bushels ? Question of sowing

Scholar. First I say, that if 30 bushels will yield 360 in one yeare, then 80 bushels will yield 960 in one yeare. Then for the second worke, I say, If one yeare yield 960, then 7 yeares will yield 6720 ; as these two figures doe shew.

M 2   Seed.

Seed.	Encrease.	Yeare.	Encrease.
30	360	1	960
80	960	7	6720

Question of Corn.

But now Sir if I set forth 30 bushels of Corne to another man for 7 yeares, agreeing so that hee shall sow every yeare the whole increase of the Corn, and I at the end of these 7 yeares to have the halfe of the whole increase : I would know how many bushels will there amount to my part, supposing the increase to be after the rate of the last question, for 30 bushels in one yeare to yield 360 ?

Master. In such a question you must have so many severall workings as there be yeares : As for Example, in the first yeare thirty bushels yield 360, then to know the yielding of the second yeare, I must say, If thirty yield 360, how many yieldeth 360 ? Worke by your Rule, and you shall finde 4320, Then say for the third yeare, If thirty yield 360, How many will 4320, yield ? you shall have 51840, and so every yeare multiplying the whole increase by 360, and dividing it by 30, the increase of the next yeare will amount, as these 7 figures following do orderly declare : where I have set 7 letters for the 7 yeeres, of which the first is set without Art, because that is the increase which you doe presuppose : and the last number of each other doth shew the increase of that yeare that it standeth for, which the letters doe declare, so that the increase of the seventh yeare is 1074954240 bushels : how many quarters that is, and also how many waies, you may by Reduction soon finde.

a
30———360

b
30 Z 360
360 Z 4320

c
30 Z 360
4320 Z 51840

d
30 Z 360
51840 Z 622080

e
30 Z 360
622080 Z 7464960

f
30 Z 360
7464960 Z 89579520

g
30 Z 360
89579520 Z 1074954240

Now with one queſtion more I will prove you. Queſtion. Iffixe *Mowers doe mow* 45 *Acres in* 5 *dayes, how* of mowing *many mowers will mow* 300 *Acres in* 6 *dayes*

Scholar. If 45 Acres require 6 Mowers, then 300 Acres require 40. Now again, if 5 daies require 40 mowers, then 6 daies need but 33 mowers.

Maſter. Why doe you not make mention of the 2 that remaineth in the laſt Diviſion? for the laſt part of the queſtion is wrought by the Backer Rule, here the firſt number 5 is multiplied into the ſecond, that is 40, whereof amounteth 200, which if you divide by the third number 6, the Quotient will be 33, as you ſaid: but then will th re remain 2, which cannot well be divided into 6 parts: howbeit you may underſtand by the 6 part of 2, the third part of one mans worke, which you muſt put

M 3   to

to the 33 ; or elſe you muſt ſay that 33 Workmen will end all the 300 Acres in 6 dayes ſave 2 mens worke for one day, or two dayes worke for one man, But ſuch broken numbers called Fractions, you ſhall hereafter better perceive, when I ſhall wholly inſtruct you of them,

Maſter. *Yet one queſtion more of field matters I will propoſe, and ſo I will make an end of this double Rule of Three.*

Scholar. With all my heart Sir I Thanke you, and I will diſpatch it as ſoon as I can, becauſe I would faine ſee the order of the next Rule of 5 numbers,

Queſtion of entrenchings.

Maſter. *If a Captain over a band of men did ſet 300 Pioners a worke, which in eight houres did caſt a trench of 200 Rods : I demand how many labourers will be able with a like trench in three houres, to intrench a camp of 3400 Rods.*

Scholar. I thinke I am now in the Backehouſe ditch : for I know not well which way to go about it.   And beſides that, truly I think I ſhal never come to preferment that way my growt is ſo ſmall,

Maſter. You know not how God may raiſe ye hereafter by knowledge and ſervice into the favor of your Prince, for the avail of your Countrey,

Example for Navigation : Sir *Francis Drake*, man greatly honoured for his knowledge, was n the talleſt man, and yet hath made as great an adve ture for the honour of his Prince and Countrey, ever Engliſhman did.

Scholar. Sir, I thanke you for your e incouragement. My minde, though I be lit is as desirous of knowledge, as any other : I have pon-dered now a little of it, and thus I set forth the worke.

Rod.	Men.
200	300
3400	Z

Saying, If 200 Rod require 300 men, what shall 3400 rods require ? I multiply 3400 by 300, and it yieldeth 1020000, which I divide by 200, and my quotient is 5100 men.

Then must I say for the second worke, if in 8 houres 5100 men, be able to discharge it, how many shall performe the same in 3 houres ? Now if I would worke by the Golden Rule of Proportion forward, I should finde a lesse number of men : because 3 houres is lesse then 8 houres : but because reason teacheth me, that the lesser the time is, wherein the trench must be made, the more Labourers I ought to have, thereupon I use now the Backer Rule, as in example. And I have in Quotient 13600. So many Pioners must I have to intrench the Campe in 3 houres.

Master. You have answered the question very officially : And truely I commend you for your diligence and apt understanding : and now according to my promise, I will (in whole numbers) give you a little taste of the Rule of Proportion, compounded in 5 numbers.

The

The Rule of Proportion, composed of 5 Numbers.

The first part of the Rule of Proportion compound, direct.

THe Rule of proportion composed, is distinct for most needfull questions into severa: parts or workings: And there belongeth unto it alwayes five numbers, whereof i: this Rule being the first par, t the secon: number & the fifth, are alwayes of one nature & lik: denomination, which Rule is to be wrought thus: yo: must multiply the first number by the second, & tha: shall be your divisor: Then again, multiply the other: numbers, the one by the other, and their Product sha: be your dividend.

And now according to my promise, we will firs: work the question of weight and carriage, whic: I delivered you in the double Rule of Three, to b: absolved by this Rule, which was this.

If the carriage of 1 C. weight, 30 miles cost 1: pence, what will the carriage of 5 C. weight stand m: in, being carried 100 miles?

C. weight.	Miles.	Pence.	C. weight.	Mil:
1	30	12	5	10C

Now marke well how these five numbers stan: Then multiply the first number by the second, : 30 by 1. which maketh but 30, that number ke: for your Divisor. Then multiply the other 3 nu: bers, the one into the other, that is to wit, 12 : 5, which maketh 60 : Lastly 60 by 100, which

V

you sée here in our Tables, ariseth to 6000, which 6000 you shall divide by the Product of the two first numbers, which here is 30.   And you sée there is found 200 pence, which is the dutie that you ought to pay for the carriage of 500 weight 100 miles, after the rate of 12 pence a hundred, and agréeth with the conclusion of the double Rule of Three.

Scholar. Sir I thank you, it is even so.

Master. *Yet note this in a generality in this Rule,* Note *looke what nature of denomination your middle number is of (which here are ponce) and of the like denomination or nature is alwayes your quotient.*

Scholar. Well now and if it please you, by your patience, I will sée how I can end the question next following of 30 Bushels of wheat sowen, which in one year yéeldeth 360,   how many then will 80 Bushels yéeld in 7 year following every year of those 7 till 80 bushels, and accor= ding to your rea= sons I set my num= bers thus.

Bush.	Year.	Bush.	Bush.	Year.
30 ——	1 ——	360—	80 —	7
		80		
		28800		
		7		
		201600		

There I multiply 30 by 1, and it maketh 30 my Divisor; then multiplying the other 3 numbers the one into the other, as here appeareth in my Tables, they make 201600, which I divide by 30: & my Quotient is 6720 bushels, my desire; for so much also it came to at two workings by the Rule of Three.

Master.

Master. Yet one question moze I will propound unto you, and so leave this Rule, till it please God hereafter, that I may make you wozke it in broken numbers.

Question of Interest. *What comes the interest of* 258 *pound, for* 5 *moneths to, after the rate of* 8 *pound, taken in the* 100 *pound for* 12 *moneths?*

Scholar. Sir, this is yet within the compasse of some reasonable usance. Therefoze to minister equitie in this case, I will see how I can wozke the same, which I set down thus, prazing you if I have not done well, to shew me mine errour.

li. *moneths*.li.li. *mon.*
100---12-8--258--5.

Master. Pzoceed, you have done very well.

Scholar. Then I doubt not by the grace of God but to end it: I multiply 100 by 12, it yieldeth 1200, and the thzee other numbers multiplied together pzoduce 10320, which I divide by 1200: and my Quotient is eight pounds. Then accozding as you have taught mee heretofoze, I turne the 720 pound that I left, into shillings: and dividing it by the first number my Quotient is 12 shillings. So I answer, that the loane of 258 pounds foz 5 moneths, after the rate of 8 pound in the 100 pound foz a yeare, comes to 8 pound 12 shillings.

Master. You say true, I commend your diligence: now behold the manner of the second part of this rule.

The

The backer Rule, or the second part of the Rule of Proportion compound.

Master.

IN the second part of this Rule of Proportion composed, the third number is like unto the first. And the Rule is to bee wrought thus : you shall now, contrary to the last Rule, multiply the third number and the fourth together, and that Product shall be your Divisor. Then multiply the fifth by the second, and the Product thereof by the first : and that is the number that shall be divided. For example I propound this question for a proofe of my last question of Interest.

A Merchant hath received 8 pound 12 shillings, for interest of certain money for 5 moneths tearme, which he received after the rate of eight pound in the 100 for a yeare. The question is now, how much money was delivered to raise this interest.

The proofe of the last question.

Behold therefore the manner, how the question is set forth.

	li.	moneths.	li.	moneths.	li.	s.
	100 —	12 —	8 —	5 —	8 ·	12

Scholar. Sir I perceive it very well : and according to the doctrine which you prescribed for the working thereof : if please you now it is set downe, I thinke I can follow the worke.

Master.

Master. Nay, stay a while, and before you worke, marke well how I deliver a reason for the perfect understanding of this Rule, which is thus:

If 8 pound in 12 moneths doe yeeld mee 100 pound, to take 8 pound 12 shillings for five moneths, must needes yeelde a great deale more.

So upon the knowledge that I have in this Art, The first part of this Rule is answerable to the Rule of Three forward : and this latter part accordeth to the Rule of Three backward.

Scholar. Sir, I yield you most heartie thanks for these your last instructions, they have given me great light into these two rules, whereby I may the better by deliberation conceive how to use them hereafter when occasion shall require.

Master. You say well, goe to now if you will, and trie your cunning in the question : But this note take with you by the way, in as much as here is mention made of shillings : turn all your monie as you work into shillings, for your more ease in working.

Scholar. If it please you to behold me a little, I will quicklie end it : for I have but my first, my second, and my last number to be multiplied together for my Dividend : And my third into my fourth for my Divisor.

(margin note: Note.)
(margin note: Note.)

li.

li.	*Moneths.*	*Moneths.*	li.	s.

```
100 ————12        8—5—8——12
 20                 20     20
————              ————   ————
2000               160     172
  12                5
————              ————
4000               800
2000

————
24000
  172

————
 48000   4128000(
168000   8     00
24000
————4128000
```

Which 4128000 I divide by 800, and my Quotient is 5160 shillings, **which in pounds yieldeth** 258 my desire.

Master. I will here for this time in whole numbers end this Rule, and I will instruct you in the Rules of fellowship. You may at your convenient leisure for your exercise worke the same by the Rule of Three at twice.   And for your aide and incouragement therein, I set down here a proffer how to apply it.

	A			B	
Moneths.	li.	s.	*pounds.*	li.	
5	8 —12		412 ⅘	100	
12	412 ⅘			258 li.	

The

The Rule of Fellowſhip.

Ut now will I ſhew you of the Rule of
Fellowſhip or Company, which hath
ſundry operations according to the
divers number of the Company. This
Rule is ſometime without difference of
time, and ſometimes there is in it difference of time.
Firſt I will ſpeak of that without difference of time,
of which let thi be an example.

Aqueſtion
of compa-
ny.

*Foure Merchants of one Company made a bank of
money diverſly : for the firſt laid in* 30 *pound, the
ſecond* 50 *pound, the third* 60 *pound, and the fourth*
100 *pound, which ſtock they occupy ſo long, till it
was increaſed to* 3000 *pound. Now I demand of you
what ſhould each receive at the parting of this money?*

Scholar. I perceive that this Rule is like the
other, but yet there is a difference which I per
ceive not.

Maſter. Then will I ſhew it to you : Firſt by
Addition, you ſhall bring all the particular ſumme
of the Merchants into one ſumme, which ſhall be the
firſt ſumme in your working by the Golden Rule
and the whole ſumme of the gaines by that ſtock
ſhall be the ſecond ſumme. Now for the thir
ſumme ou ſhall ſet the portion of each man   3
one after another, and then worke by the   5
Golden Rule, and the fourth ſumme will   6
ſhew you each mens gaines : as in exam-   10
ple.   ‾‾
24

E

The parcels of the foure Merchants make in one ſumme 240 pounds : ſet that in the firſt place, the gaines in the ſecond, and the firſt mans portion of ſtocke in the third place, thus,

240 Z 3000
30

Now multiply the ſecond by the third, and it will be 90000, which you ſhall divide by 240, and there will appeare 375 pounds, thus :

And that is the gains for the firſt man.

240 Z 3000
30   375

Now for the ſecond man ſet the 50 pound that hee brought, in the third place, and work as before : and his part will be 625 pound : as this figure ſheweth.

240 Z 3000
50   625

Likewiſe for the third man, ſet his money which was 60 pounds, and his part of gains will be 750 pounds, as here appeareth

240 Z 3000
60   750

And ſo for the fourth man ; if you ſet his ſumme which is 100 pound, his gaines will be 1250 pound, as the worke will declare.

240 Z 3000
100   1250

Scholar. This I perceive : but is there any way to examine whether I have well done or no?

Maſter. For the triall hereof, adde together all their foure portions, and if their addition make the whole ſumme of their gaines, then is the work well done.

Note this common proofe

Scholar. That will I trie by and by ; the foure parcels are thoſe, which added together make

3000,

3000, which is the juſt ſumme
of money that they gained,
whereby I know the work is
well done.

375
625
750
1250
3000

Maſter. Well, now another example will I put
to you not of gains, but of loſſe: for one reaſon ſerveth for both.

A queſtion
of loſſe.

*If three Merchants in one ſhip, and of one fellow-
ſhip, had bought Merchantdiſe, ſo that the firſt had
laid out 200 pound, the ſecond 300 pound, the third
500 pound, and it chanced by tempeſt that they did caſt
over board into the Sea Merchantdiſe of the value of
100 pound, how much ſhould each man bear in this
loſſe?*

Scholar. If I ſhall do in this as you did in the
other queſtion, then muſt I joyne their 3 portions
together, 200, 300, 500, which maketh 1000.
Then ſay I, If 1000 loſe 100 then ſhall 200 loſe
20, and 300 ſhall loſe 30, and 500 ſhall loſe 50, as
by the three figures it doth appear plain.

$$1000\; Z\; 100 \qquad 1000\; Z\; 100$$
$$200\quad\; 20 \qquad\quad 300\quad\; 30$$

$$1000\; Z\; 100$$
$$500\quad\; 50$$

Maſter. Well, ſith now you have done theſe,
will propound a queſtion of more importanc
which ſhall make you not onely the abler to u
derſtand this Rule, but alſo it will greatly ayd y
in the next Rule of fellowſhip with time, if ſu
need be that your money be of divers Denom
nations.

F

For this may not bee forgotten in all ſuch queſtions: If the number be of divers kindes, you muſt by reduction bring it into one kinde, that is to ſay, to the leaſt value that is named in the queſtion. And likewiſe ſhall you doe, if the time be of divers kindes, as ſome yeares, ſome moneths, weeks, and dayes, you ſhall make all moneths, weeks, or dayes, according as the leaſt name of time in the queſtion is, as for example.

Firſt in diverſity of money. Three companions A queſtion bought 2000 ſheep, and paid for them 241 pound of ſheepe. 13 ſhillings 4 pence, of which ſumme one payd 101 pound, 10 ſhillings. The ſecond 82 pound 17 ſhillings 10 pence. And the third paid 57 pound, 5 ſhillings 6 pence: How many ſheepe muſt each of them have? Anſwer. The firſt ſhall have 840. The ſecond 686. And the third 474. And that muſt you worke thus:

Firſt, conſidering that your money is of divers Solution. Denominations, you ſhall (by reduction) bring it all into the ſmalleſt Denomination which is in it, that is to ſay, pence; and ſo will the Totall ſumme be 58000 pence.

Now if you turne each mans money into pence alſo, the firſt mans ſumme will be 24360 pence: The ſecond mans money will bee 19894 pence; and the third mans money will be 13746 pence.

Now to know how manie ſheep everie man ſhall have, let the whole ſumme of money, that is, 58000 pence be ſet in the firſt place, and in the econd place ſet the number of ſheep, and then order-ly in the third place ſet each mans money, and then multiplying the third and the ſecond ſummes toge-ber, and dividing that that amounteth by the firſt,

P   there

there will appeare the number of ſheep that a
man ought to have : as theſe 3 figures do ſhew.

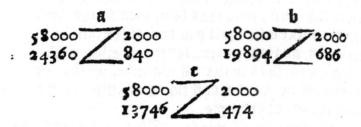

S. Why doe you ſet the money in the firſt place
ſeeing in the queſtion you ſay 2000 ſheep coſt 58000
pence, and not thus : 58000 coſt 2000 ſheep?

Maſter. You remember I taught you at the
beginning of the Golden Rule, that the firſt and the
third numbers muſt be of one name, and of like
things : & evermore the number that the queſtio
is aſked of, muſt bee ſet in the third place.

Now is the queſtion plainly this : If foure men
bought 2000 ſheep for 58000 pence, how many ſheep
ſhall each man have?

But ſeeing in this queſtion, there ought more
reſpect to be had to the ſumme of money, then
the ſumme of the perſons; (for in the ſumme of mon
is their proportion toward the ſheep, and not
the number of perſons.)

If 58000 pence bought 2000 ſheepe, how man
did 24360 buy? Againe, how many did 19894 pe
buy? And how many bought 13746 pence?

Scholar. I perceive it reaſonable, and ſo
I doe in all queſtions.

M. Even ſo. But for eaſineſſe of the wor
marke this: Whenſoever the firſt & ſecond numb

b

hath Cyphers in the firſt places, you may both in
the Multiplication and in the Diviſion leaue out
thoſe cyphers, ſo that you leaue out like manie out of
both ſummes, as in this queſtion, the firſt number
58000 hath 3 cyphers, and ſo hath the ſecond, that
is 2000 : therefoze caſt away their cyphers, and ſo
will the firſt number bée 58, and the ſecond 2. ſet
them in their places, and wozke accozding to the
Rule, and you ſhall perceiue that will bee all one,
ſauing that this is the ſhozter and eaſter way, as
theſe three figures doe ſhew.

$$ \overset{a}{\underset{\Large Z}{}} \quad {}_{58}\diagdown \diagup {}_{2} \quad {}_{14360}\diagup \diagdown {}_{840} $$

$$ \overset{b}{\underset{\Large Z}{}} \quad {}_{58}\diagdown \diagup {}_{2} \quad {}_{19894}\diagup \diagdown {}_{686} $$

$$ \overset{c}{\underset{\Large Z}{}} \quad {}_{58}\diagdown \diagup {}_{2} \quad {}_{13746}\diagup \diagdown {}_{474} $$

And this you ſée is both eaſter, and alſo the
moze certaine way to know the anſwer to this
queſtion.

Scholar. Truth it is as you ſay : But Sir me
ſéemeth I might aske a further queſtion here, not
onely ho w many ſheep each man ſhould haue, but
alſo what everie ſheep coſt.

Maſter. That queſtion doth not onelie belong
to this Rule, but may alſo be diſcuſſed by Diviſion,
eſpecially if the queſtions number bee one onely, as
thus : Divide the totall ſumme 58000 pence by 2000
(oz 58 by 2) omitting the cyphers, and the Quotient
will be 29 pence, that is 2 ſhillings 5 pence. Howe-
beit, by this Rule you may doe it, and beſt when
the number of the queſtion doth excéede 1 ; as if I

N 2   ſhould

ſhould aſke this queſtion, 2000 ſheep coſt 58000 pence, how much doe 20 coſt? Then ſhall I ſet my figures as before.   2000 Z 58000   20

And doing after the Rule, there will amount 580 pence, that is, 2 pound 8 ſhillings 4 pence, the price of one ſcore: but if you will uſe that eaſie way that I did teach you now, you may change the firſt and ſecond number thus.   2 Z 58   20

Thus doe you perceive the uſe of the Rule without time.

Scholar. All this I underſtand very well: I pray you now inſtruct mée in the Rule of Fellowſhip with Time.

The Rule of Fellowſhip with Time.

Maſter.

The Rule of Fellowſhip with time.

TO the intent you may as well perceive the ſame Rule with diverſity of time, I propoſe this example.

Feure Merchants made a common ſtocke, which at the years end was increaſed to 35145 pound. Now to know what ſhallb each mans portion of gain, you muſt know each mans ſtocke, and time of continuance.

The firſt man of theſe foure laid in 669 l. whic
h

he did take from the ſtock again at the end of 10 Queſtion
moneths.   The ſecond man laid in 810 pound, for of a Bank
8 moneths. The third man laid in 900 pound, for
ſeven moneths.   And the fourth laid in 1040, for 12
moneths.

This queſtiou ſhall you examine as you did the Note.
other befoze, ſaving that whereas in the third place A generall
of the Figure you did ſet each mans ſumme alone, Rule.
here you ſhall ſet the ſame being multiplied by the
number of their time : and likewiſe in the firſt
place of the figure you ſhall ſet the number which
amounteth of their whole ſummes ſo multiplied by
their time, and added into one whole ſumme, as
thus.

The firſt mans ſumme is 669 pounds, which I
multiply by 10 (that was the number of his time)
and it maketh 6690. The ſecond mans ſumme 810
pound, multiplied by 8 (which was his time) maketh 6480.   The third mans ſumme 900 pound,
multiplied by 7 (foz that was his time) yeeldeth
6300. The fourth mans ſumme was 1040 pound,
and his time 12 : multiplie the one by the other,
and it will be 12480.

The foure ſummes thus multiplied by their time,
muſt be ſet ozderly in the third place of the figure,
and in the firſt place muſt be ſet the whole ſumme
of all foure, which is 31950 ; and the gain muſt
et n the ſecond plcce, which is 35145. Now to end
he queſtion, J ſay firſt, Jf 31950 did get 35145,
hat did 6690 get? Anſwer.
359 pounds, as by this   31950 Z 35145
gure apeareth.   6690 Z 7359

P 3   Likewiſe,

Likewiſe, the ſecond man had to his part 7128 pound, the third muſt have 6930 pounds, and the fourth man ſhall have for his part 13728 pounds, as theſe figures doe partly declare.

b
$$3 1 9 5 0 \diagup Z \diagdown 3 5 1 4 5$$
$$6 4 8 0 \diagup \diagdown 7 1 2 8$$

c
$$3 1 9 5 0 \diagup Z \diagdown 3 5 1 4 5$$
$$6 3 0 0 \diagup \diagdown 6 9 3 0$$

d
$$3 1 9 5 0 \diagup Z \diagdown 3 5 1 4 5$$
$$1 2 4 8 0 \diagup \diagdown 1 3 7 2 8$$

Another proofe.

Scholar. This I like very well : but what proof is there of this work ?

Maſter. The ſame that I taught you for the other : howbeit, there is uſed both for this work and the other alſo, this manner of proofe, to adde all the portions together, and it may agree to the whole ſumme, then ſeemeth your worke well done: but this is no ſure proofe.

S. Yet will I prove in this example: the 4 parcels are theſe, which if I adde together, there will amount 35145, and that was the whole ſumme, whereby I perceive the work is well done.

Maſter. If it fall out otherwiſe, be ſure it is not well.

7359
7128
6930
13728
————
35145

Scholar. Then do I underſtand this worke alſo very well : But what have I now to learn ?

M. There are many other excellent parts behinde, of which I will not as now make mention, becauſe that without the knowledge of Fractions they cannot bee duely taught, and much leſſe underſtood Therefore will I propoſe to you two or three queſtion

queſtions moꝛe (that thereby you may better perceiue the uſe of this Rule and all other the like)
and ſo make an end foꝛ this Time.

¶ *Three Partners by ſome ill adventure ſuſtained* A queſtion
the loſſe of 160 *pound, whereof the firſt laid into the* of loſſe.
common ſtocke 200 *pound, for ten Moneths. The*
ſecond laid in 350 *pound, and the third* 100 *pound,*
but for how long the two latter, is unknown : But
braking off their Partnerſhip, the firſt found himſelfe
a looſer 80 *pound, the ſecond* 56 *pound, and the third*
24 *pound. The queſtion is, for how long time was*
the money of the two latter in company.

Foꝛ the ſolution hereof, and of ſuch other like,
you muſt alſo multiply the firſt mans 200 pound,
that hée put into the ſtocke by his time of continuance, which was ten moneths ; and it maketh
2000 : wherefoꝛe now I affirme, if his money
that loſt 80 pound multiplied by his time make
2000 : what ſhall his money make that loſt 56
pound, and his that loſt 24 pound, which two numbers I commit to the triall of the Rule of Three at
two woꝛkings, thus ;

If 80 giue 2000, what giueth 56 ? And again,
if 80 giue 2000, what giueth 24 ?

80	2000		80	2000
56	1400		24	600

To conclude, if you now diuide 1400, the ſecond
mans portion, by 350, which was his ſtock that he
laid into company, you ſhall find in your quotient
4 moneths, and foꝛ ſo long time did the ſecond man
put his money into the common ſtock.

Laſtly, if you diuide the third mans new laying
in, which was 600 by 100. which was his ſtock
　　that

that he put into the company, the quotient declareth his time of continuance, which was ſix moneths. And thus is the queſtion reſolved.

Scholar. Sir, I have attentively beheld your working, and the more we travell herein, the more me thinke I am in love with this excellent Art.

Maſter. Then what ſay you to this Queſtion?

A queſtion of Canons

There is in a Cathedrall Church 20 Canons, and 30 Vicars, thoſe may ſpend by year 2600 pound, but every Canon muſt have to his part 5 times ſo much as every Vicar hath : how much is every mans portion ſay you ?

Scholar. I pray you make the anſwer your ſelf alſo, ſo ſhall I perceive beſt the means to anſwer to each other like.

Maſter. In this Queſtion, you muſt doe as in thoſe beforeſaid, that have diverſitie of time, for here is diverſity of portions. Therefore ſhall you multiply the number of the perſons by their difference of portions : (as you did in the other by time:) Then muſt you multiply the 20, (which is the number of Canons (by 5.) for that is the number of their portion) ſo will it be 100. Then 30, (that is the number of Vicars) by 1, (that is the number of their portion) and it will be 30 : put theſe two ſummes together, and they make 130. Then ſay thus ; If 130 ſpend 2600 pounds, what may 100 ſpend ? The Rule ſheweth 2000 pounds.

Again for Vicars if 130 ſpend 2600 pound what may 30 ſpend? Anſwer 600 pound, as their figures ſhew.

130 Z 600   130 Z 3600
100 Z 2000   30   Z 600
                          But

But if every Canon ſhould have ſo often times 4 pound as the Vicar ſhould have 3 pound, then ſhould I multiply 20 by 4,(that were 80) & 30 by 3, (that were 90) and then both were 170. Then ſhould the figures be ſet as followeth.

	li. ſ. d.			li. ſ. d.
170 Z	2600		170 Z	2600
80	1223,--10,7		90	1376,---9,5

But this ſort is too hard for you, by reaſon of the fractions, therefore I will let it reſt to that place.

And by this rule you ſee what the 20 Canons may ſpend ; which ſumme if you divide by 20, you ſhall ſee each Canons proportion:and ſo of the Vicars, if you divide their ſumme by 30, the quotient will declare every Vicars portion.

The ſecond Dialogue.
The accounting by Counters.

Maſter.

Ow that you have learned Arithmetick with the Pen, you ſhall ſee the ſame Art in Counters : which feat doth not onely ſerve for them that cannot write and read, but alſo for them that can do both ; but have not at ſome time their pen or tables ready with them.

This ſort is in two formes commonly, The one by lines, and the other without lines. In that that

that hath lines, the lines doe ſtand for the order of places: & in that that hath no lines, there muſt be ſet in their ſtead ſo many Counters as ſhall need for each line one; and they ſhall ſupply the ſtead of lines.

    Scholar. By examples I ſhould better perceive your meaning.

    Maſter. For example of the lines; loe here you ſee ſix lines, which ſtand for ſix places, ſo that the nethermoſt ſtands for the firſt place, & the next above is for the ſecond, & ſo upward till you come to the higheſt, which is the ſixth line, and ſtandeth for the ſixth place.

```
————100000————
————10000————
   *————1000————
————100————
————10————
————1————
```

Numeration by Counters.

    Now what is the value of every place or line, you may perceive by the figure which I have ſet on them, which is according as you learned before in Numeration of figures by the Pen: for the firſt place is the place of unites or ones, and every counter ſet in that line, betokeneth but one: and the ſecond line is the place of 10, for every counter there ſtandeth for 10: the third line the place of hundreds, the fourth of thouſands, and ſo forth.

    Scholar. Sir, I doe perceive that the ſame order is here of lines, as was in the other figures by places, ſo that you ſhall not need longer to ſtand about Numeration, except there be any other difference.

                                  Maſter.

Master. If you do understand it,⸻⸻⸻

then how will you set 1543 ?   *1 ⸻⸻

    Scholar. Thus as I suppose.   5 ⸻

    Master. You have set the places 4⸻

truly, but your figures be not meet 3 ⸻

for this use : for the meetest⸻⸻

figures in this behalfe, is* ⬤⸻

the figure of a counter round,⸻⬤⸻

as you see here, where I⸻

have expressed that same⸻ ⬤⬤⬤⬤

summe.

    Scholar. So that you⸻ ⬤⬤⬤⸻

have not one figure for 2, nor 3, nor 4, & so forth, but

as many digits as you have, so many counters you

set in the lowest line, and for every 10 you set one

in the second line, and so of other. But I know

not by what reason you set that one Counter for

500 between two lines.

    Master. You shall remember this, that whenso=

ever you need to set down 5, 50, or 500, or 5000, or

set forth any number whose numerator is 5, you

shall set one counter for it in the next place above

the line that it hath his denomination of : As in

this example of that 500, because the numerator is

5, it must be set in a void space, and because the

denomination is a hundred, I know that the place

is the void place next above hundreds, that is to

say, above the third line.

    And further you shall marke, that in all work-

ing by this sort, if you shall set downe any

summe

ſumme betweeŋ 4 aŋd 10, * ───────
foʒ the firſt part of that
numbzr you ſhall ſet down ●●───────
5, ¢ then ſo maŋy counters ●
moʒe, as there reſt num- ●●●● ───<─
bers aboʋe 5. And this is ●
true both of digits ¢ articles. * ●●
And foʒ example, J will ſet ●
down this ſumme 297965, ●●●● ──────
which ſumme if pou marke ●
well, pou néd none other ● ───────
examples foʒ to learne the ●
numeration of this ſerme. ──────

But this ſhall you marke, that as you did in
other kindes of Arithmetick, ſet a pʒick in the
places of thouſands, in this woʒk you ſhall ſet a
Starre, as you ſée befoʒe. •

Scholar. Then J perceiʋe Numeration : But,
J pʒay you, how ſhall J do in this Art, to adde two
ſummes oʒ moʒe together ?

Addition.

Maſter.

THe eaſieſt way in this is to adde but two
ſummes at once together : Howbeit you
may adde more, as I will tell you
anon.

Therefoʒe when you will adde
two ſummes, you ſhall firſt ſet down
one

one of them, it forceth not which, and then by it
draw a line croſſe the other lines. And afterward
ſet downe the other ſum,
ſo that the line may be between then : as if you
would adde 2659 to 8342,
you ſet your ſummes as
you ſee here.

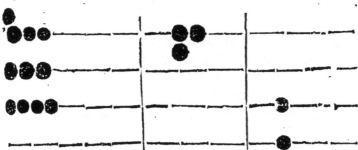

Addition
of two
ſummes.

And then if you liſt you
may then adde the one to
the other in the ſame place : or elſe you may adde
them both together in a new place : which way,
becauſe it is moſt plaine, I will ſhew you firſt.

Therefore will I beginne at the Unites. which
in the firſt ſumme is but 2, and in the ſecond ſumme
9, that maketh 11 : Thoſe do I take up, and for
them I ſet 11 in the new room, thus :

Then doe I take up all the Articles under a
hundred, which in the firſt ſumme are 40, and in
the ſecond ſumme 50, that maketh 90 : or you may
ſay better, that in the firſt ſumme there are foure
Articles of 10, and the ſecond ſumme 5, which maketh 9, but then take heed that you ſet them in their
right lines, ſee here.

Where

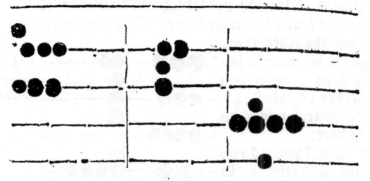

Where I have taken away 40 from the first summe, and 50, from the second, and in their stead I have set 90 in the third room, which I have set plainly, that you might well perceive: howbeit seeing that 90 with the 10 that was in the third room, already, doth make 100, I might better for those 6 Counters, set 1 in the third line, thus:

For it is all in one summe, as you may see, but it is best never to set five counters in any line, for that may be done with one counter in a higher place.

Scholar. I judge that good reason, for many are unneedfull where one will serve.

Master. Well, then will I adde forth of hundreds: I finde 3 in the first summe, and 6 in the second, which maketh 6000, them doe I take up, and set in the third room, where is 100 already, to which I put 900, and it will be 1000: therefore I set one counter in the fourth line for them all, as you see here.

Then

Then adde I the thousands together, which
in the first summe are 8000, and in the second 2000
that maketh 10000, them doe I take up for those
two places, and for them I set one counter in the
fifth line, and then it appeareth as you
sée to bée 11001, for so many doth
amount of the Addition of 8342 to
2659.

Scholar. Sir, this I do perceive :
but how shall I set one summe to an=
other, not changing them to a third
place ?

Master. Marke well how I doe it,
I will adde together 65436 and 3245,
which first I set down thus :

To adde
summes
together.

Then

Then doe I begin with the smallest Denomination, which is 1 in the second summe, and set it in his place : then doe I finde 5 in the first summe, and 5 in the second, which put together, saving the two Counters, cannot be set in a void place of 5, but for them both I must set one in the second line, which is the place of 10, therefore I take up the five of the first summe and the 5 of the second, and for them I set one in the second line, as you sée here.

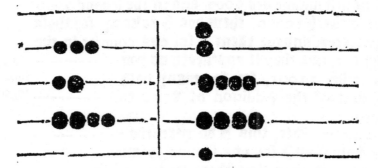

Then do I likewise take the 4 Counters of the first summe and second line (which maketh 40) and adde them to the 4 counters of the same line in the second summe, and it maketh 80 : but as I said, I may not conveniently set above 4 counters in one line, therefore to those 4. that I took up in the first summe, I take one also of the second summe, and then have I taken up 50 : for which 5 counters I set down one in the space over the second line, as here doth appear.

And

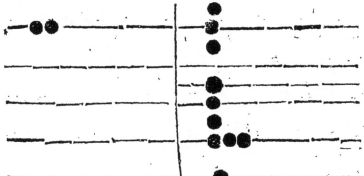

And then is there 80, as well with those 4 counters, as if you had set down the other 4 also.

Now do I take the 200 in the first summe, and adde them to the 400 in the second summe, and it maketh 600, therefore I take up the two counters in the first summe, and three of them in the second summe, and for them 5, I set 1 in the space above, thus:

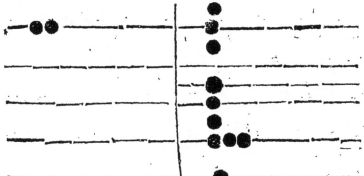

Then take I the 3000 in the first summe unto which there are none in the second summe agreeing. therefore I doe onely remove those three counters from the first summe into the second, as here doth appear.

M   And

And you fee the whole fumme that amounteth of that Addition of 65436 with 3245, to bée 68681.

And if you bé ve marked thofe two examples well you need no further inftruction in Addition of two onely fummes : but if you have moze then two fummes to adde , you may adde them thus :

First adde two of them, and then adde the third and fourth, oz moze, if there be fo many : As if I would adde 2679, with 4286, and 1391, First I adde the two first fummes thus :

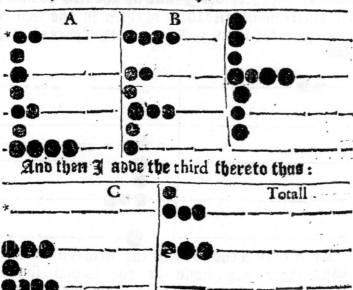

And then I adde the third thereto thus :

And so of moꝛe if you habe them.

Scholar. Now I think it beſt that you paſſe foꝛth to Subtraction, except there be any way to examine this manner of Addition, then I thinke that were gꝏd to be known next.

Maſter. There is the same pꝛoof here that is in the other Addition by the Pen, I meane Subtraction ; foꝛ that onely is a ſare way : but conſidering that Subtraction, muſt be firſt known, I will firſt teach you the Art of Subtraction; and that by this Example.

Subtraction.

*I would ſubtract 2892 out of 8746. Theſe ſummes
muſt I ſet downe as I did in Addition : but here
it is beſt to ſet the leſſer number firſt, thus :*

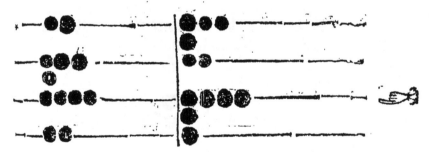

Then ſhall I begin to ſubtract the greateſt numbers firſt, (contrary to the uſe of the Pen) that is the thouſand in this example : therefore I finde among the thouſands 2, foꝛ which I withdꝛaw ſo any from the ſecond ſumme, (where are 8) and o remaineth there 6; as this example ſheweth.

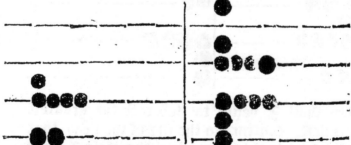

Then doe I likewise with the hundreds, of
which in the first summe I finde 8, and in the
second summe but 7, out of which I cannot take 2,
therefore this must I doe : I must look how much
my sume differeth from 10, which I find here to be 2,
then must I abate for my summe of 800, one
thousand, and set down the excesse of hundreds, that
is to say 2, for so much as 1000 is moze then I
should take up : therefore from the first summe I
take that 800, and from the second summe (which
are 6000) I take up one thousand, and leave 5000,
but then I set down the 200 unto the 700 that are
there already, and make them 900, thus :

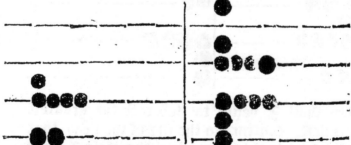

Then come I to the Articles of tennes, where in
the first summe I finde 90, and in the second summe
but onely 40, Now considering that 90 cannot b
abate

abated from 40, I looke how much that 90 doth differ from the next summe above it, that is, 100 (or else which is all to one effect) I looke how much 9 doth differ from 10, and I finde it to be 1 : then in the stead of that 90, I doe take from the second summe 100 : but considering that is 10 too much, I set down 1 in the next line beneath for it, as you see here.

Saving that here I have set 1 Counter in the space instead of 5 in the next line. And thus I have subtracted all save 2, which I must abate from 6 in the second summe, and there will remain 4 thus.

So that if I subtract 1892, from 8746 the remainer will be 5854.

And that this is truely wrought you may prove by Addition : for if you adde to this remainer the same summe that you did subtract, then will the former summe 8746 amount again.

Scholar. That will I prove, and first I set the A proof of summe that was subtracted, which was 2892, and Subtraction. then the remainer 5854. thus :

Then

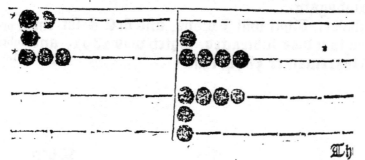

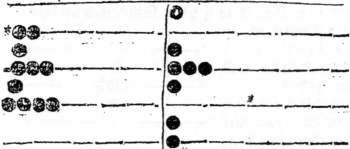

Then doe I adde the firſt 2 to 4, which maketh
6 : ſo take I up 5 of thoſe Counters, and in their
ſtead I ſet 1 in the ſpace : and one in the loweſt
line, as here appeareth.

Then do I adde the 90 next aboue to the 50
and it maketh 140. Therefore I take up thoſe
Counters, and for them I ſet 1 to the hundreds in
the third line, and 4 in the ſecond line thus :

Th

Then doe I come to the hundreds, of which I inde 8 in the first summe, and 8 in the second, that maketh 1600, therefore I take up those 8 counters, and in their stead I set 1 in the fourth line, and 1 in the space next beneath, and in the third line, as you may see here.

Then is there left in the first summe but onely 100, and in the second 5000, which is 7000, which I shall take up from thence, and set in the same line in the second summe to the one that is there already: and there will the whole summe appeare as you may well see to be 8746, which was the first of those sum: and therefore I do perceive that I had well subtracted before.

And thus may you see how Subtraction may be tried by Addition.

Scholar. I perceive the same order here with counters, that I learned before in figures.

Master. Then let me see how you can trie Addition by Subtraction.

D 4 .   Scholar.

Scholar. First I will set forth this example of
Addition, where I have added 2189 to 4988. And
the whole summe appeareth to be 7177.

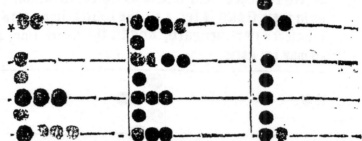

Now to try whether that summe be well added
or no, I will subtract one of the first two summes
from the third. And if I have well done, the remainer will be like that other summe: as for example, I will subtract the first summe from the third,
which I set thus in order.

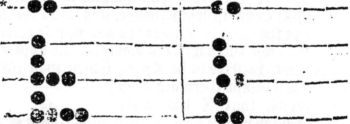

Then do I subtract 2000 of the first summe, from
the second summe, & then remains there 5000, thus

Then in the third line
I subtract the 100 of the
first from the second
summe, where is onely
100 also: and then in
the third line, resteth nothing, as you may see in
this example following.

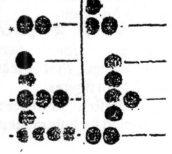

Then in the second
line with his space over
him I finde 80, which
I should subtract from
the other summe : then
seing there are but only
70, I must take it out
of some higher summe,
which is here onely

5000: therefore I take up 5000 : and seing that
is to much by 4920, I set downe so many in the
second roome, which with the 70 being there al=
ready do make 4990, and then the summes do stand
thus.

Yet remaineth therein the first summe 9, to be
abated from the second sume, where=
in that place of unites doth appear
onely 7 : then must I abate a higher
sume, that is to say 10, but seing that
10 is more then 9 (which I should
abate) by 1, therefore shall I take
up one Counter from the second, and
set downe the same in the first line,
or lowermost line, as you see here.

And

And so have I ended this work, and the summe appeareth to be the same which was the second summe of mine Addition, and therefore I perceive I have well done.

Another way of Addition.

Master. To stand longer about this, it is but folly: except that this you may also understand, that many doe beginne to subtract with Counters, not at the highest summe, as I have taught you, but at the neithermost, as they doe use to adde; and when the summe to be abated in any line appeareth greater then the other, then doe they borrow out of the next higher roome, as for example.

I should abate 1846 from 2378, they set the summes thus:

First they take 6 which is the lower line, and his space from 8 in the same roomes in the second summe, and yet there remaineth two Counters in the lowest line. Then in the second line must 4 be subtracted from 7, and so remaineth there 3. Then 800 in the third line, and his space, from 300 of the second summe cannot be, therefore doe they abate it from a higher roome, that is, from 1000, and because 1000 is too much by 200, therefore must I set downe 200 in the third line, after I have taken up 1000 from the fourth line. Then is there yet 1000 in the fourth line of the first summe, which if I withdraw from the second summe, then
doe

Doe all Figures stand in order thus : 532.

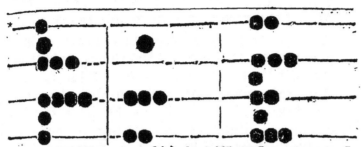

So that (as you sée) it differeth not greatly whether you begin Subtraction at the higher lines, or at the lower.

Howbeit, as some men like that one way best, so some like the other : therefore you now knowing both may use which you list.

Multiplication.

But now touching *Multiplication* : *you shall set your* numbers *into two* roomes, *(as you did in those other kindes) but so that the multiplier bee set in the first* roome , *then shall you begin with the* highest numbers *of the second* roome, *and multiply them first after this sort.*

Take the overmost line in your first working as it were the lowest line, setting on it some moveable mark (as you list) and locke how many Counters bée in him, take them up, and for them set downe the whole multiplier so many times as you tooke up Counters: reckoning (I say) that line for the Unites. And when you have done with the highest number, then

then come to the next line beneath, and doe so even with it, and so with the next, till you have done all. And if there bée any number in a space, then for it shall you take the multiplier five times, & then must you reckon that line for the Unites, which is next beneath that space. O₂ else after a ſhoꝛter way, ye ſhall take onely halfe the multiplier, but then ſhall you take the line next above the space foꝛ the line of Unites. But in each woꝛking, if by chance your multiplier bée an odde number, ſo that you cannot take the halfe of it juſtly, then muſt you take the greater halfe, and ſet down that, as if that it were the juſt halfe : and further, you ſhall ſet one Counter in the space between that line, which you reckon foꝛ the line of unites, oꝛ elſe onely remove foꝛward the same that is to be multiplied.

Scholar. If you ſet foꝛth an example hereof, I think I ſhall perceive you.

Maſter. 𝕿ake this example : I would multiply 1542 by 365. therefore I ſet my numbers thus.

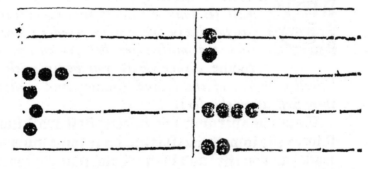

𝕿hen firſt I begin at the 1000 in the higheſt roome, as if it were the firſt place, and I take it up ſetting down foꝛ it ſo often (that is once) the
Multi-

Multiplier, which is 365, thus as you see here : where, for the one Counter taken up from the fourth line, I have set down other sixe which make the summe of the Multiplier, reckoning the fourth line, as if it were the first, which thing I have marked by the Starre set at the beginning.

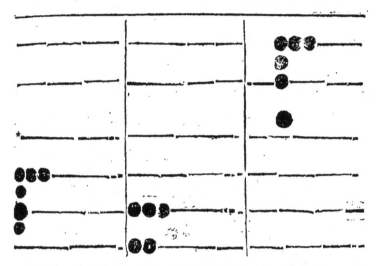

Scholar. I perceive well, for indeed this summe that you set downe, is 365000: for so much doth amount of 1000, multiplied by 365.

Master. Well then goe forth, in the next space I finde one Counter, which I remove forward, but take it not up, but (as in such a case I must) set downe the greater halfe of my Multiplier (seeing it is an odde number) which is 182, and here do I still let that fourth place stand as if it were the first, as in these examples you shall see.

Where

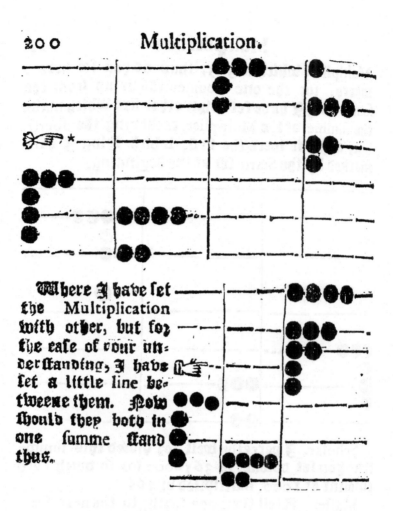

Where I have set the Multiplication with other, but for the eafe of your understanding, I have set a little line betweene them. Now should they both in one summe stand thus.

Howbeit, another forme to multiply such Counters in space is this: first to remove the finger to the next line beneath the space, and then to take up the Counter, and to set downe the Multiplier five times, as here you see.

Which summes, if you doe adde together, into one summe, you shall perceive that it will be the same that appeareth of the other working before, so that both sorts are to one intent: but as the other is shorter, so this is plainer to reason, for such as have had small exercise in this Art.

Notwithstanding you may adde them in your minde before you set them down: as in this example you might have said five times 300 is 1500, and five times 60 is 300, also five times five is 25, which all put together doe make 1825,

Another forme of Multiplication.

1825, which you may at one time set down if you lift.

But now to goe forth, I muſt remove the hand to the next Counters, which are in the ſecond line, and there muſt I take up thoſe foure Counters, ſetting downe for them my multiplier foure times ſeverally, or elſe I may gather the whole ſumme in my minde firſt ; and then ſet it down : as to ſay, 4 times 300 is 1200 : 4 times 60 are 240 : & 4 times 5 make 20, that is in all 1460 : that ſhall I ſet down alſo, as here you ſee.

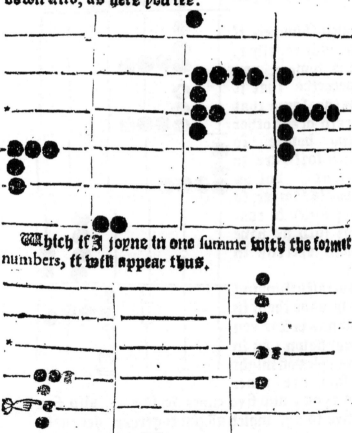

Which if I joyne in one ſumme with the former numbers, it will appear thus.

Then to end this Multiplication, I remove the finger to the lowest line where are onely 2, them doe I take up, and in their stead doe I set down twice 365, that is 730, for which I set one in the space above the 3 line for 500, & 2 more in the 3 line with that one that is there already (the rest in their order, & so have I well ended the whole sume thus:

Whereby you see, that 1542 (which is the number of years since Christ his Incarnation) being multiplied by 365 (which is the number of the daies in a year) doth amount to 562830. which declareth the number of daies since Chrifts incarnation unto the end of 1542 years, besides 385 daies, & 12 hours for leap years.

Scholar. Now will I prove by another example, as this: 40 Labourers (after 6 pence the day for each man) have wrought 28 daies. I would know what their wages doth amount unto.

In this case must I work doubly; first I must multiply the number of the Labourers, by the wages of a man for one day, so will the charge of every day amount.   P   Then

Example of wages.

Then secondly, shall I multiply the charge of one day by the whole number of dayes, and so will the whole summe appear : first therefore I shall set the summes thus.

Where in the first place is the Multiplier (that is one daies wages for one man) & in the second space is set the number of the workmen to bee multiplied.

Then say : If 6 times foure (reckoning that second line as the line of unites) maketh 24, for which summe I should set two counters in the third line, and 4 in the second, therefore do I set two in the third line, and let the 4 stand still in the second line thus.

So appeareth the whole dayes wages to bee 240 pence, that is 20 shillings.

Then do I multiply again the same summe by the number of dayes; & first I set the numbers thus : then because there are counters in divers lines I shall begin with the highest, and take them up setting for them the Multiplier so many times as I took up counters, that is twice, then will the summe stand thus.

The

Then come I to the second line, and take up those 4 counters, setting for them the Multiplier 4 times, so will the whole summe appear thus:

So is the whole wages of 40 workmen for 28 dayes after 6 pence each day for a man, 6720 pence, that is, 560 shillings, or 28 pound.

Master. Now if you would prove Multiplication, the surest way is by division: therefore will I overpasse it till I have taught you the Art of division, which you shall work thus:

Division.

First, set down the Divisor, for feare of forgetting, and then set that number that shall be divided at the right side, so farre from the Divisor, that the quotient may be set between them: as for example.

If 225 sheep cost 45 pound, what did every sheep cost? To know this, I would divide the

An example of sheep.

the whole ſumme, that is 45 pound, by 225, but
that cannot be: therefoze muſt I firſt reduce that
45 pound, into a leſſer denomination, as into ſhillings, then I multiply 44 by 20, and it is 900; that
ſumme ſhall I diuide by the number of ſheep, which
is 225, theſe two numbers therefoze I ſet thus:

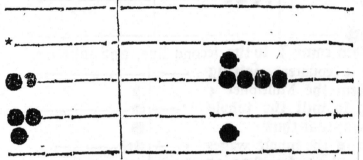

Then begin I at the higheſt line of the diuidend,
and ſéek how oft I may haue the diuiſor therein,
and that I may do four times: then ſay I, four
times 2 are 8, which if I take from 9, there reſteth
but 1, thus:

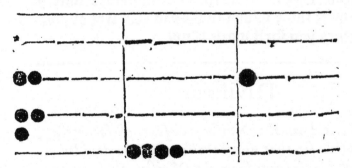

And becauſe I found the diuiſor 4 times in the
diuidend, I haue ſet as you ſée, 4 in the middle
room, which is the place of the quotient: but not
enu

muſt I take the reſt of the diviſor as often out of the remainder, therefore come I to the ſecond line of the diviſor, ſaying two times 4 make 8, take 8 from 10, and there remaineth 2, thus :

Then come I to the loweſt number, which is 5, & multiply it 4 times, ſo is it 20, that take I from 20, & there remaineth nothing, ſo that I ſee my quotient to be 4, which are in value ſhillings, for ſo was the dividend: and thereby I know that if 225 ſheep coſt 45 pound, every ſheep coſt 4 ſhillings.

Scholar. This can I do as you ſhall perceive by the example. If 160 Souldiers do ſpend every moneth 68 pound, what ſpendeth each man?

First, becauſe I cannot divide the 68 by 160, therefore I will turn the pounds into pence by Multiplication, ſo ſhall there be 16320 pence : now muſt I divide the ſumme by the number of Souldiers, therefore I ſet them in order thus :

Example of ſouldiers wages.

P 3   Then

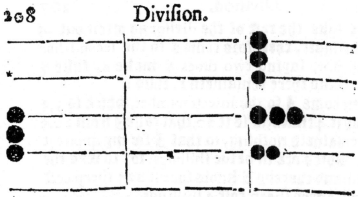

Then begin I at the higheſt place of the dividend, ſeeking my diviſion there, which I finde once, therefore I ſet 1 in the nether line.

Maſter. Not in the nether line of the whole ſumme, but in the nether line of that worke which is the third line.

Scholar. So ſtandeth it with reaſon.

Maſter. Then thus do they ſtand.

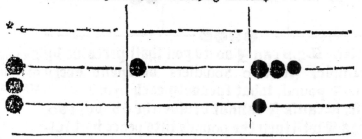

Then ſeeke I againe the reſt, how often I may finde my diviſor : and I ſee that in 300 I might finde 100 thrée times : but then the 60 will not be ſo often found in 20, therefore I take 2 for my quotient : then take I 100 twice from 300, and there reſteth 100, out of which with the 20 that maketh 120, I may take 60 alſo twice, and then ſtand the numbers thus :

Where

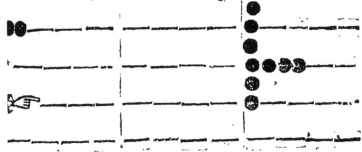

Where I haue fet the quotient 2 in the loweſt
line: ſo is euery Souldiers portion 102 pence that
is 8 ſhillings 6 pence.

Maſter. But yet becauſe you may juſtly perceiue
the reaſon of diuiſion, it ſhall be good that you fet
your diuiſor ſtill againſt thoſe numbers from which
you do take it, as by this example I will declare.

If the purchaſe of 200 acres of ground did coſt
90 pound, what did one acre coſt?

Firſt, will I turne the pounds into pence, ſo will
here be 69600 pence. Then in ſetting down theſe
numbers, I ſhall do thus:

An example of purchaſe.

Firſt fet the dividend on the right hand as it
ought, and then the diuiſor on the left hand againſt
thoſe Numbers from which I intend to take him
firſt, as here you ſee, where I haue fet the diuiſor
two lines higher then his own place.

Scholar. This is like the order of division by the Pen.

Master. Truth you say, and now must I set the quotient of this work in the third line, for that is the line of unites in respect of the divisor in this work.

When I see how often the divisor may be found in the dividend, and that I finde 3 times, then set I 3 in the third line for the quotient, and take away that 60000 from the dividend : and further I set the divisor one line lower, as you see here.

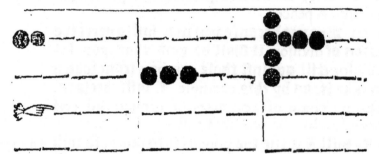

And then seek I how often the divisor will be taken from the number against it, which will be four times and 1 remaining.

Scholar. But what if it chance that when the divisor is so removed it cannot be once taken out of the dividend against it ?

Master. Then must the divisor be set in another line lower.

Scholar. So was it in division by the pen, and therefore was there a cypher set in the quotient but how shall that be noted here?

Master. Here needeth no token, for the lines do represent the places, onely look that you set your

quotient

quotient in that place which standeth for unites in
respect of the divisor. But now to return to the example: I finde the divisor four times in the dividend,
and 1 remaining : for 4 times 2 make 8, which I
take from 9, and there resteth 1, as this figure following sheweth : and in the middle space for the
quotient, I set 4 in the second line, which is in this
work the place of unites.

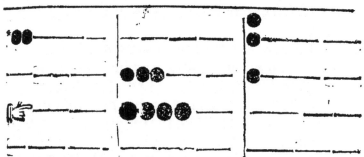

Then remove I the divisor to the next lower
line, and seek how often I may have it in the dividend, which I may do here 8 times just, and nothing remain, as in this form.

Where you may see that the whole quotient is
348 pence, that is 29 shillings, whereby I know
that so much cost the purchase of one Acre.

                                        Scholar.

Scholar. Now resteth the proofs of Multiplicati-oo, and also division.

Master. Their best proofs are each one by the other ; for Multiplication is proved by division, and division by Multiplication, as in the work by the pen you learned.

Scholar. If that be all, you shall not need to repeat again that which was sufficiently taught already : and except you will teach me any other feat, here may you make an end of this Art, I suppose.

The reason of all the former rules.

Master. So will I doe as touching whole number, and as for broken number I will not trouble your wit with it, till you have practised this so well, that you be full perfect, so that you need not to doubt in any point that I have taught you, and then may I boldly instruct you in the Art of Fractions or broken numbers : wherein I will also shew you the reasons of all that you have now learned, But yet before I make an end, I will shew you the order of common casting, wherein are both pence, shillings, and pounds, proceeding by no grounded reason, but onely by a received forme, and that diversly, of divers men : for the Merchants use one form, and Auditors another.

Merchants.

Merchants ufe.

*BVt firſt for Merchants
form, mark this example
here, in which I have ex-
preſſed this ſumme,* 198
pounds 19 *ſhillings* 11
*pence. So that you may ſee
that the loweſt line ſerveth
for pence, the next above for
ſhillings, the third for pounds,
and the fourth for ſcores of
pounds.*

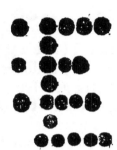

And further you may ſée that the ſpace betwéen
pence and ſhillings, may receive but one counter,
(as all other ſpaces likewiſe do) and that one ſtandeth in that place foʒ 6 pence.

Likewiſe betwéen the ſhillings and the pounds
one counter ſtands foʒ 10 ſhillings.

And betwéen the pounds and 20 pounds, one
counter ſtandeth foʒ 10 pounds.

But beſide thoſe, you may ſée at the left ſide of
ſhillings, that one number ſtandeth alone and
betokeneth 5 ſhillings.

So againſt the pounds, that one counter ſtandeth
foʒ 5 pound.   And againſt the 20 pounds, the one
counter ſtandeth foʒ five ſcore pounds, that is 100
pounds : ſo that every ſide counter is five times ſo
much as one of them againſt which he ſtandeth.

Auditors

Auditors Accompt.

*N*Ow for the Accompt of Auditors, take this ex. *ample.*

Where I have expzessed the same summe 198 pound 19 shillings 11 pence.

But here you see the pence stand towards the right hand, and the other increasing ozderly towards the left hand.

Again you may see, that Auditors will make two lines (yea and moze) foz pence, shillings, and all other values, if their summes extend thereto. Also you see that they set one Counter at the right end of each row, which so set there standeth foz five of that room, and on the left cozner of the row, it standeth foz 10 of the same row.

But now if you would adde, oz subtract after any of both these sozts, if you mark the ozder of the other feat which I taught you, you may easily do the same here without much teaching : foz in Addition you must first set down one summe, and to the same set the other ozderly, and in like manner, if you have mino ; but in Subtraction, you must set down first the greatest summe, and from it must you abate the other, every Denomination from his true place.

Scholar. I do not doubt but with a little pzactise I shall attain these both : but how shall I multiply and divide after these fozms ?

M. You cannot onely do any of both by these sozts: therefoze in such case you must resozt to your other Arts. Scholar.

Scholar. They that use such Accounts that it
exceed 200 in the summe, they set not 5 at the left
hand of the scores of pounds, but they set all the
hundreds in another farther row, & 500 at the left
hand thereof, and the thousands they set in a farther
row yet, and at the left side thereof they set the
5000, and in the space over they set the 10000, and
in a higher row 20000, which all I have expressed
in this example, which is 97869 pounds 12 shillings
9 pence ob. q. Ninetey seven thousand, eight hundred
threescore. and nine pounds, twelve
shillings and nine pence half peny farthing, for I had not told you before,
where neither how you should set
downe farthings, which (as you
see here) must be set in a voide
place sideling beneath the peace,
for a farthing one counter, ob.
two counters, for ob. farthing 3
counters, and more there cannot
bee: for 4 farthings make 1 peny,
which must be set in his due place.

And if you desire the same summe
after Auditors manner, lo here it
is.

But

But in this thing you shall take this for sufficient, and the rest you shall obserbe as you marke by the working of each sort, for the diuers wits of men haue inuented diuers and sundry wayes, almost innumerable.

THE
SECOND PART OF
ARITHMETICK,

touching Fractions, briefly set fourth.

Scholar.

Albeit I perceive your manifold businesses doth so occupie, or rather oppresse you, that you cannot as yet compleatly end the *Treatise of Fractions Arithmeticall*, which you have prepared, wherein not onely sundry works of *Geometry*, *Musick*, and *Astronomy* be largely set forth, but also divers conclusions and naturall works touching mixtures of *Metals*, and compositions of *Medicines*, with other strange examples. Yet in the meane season, I cannot stay my most earnest desire, but importunely crave of you some brief preparation toward the use of *Fractions*, whereby at the least I may be able perfectly to understand the common works of them; and the vulgar use of those rules, which without them cannot well be wrought.

Arithmeticall fractions.

Master. If my leasure were as great as my will is

is good, you should not need to use any importunate craving, for the attaining of that thing, whereby I may be perswaded that I shall any way profit the Common wealth, or help the honest studies of any good Members in the same : wherefore while mine attendance will permit me to walk and talk, I am well willing to help you as I may.

Therefore, first to begin with the explication of this name fraction, what take you it to be ?

Scholar. Marry sir, I think a fraction (as I have heard it often named) to be a broken number, that is to say, to be no whole number but part of a number.

What a Fraction is

Master. A fraction indeed is a broken number, and so consequently the part of another number, but that must be understood of such another number as cannot be divided into any other parts then fractions : for although I may take the third part of 60, or the 4 part of it, & so of other parts diversly, yet those parts be not properly, nor ought to bee called fractions, because they may be expressed by whole numbers. for the 3 part of it is 20, the 4 part is 15, the 12 part is 5, & so forth of other parts, all which bee whole numbers.

Wherefore properly a fraction expresseth the parts or part onely of a unite, that is to say, that the number which is the whole or entire summe of any fraction may not bee greater then one : and therefore it followeth, that no one fraction alone can be so great, that it shall make 1, as by example I will declare, as soon as I have taught you to know the form how a fraction is expressed or represented in writing.

What a Fraction is properly.

Nume-

Numeration.

The expressing of fractions.

But first to begin with expressing of a fraction, which is the Numeration of it: you must understand that a fraction is represented by two numbers set one over the other, and a line drawn between them, as thus $\frac{1}{3}\,\frac{3}{4}\,\frac{2}{5}\,\frac{10}{17}$, which four fractions you must pronounce, thus, $\frac{1}{3}$ one third part, $\frac{3}{4}$ three quarters, $\frac{2}{5}$ two fifth parts, $\frac{10}{17}$ ten seventeen parts.

Scholar. I understand this form of their expression and pronunciation, but their meaning or valuation seemeth more obscure. Yet I think that by the two first Fractions, I understand the valuation of the two latter Fractions, and consequently of other.

Master. Value them then, that I may perceive your taking of them.

Scholar. $\frac{2}{5}$ betokneth two fift parts, that is to say, if one be divided into 5 parts, that Fraction doth expresse two of those 5 parts: $\frac{10}{17}$ doth signifie that if one be divided into 17 parts, I must take ten of them. And this I gather of the two first examples: for $\frac{1}{3}$, that is, one third part, doth easily declare, that if one thing bée divided into thrée parts, I must take out one of them: so $\frac{3}{4}$, that is thrée quarters, doth declare that one being divided into foure quarters, I must take(for this Fraction) thrée of those quarters.

If there be no more difficulty in this Numeration, then I pray you goe forward to their Addition and Subtraction, and so to the other kindes of workes. For I understand that the same kind of wozkes be in fractions, that bée in whole numbers.

Master. There are the same kindes of wozks in both, albeit the ozder of them is divers, as I will anon declare : but yet moze in Numeration befoze we leave it. You must understand, that those two or numbers which expzesse a fraction, have severall names, the overmost which is above the line, is called the Numerator, & the other beneath the line is called the Denominator.

Scholar. And what is the reason of their divers names? Foz (in mine opinion) both bée Numerators, seeing both doe expzesse the numeration of the fraction.

Master. You are deceived; foz one onely (which is the overmost) doth expzesse the Numeration, and the Denominator doth declare the number of parts, into which the unite is divided, as in this example; when I say : divide a pound weight of gold between foure men, so that the first man shall have $\frac{2}{15}$, the second $\frac{3}{15}$, the third $\frac{4}{15}$, and the fourth $\frac{6}{15}$.

Now doe you perceive that by the Denominator (which is one in all foure fractions) it is intended that the pound weight should bee divided into so many parts, I meane 15, and by the 4 severall Numerators, is limited the divers poztion that each man should have, that is, that when the whole is parted into 15, the first man shall have two of those 15 parts : the second man thzee of them : the third man foure : and the fourth man six. And so may you sée the severall offices (as it were) of those two numbers, I meane of the Numerator and the Denominator.

And hereby you perceive that a man can have no moze parts of any thing then it was divided

T into

into, neither yet aptly fo many : fo that it were un-
aptly faid , You fhall have $\frac{15}{15}$, that is fiftéen
parts of any thing, feeing it were better faid, you
fhall have the whole thing.

Scholar. So doth it appeare reafonably, for the
labour is vain to divide any thing, and then to
apply the Divifion to no ufe. And much leffe
reafonable were it to fay $\frac{16}{15}$: for if the whole be
divided into 15 parts onely it is not poffible to take
16 of them, that is to fay, more then all together.

Mafter. This is true touching the proper and
apt ufe of the name of a Fraction ; yet improperly
(and after a vulgar acceptation for eafineffe in
worke) both thofe formes bée called Fractions,
becaufe they be written like Fractions although
they bee none indeed : for $\frac{15}{15}$ and generally in fuch
other, where the Numerator and Denominator be
equall, are not Fractions, but the whole thing with
all his parts : And fo $\frac{14}{12}$ is not to be called a Fraction, but a mixt number, of a whole number and a
Fraction, for it is as 1 $\frac{2}{12}$, that is, one whole and

An impro-
per fracti-
on of a
mixt num-
ber.

$\frac{2}{12}$ parts ; as fhall be declared in Reduction. Therefore they do abufe the names that call them
Fractions, where the Numerator is either equall,
or greater then the Denominator.

Scholar. But is there any needfull caufe, why
they fhould fo abufe the name?

Mafter. There is caufe why they fhall fometimes
for eafineffe in worke write fome numbers, after
that fort like Fractions : but they needed not to
call them Fractions, but (as they be) whole numbers, or mixt numbers, (that is, whole numbers
with Fractions) expreffed like Fractions ,

oz. as impzoper Fractions.

Now must you understand, that as no Fraction properly can be greater then one, so in smalnesse underone the nature of Fractions doth extend infinitely, as the nature of whole numbers is to increase above one infinitely, so that not onely one may be divided into infinite Fractions oz parts, but also every Fraction may be divided into infinite Fractions oz parts, which commonly be called Fractions of Fractions: and they be expzessed diversly as foz example, $\frac{1}{4}$ of $\frac{2}{3}$ of $\frac{1}{2}$, that is, thzée *Fractions of fractions.* quarters of 2 thiret parts of one halfe part. Whereby is signified, that if one be divided into 2 halfes, and the one halfe into thzée parts, and two of those thzée parts be divided joyntly into foure quarters; this fraction of fractions doth repzesent thzée of those quarters.

Scholar. I pzay you let mé pzove by an example in common money, whether I doe rightly understans you oz no. One Crowne which I take foz an unite, doth contain 60 pence; therefoze the halfe of it is 30 pence, $\frac{2}{3}$ of that half is 20 pence, whereof $\frac{1}{4}$ is fifteen pence; so then 15 pence is $\frac{1}{4}$ of $\frac{2}{3}$ of $\frac{1}{2}$. of a Crowne: and so is 3 pence $\frac{1}{4}$ of $\frac{2}{3}$ of $\frac{1}{2}$. of a shillings.

Master. You perceive this well enough: yet this note I give you by the way, that the fozm of expzessing the fractions is voluntary, and hath no other reason then the will of the Divisor, which foz many follow: foz some expzesse them thus; $\frac{1}{4}\frac{2}{3}\frac{1}{2}$ without any figure of distinction betwéen hem, which fozm also many follow. Some other oz make lines betwéene every fraction, and adde odds of distinction, after this sozt, $\frac{1}{4}$ of $\frac{2}{3}$ of $\frac{1}{2}$,

D 2   which

which form is best.

Some other expresse them thus
in slope form, to distinct them
from fractions of whole numbers,
for if they were in one right line

$$\frac{3}{4}$$

$$\frac{2}{3}$$

$$\frac{1}{2}$$

thus, $\frac{3}{4}$ $\frac{2}{3}$ $\frac{1}{2}$ then ought it to bée pronounced, three
quarters, & two third parts, & a halfe, which maketh
almost two whole unites, lacking but one twelfth
part. And so is it nothing agréeable with the other
fraction of fractions : wherefore it is a great
oversight in certain learned men, whichdoe expresse
them so confusedly with such severall fractions,
that a man cannot know the one from the other.

Therefore some men (as Stifelius *) do expresse without a line, numbers of proportion, being applied to*
Addition *or* Subtraction, *because they must be taken
as 2, where the line in fractions maketh them to be taken
for one : for of the* Numerator *and* Denominator *is
made one number.*

Schol. Then I perceive there be thrée severall
varieties in fractions : First, when one onely fraction is set for one number, as $\frac{4}{5}$, that is, foure fifth
parts. The second is, when there be set two or
moze severall fractions of one number, as $\frac{4}{9}\frac{2}{5}$, that
is, foure ninth parts, and two fifth parts. The third
fort is fractions of fractions, as $\frac{4}{9}$, of $\frac{2}{5}$, that is, foure
ninth parts of two fifth parts.

Master. You have said well, if you understand
well your own words.

Scholar. If it shall please, I will by an example in the parts of an old English Angel, expresse
my meaning.

Master. Let me hear you.

Scholar.

Three se-
verall vari-
eties.

Scholar. The old English Angel did contain 7
shillings 6 pence, that is, 90 pence : Now $\frac{4}{5}$ of it
is 72 pence : And of the same 90 pence, if I take
$\frac{4}{9}$ and $\frac{2}{5}$, that is foure ninth parts, and two fifth parts,
$\frac{4}{9}$ is 40, and $\frac{2}{5}$ is 36, which both make 76 : but if
I take $\frac{4}{9}$ of $\frac{2}{5}$, that is, foure ninth parts of two fifth
parts, seeing $\frac{2}{5}$ is but 36, then $\frac{4}{9}$ of 36 will yeeld but
16, for $\frac{1}{9}$ of 36 is but 4, and that taken foure times
maketh 16.

Master. This is plainly expressed, and truely,
and hereby (I doubt not) but you doe perceive, that
as great a difference, as is between 16 and 76,
so much difference is between those two Fractions
$\frac{4}{9}$ and $\frac{2}{5}$: and $\frac{4}{9}$ of $\frac{2}{5}$.

And now that ye understand these varieties, I
will proceed to the rest of the workes : First, admo-
nishing you, that there is another order to be followed in Fractions, then there was in whole numbers : for in whole numbers this was the order :
Numeration, Addition, Subtraction, Multiplication,
Division, & Reduction : but in Fractions, (to follow
the same aptnesse in proceeding from the easiest
workes to the harder) we must use this order of
workes, Numeration, Reduction, Addition, Subtraction, Multiplication, and Division.

The order of works in fractions.

Scholar. That Addition and Subtraction should
go together, & Division to follow Multiplication,
naturall order doth perswade : but why Reduction should be first in order here, next to Numeration,
and Addition & Subtraction, in the middle, I desire
to understand the reason.

Master. As in the Art of whole numbers, Order
would reasonably begin with the easiest, and so

M 3   goe

goe foward by degrees to the hardest : even reason teacheth in Fraction the like order. And consider that Addition or Subtraction of Fractions, can very seldome bee wrought without Reduction: & contrariwise, Reduction may bewrought without this form of Addition or Subtraction : therefore was it orderly required, that Reduction should goe before Addition and Subtraction, and this reason serveth for the placing of Reduction before the other.

Scholar. Then, if Reduction be the easiest, I pray you declare the forme of it, first by rule, and then by example.

Master. Your request is good.

Reduction of Fractions.

Herefore will I now declare the diversities of Reduction of fractions, which commonly hath five varieties, or formes.

First, when there bee sundry Fractions of one intire unite, they must be reduced to one Denomination, and also into one Fraction.

Secondly, when there be propounded fractions of fractions, they must be reduced likewise into one fraction : for otherwise they cannot bee brought into one Denomination.

Thirdly, when an improper fraction is propounded, that is to say, a fraction in form, which indeed is

greater

greater then an unite : it must be reduced into apt form, expressing the unite or unites of it, and the proper fraction distinctly. And sometimes also it shall be needfull to convert such a mixt number of unites with fractions, into the form of a fraction, that is, into an improper fraction : which two forms I esteem but as one, because they work one kinde of number.

Fourthly, there happneth sometimes fractions to be written in great numbers, which might be written in lesser numbers : therefore is there a meane to reduce such great numbers into their smallest termes. 4

Fifthly, when any fraction betokeneth the parts of a whole thing, which hath by common partition certain parts, but none of like Denomination with that fraction, then may you reduce the said fractions into another, whose Denomination shall expresse the common parts of that whole thing. 5

Scholar. This distinction in Doctrine delighteth me much, but more with hope then present fruit : for as yet I doe not understand scarcely the varieties, and much lesse the practise and use of their works.

Master. Reduction is an orderly alteration of Numbers out of one forme into another, which is never done orderly but for some needfull use, as in every of the said five severall formes, I will distinctly declare.

· First therefore, when two, or more severall fractions of any unite be propounded : as for example ⅜ and ⅔, because it is hard to tell what proportion of the intire number those two fractions do expresse, therefore was Reduction devised to be a mean whereby these severall fractions might be brought into one The first forme of Reduction

Q 4   Denomi-

Denomination   and   fraction.

And in these fractions, this is the Art for bringing them to one Denomination.

How to
reduce fra-
ctions of
divers de-
nominati-
ons into
one deno-
mination.

*Multiply first the Denominators together, and the
totall thereof you shall set twice down under two
severall lines for two new Denominators, or rather
for one common Denominator. Then multiply the
Numerator of the first fraction, by the Denominator
of the second, and set the totall thereof for the Numerator over the first line. Likewise multiply the
Numerator of the second fraction by the Denominator
of the first, and set that totall over the second line for
the Numerator of that fraction: and so are these two
first Fractions of severall Denominations, brought to
one Denomination.*

Scholar. If I understand you, as I thinke I doe
my Example shall declare the same. The Fractions,
which you propounded were these $\frac{3}{16}$,and $\frac{4}{6}$, whose
Denominators (being 16 and 6) I multiply toge-
ther, and there amounteth 96, which I set under
two lines, thus : $\frac{}{96}$ $\frac{}{96}$

Then I multiply the Numerator of the first
Fraction by the Denominator of the second, saying,
3 into 6 maketh 18, that I set over the first line
for a new Numerator, and it will be thus, $\frac{18}{96}$.

Likewise I multiply the Numerator of the second Fraction, by the Denominator of the first
saying, 4 times 16 maketh 64, that I set for th
second Numerator, and the Fraction will appea
thus, $\frac{64}{96}$.

So that both Fractions brought to one Denomination, must stand thus, $\frac{18}{96}$ and $\frac{64}{96}$.

Master. You have done well.

Schola

Scholar. I beseech you let me examine it after my accustomed forme, by common parts of coyne or other measure.

Master. Go to.

Scholar. I have a peece of Gold which is accounted worth 8 shillings, and containeth 96 pence, whereof $\frac{1}{16}$, that is, the sixteenth part, is 6^1 pence, and $\frac{3}{16}$ is 18 pence, that is $\frac{18}{96}$. Againe $\frac{1}{6}$ of the same peece of gold is 16 pence, so that $\frac{4}{6}$ parts maketh 64 pence, that is $\frac{64}{96}$. And so I finde the summes to agree with the other before.

Master. *So have you now the Art to bring two such fractions into one Denomination : And if there be more then two, then must you multiply all the Denominators together, and set the totall thereof so many times down as there be fractions ; and then to get for each one a new Numerator, multiply the Numerator of the first, by the Denominator of the second, and the totall thereof multiply by the Denominator of the third, and so forth, if there be more. Likewise multiply the Numerator of the second, by the Denominator of the first, and the totall thereof by the Denominator of the third. And in the same sort multiply the Numerator of the third into the Denominator of the first, and the totall thereof into the Denominator of the second, and so forth if there were moe. So these three fractions* $\frac{2}{5}\frac{1}{4}\frac{2}{3}$ *doe make by Reduction these other three fractions of Denomination.* $\frac{24}{60}\frac{45}{60}\frac{40}{60}$. *All which you may bring into one fraction by adding the Numerators together, and putting the totall for the totall Numerator, reserving still that same common Denominator. And those three fractions make one improper fraction, thus :* $\frac{12}{60}$.

Note the Reduction of three fractions, or more, into one.

Scholar.

Scholar. All this I perceive, and also that this last Fraction is more then an unite, and therefore you did call it an improper Fraction.

Master. There be certain other formes of working in this Reduction, which I will briefly touch also, to give you an occasion to exercise your wit therein.

The first variety of Reduction

The first variety is this : When you have made and written down your common Denominator (as I have taught before) then to get a numerator for the first, do thus : Divide the common denominator by the denominator of the first fraction, and the quotient multiplied by the numerator of the same yeeldeth a new numerator for the first new fraction. So likewise do with the second and the third, and with all the residue, if there be more.

Scholar. That will I prove in your last example of these three fractions, $\frac{2}{5}\frac{3}{4}\frac{2}{3}$. When the denominators be multiplied they make 60, for 5 into 4 maketh 20, and 20 by 3 yeeldeth 60, that I set down three times thus : $\frac{}{60}\frac{}{60}\frac{}{60}$ then to have a numerator, for the first, I must divide 60 by 5 (the denominator of the first) and the quotient is 12, which I must multiply by 2 (the numerator first) and that maketh 24, and so have I for the first Fraction, $\frac{24}{60}$.

Likewise for the second Fraction : I divide 60 by 4, and there cometh 15, which I multiply by 3, and so have I 45, for the second Fraction $\frac{45}{60}$. Then for the third in like sort will come $\frac{42}{60}$.

The second variety.

Master. *Another way is this : If it happen so, that the lesser denominator, can by any multiplication make the greater, then note the multiplier, & by it multiply the numerator over that lesser denominator,*

and

and for the lesser Denominator put the greater, as thus in these two fractions $\frac{2}{12}$ and $\frac{2}{3}$, three being the lesser Denominator multiplied by 4, will make 12, which is the greater Denominator: therefore by the same 4 I do multiply 2 which is the Numerator over 3, and that maketh 8: under which I do put 12, being the greater Denominator, which is also made by multiplication of 4 into 3, and so have I these two fractions $\frac{2}{12}$, $\frac{8}{12}$, thus shortly reduced, without altering the one Fraction.

Scholar. This I understand.

The third variety.

Master. *Then mark this third way: If the Denominators doe not happen so, that one by Multiplication may make the other, then look whether they both may be parts of any other one number, as in $\frac{2}{12}$, and $\frac{7}{18}$, although the lesser taken but twice, be too much to make 18, yet they both may be parts unto 36, therefore look how many times twelve is in 36, and that quotient being multiplied by the Numerator over 12, the totall shall be put in stead of the Numerator over 12, and for 5 put 15, thus, $\frac{15}{36}$. So likewise look how often is 18 in 36, because it is twice, therefore by 2 multiply 7, which is over 18, and it will be 14: set that for the Numerator, and in stead of 18 put 36; and then your Fractions reduced stand thus $\frac{15}{36}$ $\frac{14}{36}$, in stead of $\frac{5}{12}$, and $\frac{7}{18}$.*

And if you will prove whether you have wrought well or no, that may bee proved by Reduction of them again to their former Denominations, which Proofe. Art shall be taught in the fourth kinde of Reduction, where greater termes of Fractions be reduced into smaller in number, but no smaller in proportion. And if in such Reduction the same termes or numbers come again that were before, then

is

is the worke good, else not.

Scholar. Sir I heare your words, but I doe not understand many of them : which if it please you declare.

Master. With a good will, when convenient place serveth, but that must be in the said fourth kind of Reduction, which teacheth how to reduce fractions of fractions into one fraction, and so to one denomination.

When Fractions of Fractions be propounded, you sha'l multiply the numerators of each into other, and set the totall for the new Numerator, and then multiply all the Denominators likewise, and take their totall for the new Denominator, and so are they speedily reduced.

Scholar. If that be all, then I understand it already, as by this example I will declare. These be the fractions $\frac{3}{4}$ of $\frac{2}{3}$ of $\frac{6}{7}$ of $\frac{7}{9}$ which I would reduce to one denomination, and proper simple fraction.

Therefore begin I with the Numerators, and multiply them together, saying, 3 by 2, maketh 6 : and 6 by 6, maketh 36, which multiplyed by 7, yeeldeth 252 : that I set over a line for the Numerator thus :

$$252$$

Then I multiply the denominator, 4 by 3 maketh 12, and that by 7 bringeth 84, which multiplied by 9, yeeldeth 756, the new denominator. And so the whole reduced fraction is this, which is too hard a fraction for me to understand yet.

$$\frac{252}{756}$$

Master. You think so, and no marvell, but anon you shall learn to judge it easily, for this fraction is no more indæd then $\frac{1}{3}$, although it be in greater termes, & therefore more stranger, & more ob'cure. And

The second form of Reduction of fractions of fractions into one fraction and Denomination.

And this sufficeth for this Reduction, save that
I will shew you by a figure of measure the juft rate
and reason of this kinde of fractions, and also the
due understanding of their Reduction.

The entire meafure parted into 9.

1	2	3	4	5	6	7	8	9
1	2	3	4	5	6	7	$\frac{7}{9}$	
1	2	3	4	5	6	$\frac{6}{7}$		
1	2	3	4	$\frac{2}{3}$				
1	2	3	$\frac{3}{4}$					

Here you see the longeft meafure, (which ftandeth
for the whole & entire quantity) firft parted into 9
divifions, whereof 7 are fevered by the fecond
meafure : and thereof againe are parted out 6, and
that 6 being diftinct into three parts, two of them
are parted by the fourth meafure, of which fourth
meafure being divided into foure parts, the loweft
meafure doth contain $\frac{1}{4}$, fo that the fame $\frac{1}{4}$ muft be
named, not $\frac{1}{4}$ of the whole meafure, but indeed is
$\frac{1}{4}$ of $\frac{2}{3}$ of $\frac{6}{7}$ of $\frac{7}{9}$.

Scholar. This example is fo fenfible, that I
cannot chofe but fee it. And furthermore fee alfo,
that the fame fraction is equall to $\frac{1}{9}$ of the entire
meafure, as the lines which run up and down do
expreffly fet forth. Alfo I fee here that $\frac{2}{3}$ of $\frac{6}{7}$ is
equall to $\frac{4}{9}$, & further yet, that $\frac{6}{7}$ of $\frac{7}{9}$ is equall to $\frac{6}{9}$ or $\frac{2}{3}$

Mafter. I am glad that you fee it fo well, not
doubting but you will gather greater light of
knowledge hereby.

*But now it is time that wee come to the third forme
of Reduction, which teacheth of improper Fractions,*
                                                    that

The third
forme of
Reduction
of impro-
per fracti-
ons.

*that is to say, mixt numbers of unites and fractions
although they appear like fractions, as this $\frac{26}{5}$, which
doth conclude 5 unites wholly, and $\frac{1}{5}$ over. Wherefore
first you shall know them, by that the Numerator is
greater then the Denominator.*

Scholar **Indeed Sir, that appeareth reasonable,
that if the Numerator do expresse moze parts to be
taken of any unite then the Denominator doth signifie that unite to be divided 7 into it must needs
follow, that such a Fraction importeth moze then
the whole, that is to say, the whole with certain
parts over: but what Reduction is there in it?**

Two seve-
rall wayes
in this Re-
duction.

Master. There bee two severall kindes of Reduction, concerning such Fractions. Sometimes it shall
bee needfull to convert these fractions into unites, and
the proper fraction, that will remaine. And sometimes
contrariwise, it shall be meet to reduce mixt numbers,
that is, unites written with fractions, into the forme
of one simple fraction, & so bee there two wayes.

Scholar. **What is the mean of the first way to
turn impzoper Fractions into unites with their
pzoper Fractions?**

The fifth
way.

Reduction
of impro-
per fracti-
ons into
unites,
with their
proper fra-
ctions.

Master. *That is thus; Your* numerator *being
greater then the* denominator, *must be divided by the
same* denomination, *and the* quotient *thereof
expresseth the* unites: *the remainer shall be put for
the* numerator *of the* fraction *that resteth, and the*
denominator *must be the same that was before.*

Scholar. **Foz example, I take $\frac{17}{5}$, and dividing
17 by 5, the quotient will be 3, and there will remain 2.**

Master. **That you must wzite thus, 3 $\frac{2}{5}$, where
(you see) I have wzitten 3 without any line, as
entire**

entire numbers ought to bee written, and the 2 that remained, I have set over the former denominator, with a line, as a proper fraction. And this number doth signifie now three unites, & $\frac{2}{5}$ of one.

Scholar. Then if I would by unites here understand Crownes, so it were 3 Crownes, and $\frac{2}{5}$, that is 2 s.

Master. Even so : and therefore $\frac{17}{5}$ did signifie the same : But this happeneth sometimes that when the Reduction is so wrought, there remaineth nothing : And then it is not a mixt number, but a simple intire number, represented like a Fraction.

Scholar. As $\frac{12}{4}$ will make three 3 just, and $\frac{18}{3}$ will make even 6. This I will remember. But now, what is the second form of reduction that you spake of for these sorts of fractions?

The second way.

Master. *Whensoever you have any of these two sorts of numbers, that is to say whole numbers without fractions, or whole numbers with fractions, and you would turn them into the form of a fraction, you must multiply the whole number by that denominator which you will have to remain still, and to the totall thereof adde the numerator, which you have already, and all that you shall set for the new numerator, keeping still the former denominator : As if you have 6 $\frac{3}{4}$ which you would convert into an improper fraction, you must multiply 6 by 4, whereof cometh 24, and thereto adde the numerator, which is 3, and so have you 27 for the numerator, and 4 still for the denominator.*

Reduction of whole numbers either a-lone, or joyned with fractions into improper fractions.

Scholar. Then is $\frac{27}{4}$ equall to 6 $\frac{3}{4}$.

Master. Even just, and so backward (as appeareth by the former Reduction) 6 $\frac{3}{4}$ maketh $\frac{27}{4}$. Note.

And

And thus one of their reductions may be the proof
of the other work.

Scholar. This I perceive : But now if you
would turn whole numbers without fractions into
any fractions I see not how that may be done, because there is no Denominator to make the multiplication by.

Master. That is well marked : but this you
know, that no man intendeth to turne any whole
number into a fraction, but he hath in his minde
that Denominator by which the multiplication must
be made : for the proofe whereof I set downe 7,
which is a whole number. And if you will have
this number converted into any certain fraction,
will mée to do it.

Scholar. I pray you reduce 7 into a Fraction.

Master. Then you care not what the Fraction be,
so it be some Fraction.

Scholar. No, I passe not for the sort of the Fraction.

Master. Then how can you thinke that you require me to do any thing certain, when you leave
me to do as I list ? And séeing you stand at that
stay, whether thinke you that I must first intend
in minde what fraction I will make of it before
I can do it indéed ?

Scholar. Else you should do ignorantly.

Master. Then will I limit my self (séeing you
will not) to turne it into quarters. And therefore
I multiply 7 by 4 (which is the denomination by
quarters) and there amounteth 28 to be set for the
Numerator and the 4 must be set for the denominator, and the fraction will be thus $\frac{28}{4}$.

Scholar.

Scholar. Indeed I perceive this to be reasonable, for without much triall I understand that $\frac{35}{4}$ of any thing doth make 7. And so then if I would turn 8 into 5 parts, it will make $\frac{40}{5}$ which is all one with 8 : for 8 Crownes turned into 5 parts, (that is into shillings) will make 40 shillings, that is, $\frac{40}{5}$ of a Crown.

Master. *Seeing you understand now these three* The fourth *kinds of* Reduction, *I will declare unto you the fourth* form of *kind, that is, when fractions be written in greater* Reduction *termes then they need, how they may bee brought to lesser termes.*

Scholar. To write any thing in greater terms then needeth, seemeth to be a fault, and so this Rule seemeth to amend that fault.

Master. It were a fau't to do any thing without need, which after must bee redressed : but in this case it is not so, neither did I say absolutely (as you doe) that it needeth not to expresse those Fractions into great terms, but that the Fractions doe not need, I meane for their value, to be understood : but yet it may be needfull for the case of these workes whereto they be appl'yed ; as for example, In the first kind of Reduction this was your own example, $\frac{3}{10}$ and $\frac{2}{3}$ which when you would reduce, you were fain to turn them first into one denomination, and so appeared they thus , $\frac{18}{96}$ and $\frac{64}{96}$, where the Fractions (for their own understanding) needed not to be turned out of smaller termes into greater, but yet the easinesse of working needed it.

Scholar. Sir, I understand now, not onely the difference of this need (for the Fractions might better be understood as Fractions severall ,

R   each

each in his value, when they were in leffer termes, although they could not fo well be reduced) but alfo I underftand what you mean by greater terms,

Termes of and leffer terms, whereof before I was in doubt: for
Fractions. I fee you call the Numerator and Denominator, the terms of the fraction.

Mafter. *I am glad you underftand it fo well: now when then you would value any* fractions, *becaufe they may beft be done when the terms are fmalleft , you fhall reduce them to the fmalleft that you can, which thing you may do thus : Divide the greateft of any fuch two terms by the leffer, and if any thing remain, by that remainer divide the laft* divifor *: and if any thing remain now, by that divide the firft* divifor (*which was before the remainer of the laft* divifion) *and fo continue ftill, till nothing do remain in the* divifion : *and then marke your laft* divifor, *for it is the* number *that will eafily reduce your* fractions, *if you divide both the* numerator *and the* denominator *by the fame* number, *and put for the* numerator *the* quotient *of his* divifion, *and for the* denominator *alfo his* quotient, *that rifeth by his* divifion.

Scholar. I take for example $\frac{18}{96}$, and becaufe 96 is the greateft number, I divide it by 18, and the quotient is 5, and there refteth 6, what fhall I do with this quotient?

Mafter. Nothing in this worke, but now feing there remaineth fomewhat, by that remainer muft you divide the laft divifor.

Scholar. If I fhall divide 18, (which was the laft divifor) by 6, that was the remainer, fo is the Quotient 3, and nothing refteth.

Mafter. As for the Quotient, I omit him yet: but

but because there doth remain nothing, therefore
196 (which was your last divisor) that number by
which you may reduce the fraction propounded.

Scholar. Then as you taught me, I must divide
the Numerator 18 by 6, and the quotient is 3,
which I must put for the numerator over a   3
line thus : And then by the said 6 must I   ———
divide also the denominator 96, and the   16
Quotient will be 16, which I must take for the denominator, and so is the Fraction $\frac{3}{16}$. And so me
thinketh this rule doth prove the work of the first
Reduction.

Master. That is true, if the first Reduction
were made of fractions into their least terms, and
else not, without some help, as the second number
in that place will declare.

Scholar. The second number was $\frac{2}{3}$ which was
turned into $\frac{64}{96}$ by that Rule. Now if I shall by this
Rule reduce it againe into the least terms, I must
divide 96 by 64, and there remaineth 32, wherefore
I must take that 32, for the divisor, to reduce the
said Fractions. Then do you divide 64 by 32, and
the Quotient is 2, which I set for my numerator.
Again, I divide 96 by 32, and the Quotient will be
3, and so I have but $\frac{2}{3}$.

Master. Muse not at the matter for you have
done well enough : but you think you have not the
fraction that you looked for, that is, $\frac{2}{3}$, yet have you
one equall to it, as by the parts of a shilling you
may prove.

Scholar. Truth it is, for each of them will
bring forth 8 pence, so that $\frac{64}{96}$ and $\frac{2}{3}$ and $\frac{2}{3}$ be all
three equall. And now I perceive that because $\frac{2}{3}$

was not written in the least termes that it might
be, therefore this Reduction brought forth not it,
but that other which is written in the least terms.
Now understand I this Rule well. But is there
any other way to worke this Reduction?

Another
way to
work this
Reduction

Master. Yes: but first note this, that if you finde
no such Divisor, to reduce the fraction till you come
to 1, because one doth make no division, therefore
that fraction is already in his least termes, as by $\frac{21}{40}$
you may prove, and so $\frac{85}{98}$, and many other like.

Note that
to mediate
any num-
ber is to
divide by
two.

But now for your better aid to find the due proportion in lesse termes, with more ease for a young
learner, you shall mediate or take the halfe of the
Numerator, and also of the Denominator as long as
you may upon a line, always parting them with a
right down dash of your pen as you work, which
may easily be done, if the numbers be even , as 2.
4. 6. 8. or 10, but if they be odd (though it be but
one of them) then must you abbreviate them by 3. 5.
7, or 9. &c.

And because examples doe most instruct, I have
here set downe the manner of two or three, whose
last number at the end of the line sheweth the
least terme of valuation of that Fraction.

As for example : I would reduce $\frac{288}{576}$ into his
least terme or value, whereupon I set forth $\frac{288}{576}$ with
a long line drawn from it, thus,

288	144	72	36	18	9	3	1
576	288	144	72	36	18	6	2

And because both the Numerator and the Denominator end in even numbers, I see this may be abbreviated by 2, or 4 or 6, &c. Therefore on
the

the other side of the right down path toward the
right hand, I first take the half of the Numerator,
saying, the half of 2 is 1, the half of 8 is 4 : and
again, the half of 8 is 4 : which 144 is now a new
Numerator, and therefore I part it with a right
down path as before.

Then do I also take the half of 576, in saying,
the half of 5 is 2, and the half of 17 is 8, and the
half of 16 is 8, and so have I 288 for a new
Denominator.

Then beginning again ; saying the half of 144
is 72, and the half of 288 is 144 : thus continuing
the mediation of division by 2, untill you come to
the last worke, as appeareth here in the example,
where the same is reduced to $\frac{1}{2}$ which is equall to
$\frac{288}{576}$.

So the second example $\frac{28}{112}$ first abbreviated by
2, & again by 2, and last by 7, is reduced to $\frac{1}{4}$ which
is equall to $\frac{28}{112}$.

$$28 \mid 14 \mid 7 \mid 1$$
$$112 \mid 56 \mid 28 \mid 4$$

Again, $\frac{1465}{4395}$ abbreviated first by 5, then by 293.

1465	293	1
4395	879	3

Scholar. Sir, I thanke you much, this is very
easie and good for a young learner.

Master. So it is, but yet notwithstanding, if
you can without that division by memory, espy the
greatest number that may divide exactly both
termes of your Fractions proposed, then need you
not to use that division, as in this Fraction $\frac{63}{96}$,

R 3   I

I ſée that 12 is the greateſt number that can divide them both : and therefore without any work,
by memorie onely, I turn that into ⅛ ; but this
ability in knowledge is got by exerciſe.

Yet one other way of eaſie Reduction in this
kinde there is; when your fraction hath any cyphers
in the firſt places of both termes, then may you by
caſting away the Cyphers, make a briefe Reducti-
on as thus, $\frac{100}{400}$. Here take away the cyphers, and
it will be ¼, which is the ſame in value with $\frac{100}{400}$.

Scholar. And ſo if I have $\frac{600}{6500}$, it will be $\frac{6}{65}$.

Maſter. You are deceived, for you take away
more cyphers from the Numerator then you do from
the Denominator, which you may not do.

Scholar. I confeſſe my fault, which came of to
much haſte, I was gladder of the Rule then wiſe
in uſing it : but now I underſtand it I truſt.

Maſter. Then may I goe in hand with the fifth
or laſt kinde of Reduction, which teacheth how to
turn any fraction propoſed into any other Denomination that you liſt, or into any part of common
coynes, weights, or meaſures, or ſuch like.

The fifth *For declaration whereof, firſt you ſhall marke*
kinde of *whether your* fraction *be a ſimple* fraction, *either elſe*
Reduction *a* fraction *of ſundry parts, I mean of more terms then*
        two. And if your fraction *be a* fraction *of* fractions,
To reduce *or otherwiſe compound, you muſt reduce it to one ſimple*
fractions *fraction: And then mark well the* denomination *of*
to a deno- *that other* fraction, *into which you would turn this :*
mination *for by that* denominator *you muſt multiply the* nu-
appointed. *merator of your firſt* fraction, *and the totall* Product
        thereof ſhall you divide by the denominator *of your firſt*
        fraction, & that quotient *ſhall be the* numerator *of the*
                         deno-

denominator *proposed* : as for example, *I have this* fraction $\frac{3}{5}$, *which I would turn into ten parts : therefore I multiply this* 10 *by* 3, *that is the* numerator *of my* fraction, *and there ariseth* 30, *which I divide by* 5, *and the* quotient *is* 6, *which must be the* numerator *to* 10, *and so* $\frac{3}{5}$ *will be* $\frac{6}{10}$.

Scholar. This is easie enough to do.

Master. Then shall you see another example of the same fraction that is not so easie : as if I would turn $\frac{3}{5}$ into 8 parts, prove you that work.

Scholar. I must multiply 8 by 3, and there amounteth 24, which I divide by 5, and the quotient is 4, then is the new fraction $\frac{4}{8}$.

Master. And see you nothing doubtfull in this work?

Scholar. I see that when 24 was divided by 5, there remained 4, which I did not passe of, because ye spake nothing of any remainer, but onely of the quotient.

Master. By likelihood you remember what I said to you in Division of whole numbers, that you should not passe of the remainer there but onely note it as a summe that could not be divided without knowledge of Fractions. Wherefore now mark this, that in all divisions of whole numbers, when there is any remainer, you shall set it over a line as a Numerator, & set the divisor for the denominator, and that Fraction doth make the Division compleat, & is part of the quotient : As if I would divide 48 by 5, the quotient will be $9\frac{3}{5}$: so in your former work when 24 was divided by 5, the quotient should be $4\frac{4}{5}$, & so the new Fraction should be thus, $\frac{4}{8}$ & $\frac{4}{5}$ of $\frac{1}{8}$, that is, $\frac{4}{8}$ of the entire number, & $\frac{4}{5}$ of $\frac{1}{8}$

 part

part of any thing, which you may prove by example of some Coyne.

Scholar. When I take a Crowne, whose $\frac{3}{5}$ is 3 s. Now I would prove whether the 3 s. be $\frac{4}{8}$ and $\frac{4}{5}$ of $\frac{1}{8}$, I shall have a cumbrous work to do.

Master. Indeed for whole pence, your example is a little troublesome: yet turning the crown into halfe pence, it is easie enough.

Scholar. That will I try.

¶ First I see that $\frac{3}{5}$ of a Crown is 3 shillings which is 36 pence, or 72 halfe pence. Now if I can finde that this Fraction $\frac{4}{8}$ & $\frac{4}{5}$ of $\frac{1}{8}$ be equall unto 3 shillings, then am I fully answered.

Because I cannot take $\frac{4}{8}$ of a Crowne, I turn the Crown into halfe pence, as you willed me, which makes 120, which I divide by 8, my Quotient is 15, which taken foure times, make 60 ob. Now resteth me to have $\frac{4}{5}$ of the $\frac{1}{8}$ part of a Crown, whereof $\frac{1}{8}$ part is 15 ob. the 15 being parted into 5 parts, the Quotient is 3, which taken foure times maketh 12 ob. which with my 60 before amounteth to 72, which are then equall to $\frac{3}{5}$, my desire.

Master. I commend you for your diligence, you might have wrought it thus: either $\frac{4}{8}$ being abbreviated as before I taught, is $\frac{1}{2}$. Now halfe a Crown is 2 shillings 6 pence. Now $\frac{4}{5}$ of $\frac{1}{8}$ is a Fraction of Fractions, which if you doe reduce into one entire Fraction, as before you have learned in saying, 5 times 8 is 40, for a new Denominator, and once 4 is 4, for a new numerator: it maketh $\frac{4}{40}$, and abbreviated also make $\frac{1}{10}$. Now the tenth part of a Crown is 6 pence, which put to 2 shillings 6 pence, make also 3 shillings your desire.

But

But now one example moze for this Rule, and then we shall end it. If I have $\frac{7}{15}$ of a Soveraign (accounting the Soveraigne 20 shillings) how many shillings is that $\frac{7}{15}$?

Scholar. I must multiply 7 by 20 and that maketh 140, which I shall divide by 15, and the quotient will be $9\frac{5}{15}$, oz in lesser termes $\frac{1}{3}$.

Master. That is 9 shillings, and one third part of a shilling, that is 4 pence, as by the same Rule you may prove. And this for this time shall suffice for Reduction. And now I will proceed to Addition.

Addition.

Addition of fractions of one denomination.

Whensoever you have any Fractions to be added, you must consider whether they be of one denomination, or not, and if they be of one denomination, then adde the Numerators together, and set that that amounteth for the numerator over the common denominator, and so have you done: The reason is, because that such differ little in Addition or Subtraction from the worke of vulgar denominations where the denominators be of the number, as 3 pence and 5 pence make 8 pence, where the denomination is not altered. But if the fractions be not of one denomination or any of them be mixt of whole numbers and fractions, then must you first reduce them to one denomination, and after adde them. And if they be many, then adde first two of them, and so the summe that doth amount of the Addition, and the third, and then the 4. &c. if you have so many.

Scholar.

Scholar. This séemeth easse enough, now that I have already learned to reduce, without which I could never have wrought this : And therefore now I sée good reason why you did place Reduction before Addition.

Master. It is well considered, but yet refuse not to expresse your understanding of it by an example.

Scholar. Then would I adde first $\frac{7}{12}$ with $\frac{5}{12}$ and because the denominators are like (and so néedeth no Reduction) I adde 7 to 5, which maketh 12, and then is my summe $\frac{12}{18}$, that is in smaller numbers, being abbreviated $\frac{2}{3}$.

To adde fractions of divers Denominations.

And if you have many numbers to be added, as here $\frac{1}{8}\frac{4}{5}\frac{9}{10}$, first I must reduce them (because they have divers denominators) into one denomination, and then they will bée thus, $\frac{112}{400}\frac{280}{400}\frac{360}{400}$, or in lesser termes. $\frac{11}{40}\frac{22}{40}\frac{36}{40}$, which by Addition do make $\frac{83}{40}$, that is $2\frac{3}{40}$.

Master. Now may we go to Subtraction.

Subtraction of Fractions.

Subtraction of Fractions.

SUbtraction *hath the same precepts that* Addition *had, for if the* denominators *bee like, then must you subtract the one* numerator *from the other, and the rest is to be set over the common* denominator, *and so your subtraction is ended : but and if you have many* fractions *to be subtracted out of many, then must you reduce them to one* denomination, *and into two severall* fractions, *that is, all*

all that muſt be ſubtracted into one fraction, *and the reſidue into another* fraction, *and then work as I ſaid before.*

Scholar. **For the firſt example I take $\frac{11}{12}$ to be ſubtracted out of $\frac{17}{12}$, and the reſt will bee $\frac{3}{12}$ or $\frac{1}{4}$.**

For another example I take $\frac{3}{4}$ to be ſubtracted out of $\frac{7}{8}$, which I muſt reduce, and it will be thus $\frac{24}{32}$ and $\frac{28}{32}$.

Then doe I ſubtract 24 out of 28, and there reſteth 4, which I ſet over the common denominator **for a Remainer, thus $\frac{4}{32}$: that is $\frac{1}{8}$.**

Now for the third example, **I take $\frac{1}{4}$ and $\frac{1}{6}$ to bée ſubtracted from $\frac{7}{8}$ and $\frac{9}{10}$: and becauſe their deno-minators be divers, I doe reduce them into one de-nomination thus $\frac{48}{24}$ and $\frac{1440}{80}$ $\frac{1408}{1920}$ and $\frac{1641}{1920}$.**

Then do I adde the two firſt, and they make $\frac{1440}{1920}$. Alſo I adde the two laſt, and they yeeld $\frac{3408}{1920}$.
Then do I ſubtract 3040 out of 3408, and there reſteth 368, ſo is the remainer, $\frac{368}{1920}$, that is in ſmaller termes $\frac{23}{120}$. And thus have I done with Subtraction, except you have any more to teach me.

Maſter. **Prove one example or more out of Fractions of divers Denominations.**

Scholar. **I take the two Fractions $\frac{7}{8}$ to bée ſub-tracted from $\frac{2}{24}$ which being reduced, will ſtand thus $\frac{168}{192}$ and $\frac{72}{192}$: Now would I ſubtract 168 out of 72, but I cannot.**

168	72
$\frac{7}{8}$	$\frac{2}{24}$
192	

Maſter. **Then may you perceive that you miſtook the Fractions : for you can never ſubtract the greater out of the leſſer, although you may** adde, multiply or divide the greater with the leſſer.

And

The grea-
test of two
fractions. And albeit that $\frac{8}{}$ hath both his terms lesser then $\frac{9}{24}$
yet is $\frac{9}{24}$ the lesser Fraction : for generally if you
multiply the Numerator and the Denominators of
two Fractions crossewayes, that fraction is the
greatest of whose Numerator commeth the greatest
summe, as in this example , 7 multiplied by 24
maketh 168, and 9 being multiplied by 8 yeeldeth
but 72, therefore is the first fraction 7 the greatest
of these two, so can you not subtract it out of a
lesser fraction.

But if you should subtract a Fraction out of a
whole number, what should you doe ?

Scholar. Marry I would reduce the whole number into a Fraction of the same Denomination that
my Fraction is, and then worke by Subtraction.

Master. So may you doe, but it is much easier,
if your Fraction be a proper Fraction, that is to
say, lesse then an unite, to take an unite from the
whole number, and then turn it into an improper
Fraction, and so worke your Subtraction. As if
I would subtract 3 $\frac{2}{5}$ from 4, I may take 1 from
4, and turn it into $\frac{5}{5}$, from which I abate 3 $\frac{2}{5}$, there
will remain, $\frac{3}{5}$. And if the first be an improper Fraction, then may I take so many unites from the
whole number, that they may make an improper
Fraction, greater then that first, and then worke
by Subtraction. As if there bee proposed $\frac{12}{3}$, to bee
subtracted from 6, because $\frac{12}{3}$ is more then 3, & not so
much as 4, I must take 4 from 6, and turn them
into thirds thus, $\frac{12}{3}$ then abate $\frac{12}{3}$ from $\frac{12}{3}$, there
resteth $\frac{2}{3}$: so the whole remainer is 2 $\frac{2}{3}$.   Or else
you may at your pleasure take 3 $\frac{1}{3}$, which is $\frac{10}{3}$:
from 6 whole : then set 1 under 6, as thus $\frac{6}{1}$: And
then

then to reduce thofe two Fractions into one Denomination, as here appeareth $\frac{12}{3}$ from $\frac{4}{3}$: Then from $\frac{18}{3}$ refteth $\frac{8}{3}$, which maketh $2\frac{2}{3}$ your defire. And thus will I make an end of the work of fubtraction of Fractions, and proceed to Multiplication.

$$\frac{12}{3}$$
$$8$$
$$\frac{18}{3} \times \frac{12}{3}$$
$$3$$

Multiplication of Fractions.

Herefore when any two fractions be propofed to be multiplied together, the numerator of the one muft be multiplied by the numerator of the other : and the fumme that amounteth thereof muft be fet for a new *numerator : likewife the* Denominator *of the one muft be multiplied by the* Denominator *of the other, and that that amounteth fhall* be fet for the Denominator, and this new third *fraction* expreffeth the Product of the Multiplication *of the two firft* Fractions propofed, whereof take this Example, $\frac{3}{5}$ *multiplied by* $\frac{2}{5}$ $\frac{5}{12}$ doth make $\frac{15}{60}$.

Multiplication of Fractions

$$15$$
$$\frac{5}{12}$$
$$60$$

Scholar. I perceive then that 3 being the Numerator of the firft Fraction is multiplied by 5 being the Numerator, of the fecond Fraction, whereof amounteth 15, the Numerator of the third Fraction. And fo likewife 5 being the Denominator of the firft Fraction, is multiplied by 12 the Denomination of the fecond Fraction, whereof amounteth 60 the new denominator, fo that

that I perceiue how the worke is done, I do not perceiue how $\frac{15}{60}$ is greater then $\frac{1}{3}$, for if I shall use my former manner of examination by the parts of some coine, I see that $\frac{3}{4}$ of a Crowne is 36 pence, and $\frac{5}{12}$ of a Crowne is 25 pence, whereof the one multiplied by the other, doth make 900 pence, which is 15 Crownes, but by your multiplication there amounteth $\frac{15}{60}$, which is but 15 pence, and that is much lesse then any other of both the first Fractions.

Master. That difference is between multiplication in whole numbers, and Multiplication in broken numbers, that in whole numbers, the summe that amounteth is greater then both the other whereof it came: but in fractions it is contrariwise : for the summe that amounteth is lesser then any of the other two fractions whereof it is produced.

S. I desire much to vnderstand the reason thereof.

Master. Although I purposed to reserue the reasons of workes Arithmeticall for the perfect Booke of Arithmetick, yet I will shew you this, because of the strangenesse of the worke.

You see in whole numbers, that of two numbers, being multiplyed together, is made the third number, which third number doth beare the same proportion to the number multiplied, that the multiplier doth beare to an vnite. And so in Fractions the third number which amounteth of Multiplication, beareth the same proportion to each of the two first Fractions, that the other of these two fractions doth beare to an vnite.

Scholar. Sir, I vnderstand your words thus : when 40 is multiplyed by 12, there doth amount

480

480, which 480 doth contain 40 so many times in it, as 12 doth containe Unites, that is to say, twelve times. And so it appeareth that 480 doth containe twelve so many times also as 40 doth containe unites, that is 40 times. But now I see not how the third number in this example of Fractions can containe any of the two former (as it happened in whole numbers) seeing it is lesser then either of them.

Master. No marvell if you cannot see that thing which is not possible to be seen of any man, how the third number in Multiplication of Fractions should be greater then any of the two former Fractions: but yet this may you see (which I said) that the third number in fractions so multiplyed doth beare the same proportion to any of the two former Fractions that the other of those two Fractions doth bear to an unite, as in your example, $\frac{3}{5}$ being multiplyed by $\frac{1}{12}$ doth make $\frac{11}{60}$. Now I say that $\frac{11}{60}$ doth beare the same proportion to $\frac{3}{5}$ that $\frac{1}{12}$ doth bear to a unite, as you may in your owne forme of examination by Coine, try it: for in an old angell (which in times past was, currant for 7 shilling 6 pence) are 180 halfe pence which I set for the intire unite, whose parts (according to the Fractions aforesaid) are these, for $\frac{11}{60}$ set 45 halfe pence, for $\frac{3}{5}$ take 108 halfe pence, and for $\frac{1}{12}$ put 75 halfe pence. Now doth 45 bear the same proportion to 108, that 75 doth bear to 180, for 45 is $\frac{1}{12}$ of 108, and so is 75 also $\frac{1}{12}$ of 180.

But these reasons may be better reserved till another time, when the knowledge of proportions in due order shall be taught: yet in the mean season

I

I will shew you how it commeth to passe, that in Fractions the third summe must needs be lesser then any of the other two.

Consider this, that when a Fraction is proposed, as in the former example ⅗ if it be multiplied by more then 1, it will make more then one entire number. As if I multiply ⅗ by 5, that is to say, if I take it 5 times, it will make three entire unites. Example : in a Crown ⅗ of it maketh 3 shillings, which if I take five times, it will amount to 15 shillings that is, three entire Crownes ; so if I take the same ⅗ but twice, it will yeeld 6 shillings, that is, one entire Crowne, and ⅕. Now if I take it but once, it cannot be more then it was before, that is 5 shillings. And if I take it lesse then once, it cannot bée so much as it was before. Then seeing that a Fraction is lesse then one, if I multiply a Fraction by another Fraction, it followeth that I doe take the first Fraction lesse then once, and therefore the somme that amounteth, must néeds be lesse then the first Fraction.

Scholar. Sir , I thank you much for this reason. And I trust I doe perceive the thing, as by example of this same Fraction ⅗ I will expresse. If I take ⅗ of a Crowne once, that is to say, if I multiply ⅗ by 1, it will be as it was before, but 3 shillings : so if I doe multiply it by ½, that is, if I take but halfe one time, then will it be but halfe so much : likewise if I multiply it by ⅓ that is, if I take but the third part of one, it will yeeld but 12 pence, that is, the third part of the first Fraction.

And

And so to make an end: if I take but the twelfth part of one, that is, if I doe multiply it by $\frac{1}{12}$ it will yeeld but the twelfth part of the first Fraction, which is but 3 pence. And it followeth, that if $\frac{1}{12}$ make 3 pence, then $\frac{5}{12}$ must needs make five times so much, that is, 15 pence, which was the summe that hath given the occasion of all this doubt.

Master. When I perceive you have sufficient understanding in this sort of Multiplication for this time, wherefore I will proceed to the rest.

In Multiplication it happeneth sometime, that there bee whole numbers to be multiplyed with Fractions, and may be in two sorts: for either the whole number is severall from the fraction, & is the multiplier, or else the whole number is joyned with one, or both of the fractions, and so maketh a mixt number thereof. If it bee in the first sort, then needeth there no Reduction, but onely multiply the numerator of the fraction by that whole number, and the totall thereof set for the new numerator.

To multiply a whole number into a fraction.

Scholar. I understand you thus. If I have $\frac{6}{23}$ to be multiplyed by 16 then must I multiply that 16 with 6, which is the Numerator, whereof commeth 96, and that must I set for the new Numerator: keeping still 23 for the Denominator, and so the Fraction will be $\frac{96}{23}$ that is $4\frac{4}{23}$.

Mast. And in this sort of work you may abridge the labour thus. If it happen the denominator to be such a number as may evenly be divided by the said whole number proposed, then divide it thereby, and set the Quotient of that division for the former denominator, but reserve still the numerator, and so is the multiplication ended.

S   Scholar.

Scholar. Then fain this example $\frac{7}{20}$ to be multiplied by 5, and becaufe 5 will juftly divide 20, therefore I take the Quotient of that divifion, which is 4, and fet it in ftead of 20, and fo the Fraction will be $\frac{7}{4}$ that is 1 $\frac{3}{4}$.

Mafter. Which is all one with $\frac{35}{20}$ that would have followed of the other fort of work.

Scholar. I perceive it very well.

How to multiply mixt numbers.

Mafter. *Now then for the other fort, where the number is mixt, take this way: firft to reduce the faid whole number and Fraction into one improper Fraction (as I fhewed you in* Reduction *) and then multiply them together, as if they were proper Fractions.*

Scholar. 13 $\frac{3}{5}$ being fet to be multiplyed by $\frac{5}{8}$ firft I muft reduce the mixt number, as in this example appeareth, by multiplying 13 by 5, and that maketh 65, whereto I muft adde the numerator 3, and fo the Fraction will be $\frac{68}{5}$ which two Fractions now I fhall multiply after the accuftomed forme, and it will be $\frac{340}{40}$, or $1\frac{3}{4}$.

$$13\frac{3}{5} \qquad 340$$
$$\overline{68\ by} \qquad \overline{\ }$$
$$\underline{} \qquad 5$$
$$5\ 40 \qquad \overline{8}$$

Mafter. You have done well: and fo may you fée, that although moft part of the formes of multiplication may be wrought without Reduction, yet fome cannot, as namely, mixed numbers.

Duplation.

And yet one note more I will tell you of Multiplication before we leave it : That is ; whenfoever you would multiply any Fraction by 2, *which commonly is called* duplation, *you may doe it not only by doubling the numerator, but alfo by parting the denominator into half, if it be eaven.*

Scholar. Then if I would double $\frac{7}{12}$ I may chufe chuſe whether I will make it $\frac{10}{12}$ oʒ elſe $\frac{5}{6}$. And indeed I ſée that is all one, but the dividing of the Denominator ſéemeth the better way to make ſmaller termes of the Fraction, and ſo they ſhall néed the leſſe Reduction.

Maſter. It is ſo: and now I ſhall not néed to tell you that Multiplication is pʒoved by Diviſion, and Diviſion likewiſe by multiplication: but the like woʒk that I ſhewed you in multiplication, will I ſhew you in diviſion.

Diviſion of Fractions.

Henſoever two fractions bee propoſed, that one ſhould be divided by the other, I muſt ſet down firſt the fraction that ſhall be divided (which is called the dividend) and then after it the other which is the diviſor: Then ſhall I multiply the numerator of the dividend by the denominator of the diviſor, and that which amounteth I muſt put for a new numerator. Againe I ſhall multiply the denominator of the dividend by the numerator of the diviſor, and the number that amounteth thereof I muſt put for the new denominator. And this third fraction is the Quotient of the ſaid diviſion.

Diviſion of Fraƈtions.

Scholar. This ſéemeth eaſie in foʒme, as by example thus: If I would divide $\frac{5}{8}$ by $\frac{2}{6}$, firſt I multiply 5, (being the numerator of the dividend) bʒ 6, which is the denominator of the diviſor,

and

and thereof riseth 30 : then I
multiply 8 (being the denomina-   30
tor of the dividend) by 2 being the   5   2
numerator in the divisor : and so   8   6
riseth 16, the which I must make   16
a third Fraction thus $\frac{30}{16}$.

Master. Me seemeth you are quicker in understanding now, then you were when I taught you
the Art of whole numbers, but that is no marvell :
for the more knowledge that a man getteth the
readier shall be finde his wit, and be quicker in
understanding : but yet of two things I will ad-
monish you, which you might have observed here
for the ease of worke and lightnesse of understanding the nature of the Quotient.

Whensoever you divide one Fraction by another, either they be both equall together, or else the
one is greater then the other : if they be equall,
their Quotient shall be such, that the Numerator
and the Denominator of it shall be equall also. And
if the two first Fractions be unequall their Quotient
shall declare the same by the inequality of the Numerator and Denominator, as in these examples following shall appear.

First, if equall Fractions $\frac{4}{9}$ and $\frac{12}{27}$ be equall to-
gether, and if the one be divided by the other, the
Quotient will be $\frac{108}{108}$, as you may perceive by the
Rule aforesaid.

Now in the unequall Fractions, as $\frac{4}{9}$ and $\frac{3}{15}$ the
Quotient will be $\frac{60}{27}$, where the Numerator is greater then the Denominator.

Scholar. I see it is so: but I see not the reason
why it should be so.

Master

Note how to know the proportion between to numbers.

Master. The reason is this; when any Fraction is divided by another, the Quotient declareth what proportion the Dividend beareth to the Divisor. So $\frac{1}{2}$ divided by $\frac{1}{4}$, maketh 2, which must be sounded, not 2, but twice, declaring that $\frac{1}{4}$ is contained twice in $\frac{1}{2}$.

And note this, that the Numerator in the Quotient representeth the Dividend, and the Denominator representeth the Divisor. And this is alwayes true, whether the greater Fraction be divided by the lesser, or the lesser by the greater. But this proportion will not be exactly known, till you have learned the Art of proportions: notwithstanding somewhat of it I have declared in the Rule of Reduction. But now for the easie remembrance of the Quotient in Division, as soone as you have set downe your two Fractions the one against the other, then make a straight line for the Quotient: and as soone as you have multiplied the Numerator of the Dividend, by the Denominator of the Divisor, set the Number that amounteth over the said line, and then multiply the other two Numbers, and set their totall under the same line.

Scholar. I perceive you would not have me trust to memory till I were better expert, lest oftentimes I happen by misse-remembrance to be abused. This Example I take for that declaration.

If I would divide $\frac{2}{3}$ by $\frac{3}{4}$ I must set the numbers one against the other (as here doth appear) and then make another line for the Quotient in some good

$$2 \quad\underline{\quad by \quad}\quad 3$$
$$3 \qquad\qquad\quad 4$$

distance

S 3

diſtance, where I may ſet the numbers of the Quotient, as ſoon as any of them is multiplied. So then as ſoone as I have multiplied 2 by 4 which maketh 8, I ſhall ſet that 8 over that line, thus: 8
And then multiply 3 by 3, which yeeldeth   ———
9 : and that 9 muſt be ſet under the ſame   9
line, and then will the whole Quotient appeare thus $\frac{8}{9}$: whereby it appeareth (as I remember your words) that $\frac{2}{3}$ is in proportion to $\frac{3}{4}$ as 8 is to 9, but how may I perceive that?

Maſter. Although you might better perceive it by the Rule of Reduction, yet this example may be declared in common coines, as in a common ſhilling of 12 pence, of which $\frac{2}{3}$ maketh 8 pence, and $\frac{3}{4}$ doth make 9 pence, an ſo you may eaſily ſee that their proportions doe agrée. And if you had taken this example before when you tooke the example of $\frac{2}{3}$ & $\frac{1}{2}$, your Quotient would appeare (as this doth) more eaſie to underſtand; whereas that Quotient being $\frac{30}{16}$, is not an eaſie proportion for you to perceive, being yet little acquainted with proportions.

Scholar. If there be whole Numbers to bée divided by a Fraction, how ſhall I performe it?

Maſter. *When any whole number ſhall be divided*
To divide a whole number by a fraction. *by a fraction, you muſt multiply the ſaid whole number with the Denominator of the Fraction, and ſet the totall thereof for the new Numerator, and for the Denominator ſet the Numerator of the fraction.*   80

Scholar. Then 20 divided by $\frac{3}{4}$ will   20 by
To divide the fraction by the whole number. make $\frac{80}{3}$, as here appeareth $\frac{20}{3}$.   ———
   3

Maſter. *Even ſo: but if you would divide the Fraction by the whole number, then multiply the Denominator by the ſame whole number, and ſet the total*
                                           f

for the Denominator, without changing the Numera-
tor.

Scholar. Then to divide $\frac{20}{23}$ 20
by 4, it will be $\frac{20}{92}$, as here   20 by 4
appeareth $\frac{20}{23}$ by $\frac{4}{1}$ in this Ex-   ————
ample $\frac{20}{92}$.   23 1

     92

Master. You say well. And by the same Example you give me cause to remember another
briefe way to doe the same : for if you had divided
the said Numerator by 4, and set the Quotient for
the Numerator, keeping still the old Denominator,
it would have been not only as well done, but also
in a Fraction of lesser terms.

Scholar. I guesse it to be even so, by a like work
that you taught me in multiplication : And for
prof thereof $\frac{20}{23}$ being the Dividend, and 4 the
Divisor, I divide the Numerator 20 by 4, and the
Quotient is 5, which I set for 20 over 23, thus $\frac{5}{23}$:
And I see that it is all one with $\frac{20}{92}$, as by dividing
or abbreviating both these termes by 4, and so
reducing them to their least Denomination, I may
easily prove : as appeareth by this example $\frac{20}{2}\frac{5}{1}$.

Master. You conceive it well. And if there be
mixt numbers, (either one or both) you must first
reduce that mixt number into an improper Fraction,
and then work as you have learned.

Scholar. That was sufficiently taught in Multiplication. Therefore I pray you go forward to
some other thing.

Master. Then take this note yet for Division: if
the denominators be like, then divide the numerators,
as it were in whole numbers, and the Quotient,
whether it be Fraction, whole number or mixt, is a

S 4   gvo

good Quotient for that Division. And generally, if one of the numerators may justly divide the other by that Quotient, multiply the Denominator of the lesser numerator, and set it that doth amount in the roome of the same denominator, and then for a numerator to it, set the denominator of the other Fraction.

Scholar. Then if I would divide $\frac{3}{4}$ by $\frac{12}{17}$ I see that 3 will divide 12, and the Quotient will be 4, by which I must multiply the other 4, that is the denominator under 3, and then it is 16, which is set for the denominator 4, and over it in stead of 3 I must set 17 the other denominator, and so it is thus $\frac{17}{16}$.

Master. And so is $\frac{17}{16}$ in stead of $\frac{51}{48}$, which would have rs'en by the common worke, as here appeareth.

$$\frac{3}{4} \; by \; \frac{12}{17} \quad \frac{51}{48}$$

And now for Mediation (which is to divide by 2) marke this, if the Numerator be an even number, set the half of it in his place without the divisor, and so have you done : and if the numerator be not even, then double the denominator.

Scholar. That is, if I would mediate $\frac{16}{11}$, I may make the Quotient $\frac{8}{11}$, and if I would mediate $\frac{7}{11}$, I must make it $\frac{7}{22}$.

Master. And thus will I make an end of the workes of common fractions for this time, not doubting but you can apply them both to the Rule of progression, and also to the Golden Rule, without any other teaching then you have learned before which might seem tedious to repeat, in regar you have sufficient knowledge in Reduction

Addition

Addition, Subtraction, Multiplication, and Division: And therefore will I goe in hand with the Rule of Proportion, or Golden Rule, which now will appeare easie enough.

The Golden Rule direct in Fractions.

Master.

Herefore as touching the *Golden Rule* for the placing of the three numbers proposed in the question whereby to finde the fourth and for the form of their worke, with other like notes, I referre you to that which you have already learned.

The rule of proportion in Fractions.

But this easie forme of working by fractions shall you note, that if your three numbers be fractions, for an apt worke and certaine, multiply the numerator of the first number in the question, by the denominator of the second : And all that againe multiply by the denominator of the third number, and the totall thereof shall you keepe for to bee the divisor. Then multiply the Denominator of the first number by the numerator of the second , and the whole thereof by the numerator of the third, & the totall thereof shall be your dividend.

Note this for a generall Rule.

Now divide this dividend by the divisor which you found out before, and that number shall be the fourth number of the question which you seek for, as in this example.

If $\frac{1}{4}$ of a yard of Velvet cost $\frac{2}{3}$ of a soveraign, esteemed at 20 shillings. what shall $\frac{3}{6}$ cost ?

A question of velvet.

Scholar. If it please you to let me make the answer, I would first place these three numbers as I learned in whole numbers, thus:

$$\frac{1}{4} \quad \frac{2}{3}$$
$$\frac{3}{6} \quad Z$$

And

And then according to your new rule, I muſt multiply 3, being numerator in the firſt number, by 3, the denominator of the ſecond: & thereto commeth 9, which I multiply again by 6, the denominator of the third number, and ſo have I 54, which I keep for the diviſor. Then multiply I 4 the denominator of the firſt, by 2 the numerator of the ſecond, and there ariſeth 8, which againe I multiply by 5 the numerator of the third, and it maketh 40. Then muſt I divide 40 by 54, and it will be $\frac{40}{54}$ that is $\frac{20}{27}$ in leſſer termes, and then the figure will ſtand thus:

$$\begin{array}{ccc} \frac{2}{4} & & \frac{3}{3} \\ & Z & \\ \frac{5}{6} & & \frac{20}{27} \end{array}$$

But what that is in money I cannot tell, except I ſhall worke it by Reduction, as you taught me.

Maſter. It forceth not now, you may reduce it when you liſt, but it were diſorderly done here to mingle divers workes together, where we do not ſeek the value of the thing in common money, but in apt number, which you have well done: and therefore will I yet ſhew you another like way of eaſtneſſe in worke, how you may change your three Fractions into three whole numbers, by which you ſhall worke, as if the queſtion were propoſed in whole numbers. The firſt number you ſhall finde as I taught you: now to finde the diviſor of the ſecond number, take the numerator for the ſecond fraction: and for the third number, take that that ariſeth of the multiplication of the denominator of the firſt, by the numerator of the third, and then worke your queſtion.

A queſtion of Silver.

Scholar. *For example hereof, I put this queſtion, If $\frac{11}{12}$ of 1 pound weight of ſilver, be worth $\frac{12}{4}$ of a Soveraigne, what is, $\frac{1}{2}$ of 1 pound weight worth?*

For

For the answer, first I place the $\frac{11}{12}$ **Z** $\frac{12}{4}$
Fractions in order thus : $\frac{1}{2}$

Then to turne these Fractions into whole numbers, I multiply 11, which is the numerator of the first by 4 (the denominator of the second) and there commeth 44, which I multiply by 2 the denominator of the third, and so amounteth 88, which I set for the Divisor in the first place. Then in the second place I set 12, which is the numerator of the second fraction, and in the third place I set the summe that amounteth of 12, being the denominator of the first number, multiplied by one, being numerator in the third number, and so   88 **Z** 12 the figure will stand as here you see.   12

Then to work it forth, I multiply 12 by 12, and there amounteth 144, which I divide by 88, and the quotient will be 1 $\frac{56}{88}$, or in lesser termes, 1 $\frac{7}{11}$, and then the figures will stand $\frac{11}{12}$ **Z** $\frac{12}{4}$ thus : $\frac{1}{3}$ 1 $\frac{7}{11}$

Master. These two formes now you understand The proof well enough, and as for any other at this time I of the golwill not repeat, onely this shall you mark for the den rule. proof of this Rule, whether your work be well wrought or no.   Multiply the first number by the fourth, and note what amounteth; then multiply the second by the third, and mark what amounteth also.   Now if these two numbers so amounting be equall, then is your work well done, else you have erred.   And this shall suffice for the former Rule.

The

The Backer Rule, or Reverſe Rule in Fractions.

Ut in the Backer Rule, this ſhall you note for your eaſe of worke, that you mul-tiply the Numerator of the firſt by the Numerator of the ſecond, and the whole thereof by the Denominator of the third, and that amounteth thereof, ſhall be the divi-dend. Then multiply the Denominator of the firſt, by the Denominator of the ſecond, and that whole by the Numerator of the third, and that that ariſeth thereof, ſhall be the Diviſor. Example of this.

I did lend my friend $\frac{3}{4}$ of a Porteguiſe, ſeven Moneths upon promiſe that he ſhould do as much for me againe; and when I ſhould borrow of him he could lend me but $\frac{1}{12}$ of a Porteguiſe: now I demand how long time I muſt keep his money in juſt recompence of my loane, accounting 13 Months in the year?

Scholar. The firſt number muſt be the firſt mo-ney boꝛꝛoweꝺ, that is $\frac{3}{4}$ of the Porteguiſe: the ſe-conꝺ number the 7 moneths, that is $\frac{7}{13}$ of a yeare: anꝺ the thirꝺ number the money that was lent in recompence, that is $\frac{1}{12}$ of a Porteguiſe:

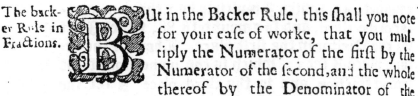

then J ſet the numbers thus:

Then (as you taught me) J multiply 3 (being Numerator in the firſt number) by 7, the Numerator of the ſecond number, anꝺ it maketh 21, which J multiply by 12 the Denominator of the thirꝺ, anꝺ ſo have J 252 foꝛ the dividend: then J multiply 4, the Denominator of the firſt, by 13 the Denomina-

tor

tor of the 2^d, & it yeeldeth 52, which I multiply again by 5 the Numerator of the third, & it will make 260, that is the divisor. Then must I divide 252 by 260 so it will be in the small Fraction $\frac{63}{65}$ of a year.

Master. And thus do you see some ease in working, better then to multiply and divide tediously so many Fractions.

Another question yet I will propose, to the intent you may see thereby the reason of the Statute of Assise of Bread and Ale, which in all statute Bookes, in Latine, French and English, is much corrupted for want of knowledge in this Art; for the right understanding whereof I propose this question.

When the price of a quarter of wheat is 2 *shillings, the farthing white loafe shall weigh* 68 *shillings; then I demand what shall such a loafe weigh, when a quarter of wheat is sold for* 3 *shillings.*

Scholar. This question must be wrought as it is proposed in whole numbers: and not in Fractions.

Master. You seem to say reasonably, howbeit in the Statute of Assise, the rate is made by the proportion of parts in a pound weight Troy else could it not be a Statute of any long continuance, seeing the shillings doe change often as all other moneys doe: but this Statute being well understood, is a continuall Rule for ever, as I will anon declare by a new Table of Assise, converting the shillings into ounces, and parts of ounces.

Wherfore here by a shilling you must understand $\frac{1}{20}$ of a pound weight, and so by a penny $\frac{1}{20}$ of an ounce: wherefore although you might worke this question proposed by whole numbers well enough, for that time when the Statute was made, yet

Statute of Assise of bread and ale.

Question of bread.

yet to apply it to your time, and make it ferve for all times generally, it is beft to worke it by fracti-ons, fetting for 2 fhillings $\frac{2}{20}$, and for 68 fhillings $\frac{68}{20}$, and fo for three fhillings $\frac{3}{200}$, and then $\frac{2}{20}$ $\frac{68}{20}$ $\frac{3}{20}$ will the figure of the queftion ftand thus. Z

In which queftion, becaufe all the Denominators be like, you fhall work onely with the Numerators.

Scholar. Then fhall I multiply 68 by 2, where-of commeth 136, which if I divide by 3, the quoti-ent will be 45 $\frac{1}{3}$: but how fhall I make a fraction of that, to ftand with the other ?

Mafter. Have you fo foon forgotten what was taught you fo lately? this is his forme. 45 $\frac{1}{3}$ 20

Scholar. I remember it now, and then it figni-fieth 45 twenty parts, and the third deale of one twenty part.

Note what a fhilling is.

Mafter. So is it that maketh in fhillings 45 fhil-lings 4 pence, whereby you may note one great er-rour in the Statute Books ; which have conftantly 48 fhillings in that Afsife. And by this Rule, if you examine the Statute, you fhall finde many fummes falfe. Wherefore for the true underftanding of that Statute, and fuch like, as I have made mention of it, and fomewhat recognized it, fo doe I wifh that all Gentlemen and other ftudents of the Lawes would not neglect this Art of Arithmetick, as un-needfull to their ftudies. Wherefore to encourage them thereto, and to gratifie both them & all other in generall I will exhibite a Table of that part of the Statutes in two Columnes and in a third Co-lumne, I will adde the correction of thofe errours which have crept into it.

Here followeth the Table.

The price of a quarter of Wheat.		The weight of a farthing white loafe by the Statute Bookes.			The Correction by just Assise.		
S.	D.	l.	s.	d.	l.	s.	d.
1	0	6	16	0	6	16	0
1	6	4	10	8	4	10	8
2	0	2	8	0	3	8	0
2	6	2	14	$4\frac{1}{5}$	2	14	$4\frac{4}{5}$
3	0	2	8	0	2	5	4
3	6	2	2	0	1	18	$10\frac{2}{7}$
4	0	1	16	0	1	14	0
4	6	1	10	0	1	10	$2\frac{2}{3}$
5	0	1	8	$\frac{1}{2}$	1	7	$2\frac{2}{5}$
5	6	1	4	$8\frac{1}{4}$	1	4	$8\frac{8}{11}$
6	0	1	2	8	1	2	8
6	6	0	16	11	1	0	$11\frac{1}{13}$
7	0	0	19	1	0	19	$5\frac{1}{7}$
7	6	0	18	$1\frac{1}{2}$	0	18	$1\frac{1}{3}$
8	0	0	7	0	0	17	0
8	6	0	16	0	0	16	0
9	0	0	15	$0\frac{1}{4}$	0	15	$1\frac{1}{3}$
9	6	0	14	$0\frac{1}{4}$	0	14	$3\frac{13}{19}$
10	0	0	13	$7\frac{1}{2}$	0	13	$7\frac{2}{5}$
10	6	0	12	$11\frac{1}{2}$	0	12	$11\frac{2}{7}$
11	0	0	12	$4\frac{1}{4}$	0	12	$4\frac{4}{11}$
11	6	0	11	10	0	11	$9\frac{21}{23}$
12	0	0	11	4	0	11	4

In the common Bookes there is no further rate of Assise made, then unto 12 s. the quarter of wheat, but in an ancient copy of 200 years old (which I have) there is added the rate of Assise unto 20 s. the quarter, but yet was that Assise also either wrong cast at the first penning, or else corrupt since that time, for lack of just knowledge in the Rule of Proportion, which I will adde here also to gratifie such as be desirous to understand truth exactly.

The price of a quarter of wheat.		The weight of a farthing white loafe by the Statute Bookes.			The Correction by just Assise.		
s.	d.	l.	s.	d.	l.	s.	d.
12	6	0	11	0	0	10	10 $\frac{14}{23}$
13	0	0	11	0 $\frac{2}{7}$	0	10	5 $\frac{2}{13}$
13	6	0	10	1 $\frac{1}{2}$	0	10	0 $\frac{8}{9}$
14	0	0	9	7	0	9	8 $\frac{4}{7}$
14	6	0	9	2 $\frac{1}{3}$	0	9	4 $\frac{16}{15}$
15	0	0	9	1 $\frac{1}{2}$	0	9	0 $\frac{4}{5}$
15	6	0	9	1 $\frac{1}{4}$	0	8	9 $\frac{2}{31}$
16	0	0	9	0	0	8	6
16	6	0	8	6	0	8	2 $\frac{10}{11}$
17	0	0	8	3	0	8	0
17	6	0	7	10	0	7	9 $\frac{2}{33}$
18	0	0	7	6	0	7	6 $\frac{2}{3}$
18	6	0	7	3	0	7	4 $\frac{8}{37}$
19	0	0	7	2	0	7	1 $\frac{11}{19}$
19	6	0	5	10	0	6	11 $\frac{2}{13}$
20	0	0	5	6	0	6	9 $\frac{1}{5}$

These

These two Tables I have set severall, because no man should think that I would either add or take away from any Law those parts which might of right séem either superfluous, either diminute: but yet I may not be so curious as to negect manifest errours, which is not onely my part, but every good Subjects duty with sobriety to correct. And for avoiding of offence, I have rather done it in this private Booke, then in any Booke of the Statutes it selfe, trusting that all men will take it in good part.

Scholar. I would with so, but I dare not so hope, sith never good man that would reform errour, could reforme the venemous tongues of envious detractors, which because they either cannot or list not to doe any good themselves, doe delight to bark at the doings of others, but I beséech you to stay nothing for their perverse behaviour.

Master. I consider many things that some may object, whereunto I am not unprovided of just answers, but I will not séem so hasty to make the answers before I heare their Objections, but as I trust that men are of a better nature, and more gratefull now then some have béen in times past. As I have done in the Statute of Assise of bread in rate of shillings, so will I set forth the like Table in pounds and ounces, and the parts thereof, that it may be easily applyed to all times : But I meane not by this to alter any word of the Statute, (being so good an Ordinance, and of so great continuance) but onely to make it as a kinde of exposition and declaration of the said Statute, trusting that thereby the Statute may be better understood and con-

Concerning the following Tables.

T   sequently

sequently better put in execution. And here you
shall note, that I have accounted the shillings after
the rate of 60 shillings to the pound weight, because
I esteem it the most apt for our time. Wherefore in the first Columne you finde the price of
Wheat directly against it; in the second Columne,
you may finde the weight of a farthing white loafe,
in this our time; and if you double the number (as
I have done in the third Columne) then have you
the weight of the halfe penny white loafe; and so in
the fourth Columne) is set the weight of a penny
white loafe; It needeth not to tell, that the sight
doth testifie how that every Columne is parted into three smaller pillers, whereof the first Columne
hath these three titles; pounds, ounces and penny
weights. And as in the first Columne 12 pence
make a shilling, and 20 shillings make a pound, so in
the other three Columnes 20 pence weight maketh
an ounce, and 12 ounces make a pound.

Gentle Reader, touching the understanding of the *Table* following, wherein according to our time, *Master Record* alloweth 60 *pence* to the *ounce*, and 3 *pound* or 60 *shillings* to the
pound, and thereupon after the rate of 60 *shillings* to the *pound*
Troy, doth he frame or produce this his *Table*, beginning at 3
shillings the quarter till it come to 40 *shillings* 6 *pence the quarter*,
and this is his proportion (for that he hath not set downe any one
Example to continue the worke) hath been hard for many to conceive or comprehend, and therefore the onely chief cause why I
have written this *digression* for the better understanding of him
therein.

The first thing therefore that is sought for in this *Table*, as in
the other aforesaid, is a *Maxime* grounded upon the *Statute*, which
is this. When the *quarter of wheat* is sold for *two shillings*, then the
farthing white loafe shall weigh 68 *shillings*, whereby a *shilling* is
$\frac{s}{1}$ meant of a *pound*, and by a *penny* $\frac{1}{1}$ of an *ounce*. Now therefore

fore for a generall Rule, to finde what weight the *farthing white loafe* shall weigh at 3 *shillings the quarter*, till you come to 40 *shillings* 6 *pence the quarter*, is thus to be wrought. Comming to the *first ground*, and working by the *Backer Rule*, say ; If *two shillings the quarter* give, or allow the *farthing white loafe* to weigh 68 *shillings*, what weight ought the *farthing white loafe* to weigh at 3 *shillings the quarter* ? Worke, and you shall find 45 *shillings* 4 *pence*, as before in the correction of the *first table* is noted. Then for the *second work*, say by the *Rule of Three Direct*, if 20 *pence* give one *ounce*, what giveth 45 *shillings* 4 *pence* ? multiply and divide, and you shall finde 544 *ounces*, which 544 *ounces* being multiplied by 3, for 3 *pounds*, or 60 *shillings*, yeeldeth 1632 *ounces*, which divided by 20, produceth 81 *ounces*, & $\frac{12}{20}$ or rather $\frac{3}{5}$ of an *ounce*, equall unto 12 *penny weight*, which is *halfe an ounce*, & 2 *penny weight*, & so maketh in all 6 *pounds*, 9 $\frac{1}{2}$ *ounces* & 2 *penny weight*. Now the next way to continue this *table*, to know the weight of the halfe *penny white loafe*, is thus, multiply 1632 *ounces* by 2, & it bringeth forth 3264 *ounces*, & divided by 20, it yeeldeth 163 *ounces* and $\frac{1}{5}$, which is equall to 13 *pounds* 7 *ounces*, and 4 *penny weight*, as M. *Record his Table noteth*.

Thirdly, for the weight of the penny white loafe, multiply 1632 *ounces* by 4, and divide by 20, and after by 12 as before, and you shall finde 27 *pounds* 2 *ounces*, and 8 *penny weight*, &c. This *Method*, or else by doubling the *farthing white loafe*, for the *weight of the halfe penny white loafe*, and so doubling the *halfe penny white loafe*, for the *weight of the penny white loafe*, is the order to continue the *table* to the end thereof.

The price of a quarter of Wheat.			The weight of a farthing white loafe.		
li.	s.	d.	Po.	Ounces.	Penny weight.
0	3	0	6	9	12
0	4	6	4	6	8
0	6	0	3	4	16
0	7	6	2	8	$12\frac{4}{5}$
0	9	0	2	3	4
0	10	6	1	11	$6\frac{8}{7}$
0	12	0	1	8	8
0	13	6	1	6	$2\frac{2}{3}$
0	15	0	1	4	$6\frac{4}{5}$
0	16	6	1	2	$16\frac{4}{11}$
0	18	0	1	1	12
0	19	6	1	0	$11\frac{4}{13}$
1	1	0		11	$13\frac{5}{2}$
1	2	6		10	$17\frac{2}{5}$
1	4	0		10	4
1	5	6		9	12
1	7	0		9	$1\frac{1}{3}$
1	8	6		8	$11\frac{15}{19}$
1	10	0		8	$3\frac{1}{5}$
1	11	6		7	$15\frac{1}{7}$
1	13	0		7	$8\frac{4}{11}$
1	14	6		7	$1\frac{21}{23}$
1	16	0		6	16
1	17	6		6	$10\frac{14}{25}$
1	19	0		6	$5\frac{2}{13}$
,	0	6		6	$0\frac{8}{0}$

| The price of a quarter of Wheat | | | The weight of a half penny white loafe | | | The weight of a half penny white loafe | | |
li	s.	l.	Po.	Ounces	Penny weight	Po.	ounces	Penny weight
0	3	0	13	7	4	27	2	8
0	4	6	9	0	16	18	1	12
0	6	0	6	9	12	13	7	4
0	7	6	5	5	$5\frac{3}{5}$	10	10	$11\frac{1}{5}$
0	9	6	4	6	8	9	0	16
0	10	6	3	10	$12\frac{4}{7}$	7	9	$5\frac{1}{7}$
0	12	6	3	4	16	6	9	12
0	13	0	3	8	$5\frac{1}{3}$	6	0	$10\frac{2}{3}$
0	15	0	2	8	$12\frac{4}{5}$	4	5	$5\frac{2}{5}$
0	16	6	2	5	$13\frac{5}{11}$	4	11	$6\frac{1}{11}$
0	18	0	2	3	4	4	6	8
0	19	6	2	1	$2\frac{2}{13}$	4	2	$4\frac{4}{13}$
1	1	0	1	1	$16\frac{2}{7}$	3	10	$12\frac{4}{7}$
1	2	6	1	9	$15\frac{1}{5}$	3	7	$10\frac{2}{5}$
1	4	0	1	8	8	3	4	16
1	5	6	1	7	4	3	2	8
1	7	0	1	6	$2\frac{2}{3}$	3	0	$5\frac{1}{3}$
1	8	6	1	5	$3\frac{11}{19}$	2	10	$7\frac{1}{19}$
1	10	0	1	4	$6\frac{2}{5}$	2	8	$12\frac{4}{5}$
1	11	6	1	3	$10\frac{4}{7}$	2	7	$1\frac{1}{7}$
1	13	0	1	2	$16\frac{8}{11}$	2	5	$13\frac{5}{11}$
1	14	6	1	2	$3\frac{12}{23}$	2	4	$7\frac{1}{23}$
1	16	0	1	1	12	2	3	4
1	17	6	1	1	$1\frac{8}{25}$	2	2	$2\frac{6}{23}$
1	19	0	1	0	$11\frac{1}{13}$	2	1	$2\frac{4}{13}$
2	0	9	0	0	$1\frac{6}{7}$	2	0	$3\frac{1}{9}$

HAving spoken before for the understanding of the *Table* placed by M. *Record*, a man indued with rare knowledge in *Arithmeticall* and *Geometricall* proportions, touching the *Statute* of *Coynage*, and the *Standard* thereof, as appeareth in his Epistle of this *Book dedicated* to *King Edward the sixth*, insinuating unto his Highneſſe that the *Standard of Coyne* is much altered from the 14. yeare of *King Edward the third* (when this *Statute and Aſſiſe* was confirmed) to the *Standard of this our time*. For it appeareth that in *King Edward the thirds* time, when the *Aſſiſe* of *Bread and Drink* was eſtabliſhed, that a *Sterling penny*, round without clipping, did then weigh 32 *cornes of wheat dried*, and taken out of the middle of the *eare*, and 20 of theſe *pence* made an *ounce*, and 12 *ounces* made a pound *Troy*. And ſo from the *weight of a penny* to 20 *ſhillings ſterling*, which then weighed 12 *ounces*, tooke *Bread his weight* and proportion : And now finding 60 *pence* is an *ounce*. That onely cauſe (I perceive, for the zeale of a Common-wealth) moved him to ſet downe the ſame *Table in this private Book* ; meaning not thereby to alter any word of the *Statute* being ſo good an ordinance & of ſo long continuance, but as a kinde of expoſition by the way, that thereby the *Statute* may be better underſtood, and ſo conſequently better put in execution. Which *Aſſiſe of his*, is three times greater then the *Statute* now alloweth. Therefore alſo, to gratifie ſuch as are deſirous of knowledge, according to theſe *prices of a quarter of wheat* I have added to this *Author theſe three other new Tables following* and reduced their *prices* into their juſt proportions of *Sterling money*, and alſo reduced the *money* into known *weight Troy*, according to the Statute. And thereafter according to proportion in my other three *Tables*, have I noted the *juſt weight* that a *Farthing, Half penny, and Peny white-loafe*, ought to weigh by the *Statute*.

The

The price of a Quarter of Wheat.

l.	s.	d.
0	3	0
0	4	6
0	6	6
0	7	6
0	9	6
0	10	6
0	12	6
0	13	6
0	15	6
0	16	6
0	18	6
0	19	6
1	1	6
1	2	6
1	4	6
1	5	6
1	7	6
1	8	6
1	10	6
1	11	6
1	13	6
1	14	6
1	16	6
1	17	6
1	19	6
2	0	6

☞ The weight of a Farthing white loaf in Sterling money by Assise.

l.	s.	d.
2	5	4
1	10	$2\frac{2}{3}$
1	1	8
1	18	$1\frac{2}{5}$
1	15	$1\frac{1}{5}$
1	12	$15\frac{3}{7}$
1	11	4
1	10	$0\frac{8}{9}$
0	9	$0\frac{4}{5}$
0	8	$2\frac{11}{10}$
0	7	$6\frac{2}{3}$
0	6	$11\frac{2}{13}$
0	6	$9\frac{3}{5}$
0	6	$0\frac{8}{13}$
0	5	8
0	5	4
0	4	$0\frac{4}{9}$
0	4	$9\frac{2}{19}$
0	4	$6\frac{4}{5}$
0	3	$3\frac{17}{21}$
0	3	$2\frac{1}{11}$
0	3	$11\frac{7}{23}$
0	3	$9\frac{2}{3}$
0	3	$7\frac{12}{25}$
0	3	$5\frac{11}{13}$
0	3	$4\frac{8}{17}$

The weight of a Farthing white loaf in Troy weight by Assise.

Oun-ces.	Po.	penny weight.
2	3	4
1	6	$2\frac{2}{3}$
1	1	12
0	10	$17\frac{2}{5}$
	9	$1\frac{1}{3}$
	7	$15\frac{1}{7}$
	6	16
	6	$0\frac{8}{9}$
	5	$8\frac{4}{5}$
	4	$18\frac{10}{11}$
	4	$10\frac{2}{3}$
	4	$3\frac{7}{13}$
	4	$1\frac{3}{2}$
	3	$2\frac{8}{15}$
	3	8
	3	4
	3	$0\frac{4}{9}$
	3	$17\frac{1}{19}$
	2	$14\frac{4}{5}$
	2	$11\frac{17}{21}$
	2	$9\frac{1}{2}$
	2	$7\frac{2}{3}$
	2	$5\frac{2}{3}$
	2	$3\frac{11}{25}$
	2	$1\frac{1}{13}$
	2	$0\frac{8}{27}$

The price of a Quarter of Wheat.

li.	s.	d.
0	3	0
0	4	6
0	6	6
0	7	6
0	9	6
0	10	6
0	12	6
0	13	6
0	15	6
0	16	6
0	18	6
0	19	6
1	1	6
1	2	6
1	4	6
1	5	6
1	7	6
1	8	6
1	10	6
1	11	6
1	13	6
1	14	6
1	16	0
1	17	6
1	19	0
1	0	6

The weight of a half penny white loaf in Troy weight by Assise.

Pou.	Ounces	Penny weight.
4	6	8
3	0	$5\frac{1}{3}$
2	3	4
1	9	$15\frac{1}{5}$
1	6	$2\frac{2}{3}$
1	3	$10\frac{6}{7}$
1	1	12
1	1	$1\frac{1}{9}$
1	0	$17\frac{1}{5}$
1	0	$17\frac{1}{11}$
0	9	$1\frac{1}{3}$
0	8	$7\frac{4}{13}$
0	8	$3\frac{1}{3}$
0	7	$5\frac{1}{15}$
0	6	16
0	6	8
0	6	$0\frac{8}{9}$
0	5	$14\frac{10}{19}$
0	5	$8\frac{4}{5}$
0	5	$03\frac{14}{21}$
0	4	19
0	4	$14\frac{14}{23}$
0	4	$11\frac{1}{3}$
0	4	$7\frac{1}{25}$
0	4	$3\frac{2}{3}$
0	4	$0\frac{16}{2}$

The weight of a half penny white loaf in Troy weight by Assise.

Po.	Ounces	ny weight.
9	0	16
6	0	$10\frac{2}{3}$
4	6	8
3	7	$10\frac{4}{5}$
3	0	$5\frac{1}{3}$
2	7	$1\frac{2}{5}$
2	3	4
2	0	$3\frac{4}{9}$
1	9	$15\frac{3}{5}$
1	7	$15\frac{2}{9}$
1	6	$3\frac{2}{3}$
1	4	$14\frac{2}{11}$
1	4	$0\frac{4}{5}$
1	2	$10\frac{2}{15}$
1	1	12
1	0	16
1	0	$1\frac{7}{9}$
0	11	$9\frac{1}{10}$
0	10	$17\frac{2}{5}$
0	10	$7\frac{1}{12}$
0	9	18
0	9	$9\frac{5}{23}$
0	9	$2\frac{1}{3}$
0	8	$14\frac{2}{25}$
0	8	$7\frac{1}{28}$
0	8	$1\frac{1}{2}$

Scholar. Sir I do thanke you most heartily for this, not onely in mine owne name, and in the name of all students, but also in the name of the whole Commons, to whom the restitution of this Assise (I trust) shall bring restituton of the weight in Bread, which long time hath béen abused. And if you know any thing more, wherein you would vouchsafe to declare the errours, and set forth the truth, you cannot but obtain great thankes of all good hearted men that love the Commonwealth.

Mast. I have sundry things to declare, but I have reserved them for a private Booke by it self, yet notwithstanding because the statute of the rate of measuring of grounds is so common, that it touch= eth all men, and yet no more common then néedfull, but so much corrupt, that is, to farre out of all good rate; not onely in the English Bookes of statutes, commonly Printed, but also in the Latin Bookes, and in the French also, (for I have read of each sort, and conferred them diligently) I will give you a Table for the restitution of those errours, as may suffice for this present time. And first I will pro= pose one question to you touching the use of that statute, whereby you may perceive the order how to examine the whole statute, and every parcell thereof, and the question is this.

When the Acre of ground doth containe foure A question *Perches in breadth, then must it contain* 40 *Perches* of measu= *in length. Then do I demand of you, how much shall* ring of *the length of an Acre be, when there is in the breadth* ground. *of it* 13 *Perches. But before you shall answer to this question, I will declare unto you another Statute, which is the ground of the former Statute. And that Statute is this.*   It

Statute measure.

It is ordained, that three Barly-cornes dry and round, shall make up the measure of an inch: 12 inches shall make a foot, & 3 foot a yard, (the common English books have an Elne) five yards and a halfe shall make a Perch, and 40 Perches in length, and 4 in breadth shall make an Acre. This is that **An Acre.** Statute, whereby you may perceive, that the intent of the statute is, that one Acre should containe 160 square Perches. Now let me hear you answer to the question.

Scholar. As I perceive by the words of the Statute, a Perch to be the $\frac{1}{160}$ part of an Acre, so could I make those numbers all in fractions, and so worke the question: but seeing I may doe it also in whole numbers, I take that forme for the most ease; therefore thus I set the question in forme. Then doe I multiply 40 by 4, and it maketh 160, which I divide by 13, and the Quotient is 12 $\frac{4}{13}$.

$$13 \underset{12}{\overset{40}{Z}} 4$$

Master. Now turne that $\frac{4}{13}$ into the common parts of a Perch, as they be named in the former statute: Howbeit it shall be best to take one of the least parts in Denomination for avoyding of much labour, as Feete, whereof the Perch containeth 16 $\frac{1}{2}$.

Scholar. Then to return $\frac{4}{13}$ into Feet, I multiply 16 $\frac{1}{2}$ by 4, and it maketh 66. which I must divide by 13, and the Quotient is 5 $\frac{1}{13}$.

Master. So I find, that if the Acre hold in breadth 13 Perches, it shall containe in length 12 Perches 5 Foot, & $\frac{1}{13}$ of a foote, which is not fully an **Note this** Inch, for the Inches is $\frac{1}{12}$ of a foot. But here all the **errour** Statute Bookes in Latine and English (that I have seene) dos note it to be 13 Perches, 5 foote, and one

one Inch, which make above 13 Perches to many in the Acre: ſo that I would have thought the errour to have crept into the Printed Bookes, by the great negligence that Printers in our time doe uſe, ſave that in written copies of great antiquity, I doe finde the ſame; yet have I one French copy which hath 12 Perches ¼ and one foote, and that miſſeth very little of the truth.

Scholar. Then I ſée it is true that I have often heard ſay, that the trueſt Copies of the Statutes, be the French copies.

Maſter. That is often true, but not generally, as I have by conference tried diverſity: but in this Statute the French Booke is moſt coʒrupt: in all other places lightly.

But now to perfoʒme my pʒomiſe, I will ſet foʒth the Table foʒ meaſuring of an Acre of ground, onely by ſuch parts as the Statute doth mention, becauſe at this time I doe of purpoſe wʒite it foʒ the better underſtanding of that Statute, and hereafter with other things intend to ſet foʒth this ſame moʒe at large.

In this Table *following, I have not done as in the other* Statute *before compared by reſtitution with the faults crept into the* Statute, *but only have written that true meaſure, which the equity of the* Statute *doth pretend. For it were vile to judge of ſo noble Princes and worthy Counſellors, as have authoriſed and ſet forth this* Statute, *that they would make one* acre *in any forme greater then another, but every one to be juſt and equall with each other, which is the ground alſo of my worke: & hereby may all men perceive how needfull* Arithmetick *is to the* Students of Law. *But*

now

now I think best to make an end of these matters for this present time, sith the Table hath in it no obscurity that I should need to declare.

The breadth of the Acre.	The length of the Acre.			
Perches	Perches	Feet.	Inches.	Parts of an Inch
10	16	0	0	0
11	14	9	0	0
12	13	5	6	0
13	12	5	0	$\frac{12}{13}$
14	11	7	0	$\frac{6}{7}$
15	10	11	0	0
16	10	0	0	0
17	9	6	9	$\frac{2}{17}$
18	8	14	8	0
19	8	6	11	$\frac{7}{19}$
20	8	0	0	0
21	7	10	2	$\frac{4}{7}$
22	7	4	6	0
23	6	15	9	$\frac{2}{23}$
24	6	11	0	0
25	6	6	7	$\frac{1}{5}$
26	6	2	6	$\frac{6}{13}$
27	5	15	3	$\frac{1}{3}$

The

The breadth of the Acre. Perches	The length of the Acre.			
	Perches	Feet.	Inches.	Parts of an Inch.
28	5	11	9	$\frac{1}{7}$
29	5	8	6	$\frac{18}{29}$
30	5	5	6	0
31	5	2	7	$\frac{22}{31}$
32	5	0	0	0
33	4	14	0	0
34	4	11	7	$\frac{11}{17}$
35	4	9	5	$\frac{1}{5}$
36	4	7	4	0
37	4	5	4	$\frac{1}{37}$
38	4	3	5	$\frac{1}{19}$
39	4	1	8	$\frac{4}{13}$
40	4	0	0	0
41	3	14	10	$\frac{24}{41}$
42	3	13	4	$\frac{2}{7}$
43	3	11	10	$\frac{12}{43}$
44	3	10	6	0
45	3	9	2	0

Scholar. Indeed Sir, I understand the Table
(as I thinke) by those other which you set forth
before. For in the first columne is set the Perches
of the breadth of an Acre, and then in the two Columnes following appeareth how many Perches,
and how many foote that same Acre must have for
his length.

Master.

Master, ₽ou take it well : howbeit to speake exactly of breadth and length, and the first Columne doth sometime betoken the breadth, and sometime the length : for properly the longest side of any square doth limite his length, and the shorter side doth betoken the breadth, yet it is no great abuse in such Tables, where a man cannot well change the Title, to let the name remaine, although the proportions of the numbers doe change : for still by the first Columne is expressed the meásure of the one side, and by the two other pillars in one Columne, is set forth the measure of the other side. And this shall be sufficient now for the use of the Golden Rule.

The Rule of Fellowſhip.

NOw will I touch certaine other Rules, which for their ſeverall names may ſeeme divers Rules, and diſtinct from this, but indeed they are but branches of it : yet becauſe they have ſeverall workings in appearance, but alſo pleaſant in uſe, I will give you a taſte of each of them. As for the Rule of Fellowſhip, both ſingle and double, with time and without time, I ſhall need to ſay little more then I have already ſaid in teaching the workes of whole numbers : yet an example or two will we have to refreſh the remembrance of the ſame, and to declare certain proper uſes and applications of it, as this for one.

The Rule of Fellow-
ſhip with-
out time.

A queſtion
of inequall
ſociety.

Foure men got a booty or prize in time of warre, the prize is in value of money 8190 *pound, and becauſe*
                                        the

the men be not of like degree, therefore their shares may
not be equall ; but the chiefest person will have of the
booty the third part, and the tenth part over ; the se-
cond will have a quarter, and the tenth part over :
the third will have the sixth part : and so there is left
for the fourth man a very small portion, but such as
his lot (whether he be pleased or wroth) he must be
content with one 20 part of the prey : Now I demand
of you what shall every man have to his share ?

Scholar. You muſt be fain to anſwer to your
owne queſtion, elſe it is not like to be anſwered at
this time.

Maſter. The forme to underſtand the ſolution of
this queſtion, and all ſuch like, is this : Reduce all
the Denominators into one number by Multiplication, except that any of them be parts of ſome other
of them, for all ſuch parts you may overpaſſe and
take for them all thoſe numbers, whoſe parts they
be : as in this example the ſhares be theſe $\frac{1}{3} \frac{1}{10} \frac{1}{4} \frac{1}{10} \frac{1}{6} \frac{1}{20}$,
if I multiply all the Denominators together, begin-
ning with 3, and ſo goe on unto 20, it will make
144000 : but conſidering that 3 is a part of 6, I will
omit that 3, and likewiſe 10, which is a part of 20,
I may overpaſſe alſo, & then is there but 3 Denominators to multiply, that is, 4, 6, and 20 which
make 480, which ſumme I take for my worke,
becauſe all the Denominators will be found in it.
Then I take ſuch parts of it as the queſtion im-
porteth, that is, for the firſt man $\frac{1}{3}$ and $\frac{1}{10}$, the $\frac{1}{3}$ is
160, the $\frac{1}{10}$ is 48, which I put in one ſumme for
the firſt mans ſhare, and it maketh 2o8. Then
for the ſecond mans ſhare, I take $\frac{1}{4}$, which is 120,
and $\frac{1}{10}$ which is 48, and that maketh in the
whole

whole 168.   Now for the third man which must
have ⅙, I take 80.   And for the fourth man there
remaineth but 24, which is 1/20 of the whole summe :
so that if the whole prey had béen but 480 pound,
then were the question answered : but because the
summe was of greater value, by this meanes now
shall I know the partition of it.   I must set my
numbers by the order of the Golden Rule, putting
in the first place the number of that I found by

The rea-
son of this
rule.

multiplying the Denominators, and in the second
place the summe of the booty,   And look what proportion is betweene the first number and the second,
the same proportion shall be between the parts of that
first number, and the parts of the second, comparing
each to his like.   Therefore I must put in the third
place, one of the parts or shares, and then worke
by the former Rule of proportion or Golden Rule.
And because I have 4 severall parts of the first
number, by which I would find out four like parts
of the second number, therefore must I make foure
severall figures.

Scholar.   Now I trust I can answer to
your question , as by your favor I will
prove :

And to trie it , I set the foure figures
thus, marked with A, B, C, D, to shew their
order.

A

A
480 Z 8190
208

B
480 Z 8190
168

C
480 Z 81190
80

D
480 Z 8190
24

And then in each of them I multiply the ſecond number by the third, and divide their totall by the firſt, and ſo amounteth the fourth ſumme which I ſæk for: for if I doe multiply 8190 by 208, it maketh 1703520, which being divided by 480, maketh in the Quotient 3549 for the firſt mans portion.

And ſo working with the other three figures, I finde for the ſecond man 2866½. and for the third man 1365, and then for the fourth man 409½, and ſo every mans ſhare is ſet forth in the figure here annexed.

A
480 Z 8190
208   3549

B
480 Z 8190
168   2866½

C
480 Z 8190
80   1365

D
420 Z 8190
24   490½

And thus I think I have done well.

Maſter, If you miſdoubt your working, and liſt to prove it, adde all the ſhares together; and if they make the totall, then ſeemeth it well done.

Scholar. I may ſet them thus: and then by Addition the juſt ſumme doth amount, that is, 8190, and therefore (as you ſay) it ſæmeth to be well wrought.

The proof of Addition.

3549
2866½
1365
409½
8190

But I beſéech you, is there any doubt in this triall, that you uſe that woꝛd ſeemeth?

Maſter. You may eaſily conjecture, that if you did aſſigne the firſt mans ſhare to the laſt, and ſo change all the reſt, and one had anothers ſhare, yet would the Addition appeare all one, and therefoꝛe is not the pꝛoofe exact.

The juſt Proof. But if you will make a juſt pꝛoofe foꝛ the firſt mans part, take $\frac{1}{3}$ and $\frac{1}{10}$ of the whole ſumme, and if it agrée with the number in the figure, then it is well done. And ſo do foꝛ the ſecond, third, and fourth ſummes, and this pꝛoofe faileth not. Now will I pꝛopound certaine other queſtions, which have bén ſet foꝛth by certain learned men, albeit not without ſome overſight, which queſtions I pꝛo-teſt heartily, I doe not repeate to depꝛave thoſe good men, whoſe labours and ſtudies I much pꝛaiſe and greatly delight in. But onely accoꝛding to my pꝛofeſſion, to ſéeke out truth in all things, and to remove all occaſions of errour as muchas in me lieth: and foꝛ that cauſe I will onely name the queſtions whithout hurting the Authors name.

A queſtion of building The firſt queſtion is this.

Foure men did build an houſe, which coſt them 3000 crownes, their ſhares were ſuch that one man ſhould pay $\frac{1}{2}$ of the ſumme & ſix crownes over: the ſecond ſhould pay $\frac{1}{3}$ and 12 crownes over: the third man muſt lay out $\frac{2}{3}$ abating 8 crownes: and the fourth man ſhould pay $\frac{1}{4}$ and 20 crownes more. Can you anſwer to this queſtion?

Scholar. No, I cannot ſir, and that you know beſt of any man, foꝛ I know no moꝛe then you have taught me.

Maſter

Maſter. Then I dare ſay, you cannot doe it, neither yet the beſt learned man that ever did propoſe
it : for the queſtion is impoſſible. For declaration
whereof, I will be bold to uſe firſt the repreſenta-
tion of the Numbers in their apteſt forme, (although I have not yet taught that manner of
worke) becauſe it may appeare plainly that the
queſtion is not poſſible. For here I have ſet the
parts, and added them, and they make the whole

$$\sum, \text{ and } \tfrac{1}{4} \& \text{ 3 more. Now how is it}
\begin{cases}
\tfrac{1}{2} & + & 6 \\
\tfrac{1}{3} & + & 12 \\
\tfrac{2}{3} & - & 8 \\
\tfrac{3}{4} & + & 20 \\
\tfrac{1}{4} & + & 30
\end{cases}$$

poſſible to diviue truely either gaines,
either charges, ſo that the particulars
1 ¼ ſhall be more then the totall?

Scholar. It is againſt the forme
of proofe by addition of parts.

Maſter. You ſay truth. And (becauſe you ſhall
perceive it the better) I will trie it after the vulgar forme, as in this figure you ſee where   1505
the ½ with 6 over, is 1506, for the totall as   1012
you heard before, is 3000, the ⅓ and the 12   1992
more is 1012: the ⅔ would be 2000, but   770
then abating 8, it is but 1992, and then laſt   ‾‾‾‾
of all, the ¼ is 750, and the 20 more maketh 770:   5280
which all being added in one ſumme doe make
5280 where the totall ſumme ſhould be but
3000, which ſumme of 3000 if you divide by ¼, you
ſhall have ¼ of it, that is 1250, and thereto adde 30
more, then will thoſe three ſums make 5280:   2000
whereby you may ſee how this forme   1250
(as well as the other) doth declare that the   ‾‾‾‾
particulars in that queſtion would make   5280
more then the whole ſumme by ¼ and 30 more, and
therefore can that queſtion not be accepted as a

possible thing, but yet dos certaine learned men propound such questions, and answer to them: Therefore somewhat to say to their excuse (rather of
their good meaning, then for their doing) I will a-
non declare what may be said for their defence:
but in the meane season, I will propound the question as it may be wrought by good possibility.

The for-
mer que-
stion of
building
now possi-
ble.

*As if foure men build a house together, and it cost
them 3000 Crownes, and then for the partition they
agree thus: that as often as the first man doth pay 6
Crownes, so often the second man shall pay 4, the third
man 8, and the fourth man 3. Or else thus, that the
first man shall pay double so much as the fourth, and
the second man shall pay $\frac{1}{3}$ of the first mans charge: the
third man shall pay double so much as the second: (and
these two ways are to one end) but further for their a-
greement it is appointed also, that the first shall give 6
Crownes overplus, and the second 12, and the fourth
shall give 20, but the third man shall give no overplus,
but shall have 8 Crownes abated of his charge.*

Now is the question possible to be assoiled, and
this is the way to doe it. Marke the proportion of
the severall charges, and set out small numbers in
that Rate, by which you may reduce the worke to
the Golden Rule, as here in the first forme the
numbers are already named, 6, 4, 8, 3: and in the
second forme, although they be but plainly named,
yet they may be the same numbers: for 6 is double to
3, and 4 is $\frac{1}{3}$ of 6: and againe 8 is double to 4.
Now adde these together, and they make 21,
which 21 must be set for the first number in the
Golden Rule: for if it with the overplus of each
mans charge would make the totall summe of the
charges

charges, then were those severall summes the charges of each man, besides his overplus: but now it is not so.

But yet this is true: (so excellent are conclusions Arithmeticall) that look what proportion each of their severall summes doth bear to 21, the same proportion doth the just charges of every man (besides his overplus) beare to the totall of the charges, the overplus being deducted: wherefore this may you note, that before you doe apply the totall of his charges to the golden rule, you must deduct the overplus, which is 6, 12, and 20, that is in the whole 38 : but then 8 must be restored for the abatement of the third man, and then remaineth to be deducted 30 : take 30 therefore out of 3000, and there will rest 2970, which I must set in the Golden Rule, for the second summe : and for the third summe, I must put each of the small numbers before mentioned, which although they be not severall charges, yet they represent them in proportion. And so making for every mans charge a severall question, the figures will be 4, which I marke with four letters, A, B, C, D, thus.

$$\text{A} \qquad 21\ \diagup\mathbf{Z}\diagdown\ 2970 \qquad\qquad 6\ \diagup\mathbf{Z}\diagdown\ 848\tfrac{4}{7}$$

$$\text{B} \qquad 21\ \diagup\mathbf{Z}\diagdown\ 2970 \qquad\qquad 4\ \diagup\mathbf{Z}\diagdown\ 565\tfrac{5}{7}$$

$$\text{C} \qquad 21\ \diagup\mathbf{Z}\diagdown\ 2970 \qquad\qquad 8\ \diagup\mathbf{Z}\diagdown\ 1131\tfrac{3}{7}$$

$$\text{D} \qquad 21\ \diagup\mathbf{Z}\diagdown\ 2970 \qquad\qquad 3\ \diagup\mathbf{Z}\diagdown\ 424\tfrac{2}{7}$$

Where I have set for breifnesse the summe of every mans charge in the fourth place, presupposing that you can tell how to try out the fourth summe by so many Examples as ye have had.

Scholar. As I trust that I understand this forme, so I desire much to know what may bee said for them that mistoke this Question.

Master. You seem so desirous to know this errour, that you have forgotten to examine, whether this work be without fault.

Scholar. Me seemeth this worke to be well done, because the Addition of the foure severall numbers doth make the totall summe of 2970, which was to be divided into such foure parts.

Master. But then have you forgotten that the first man must pay sixe crownes more besides his share, and the second man 12 Crownes more, the third man 8 crownes lesse, and the fourth man 20 crownes more ; for without these, your first totall of 3000 crownes will not be made.

Scholar. Then must I adde to the first mans summe 6 more, and it will bee $854\frac{4}{7}$ and to the seconds summe, I must adde 12, and it will be $577\frac{4}{7}$: from the thirds summe I must abate 8, and then will the summe bee $1131\frac{3}{7}$: then adding unto the fourths summe 20, it will be $444\frac{2}{7}$, and these foure summes will make 3000, which is the whole charge, as in this example it may appear, where first I gather the $\frac{14}{7}$, that maketh 2, and so proceed I in the Addition to the end.

$$854\frac{4}{7}$$
$$577\frac{4}{7}$$
$$1123\frac{3}{7}$$
$$444\frac{2}{7}$$
$$\overline{\quad\quad}$$
$$3000$$

Master. Now have you well done, and this worke in the same summes is brought of other Learned men for the true solution of the question as it was proposed, which (as I said) was im-

possib

poſſible : and now examine by theſe ſeverall ſummes, and ſée whether it doth agrée with the ſummes in the Queſtion propoſed.

The firſt man muſt pay ½ and 6 over of the totall ſumme : how think you, is 854 ⅔ the halfe and 6 more of 3000 ?

Scholar. No that it is not, for it ſhould bée 1506: and for the ſecond man 1012 : and for the third man 1992, and for the fourth man 770 : whereof not one ſumme agréeth to this worke. But I marvell, that ſo wiſe men could bée ſo much over-ſéen.

Maſter. It is commonly ſéen, that when men will receive things from elder writers & will not examine the thing, they ſéeme rather willing to erre with their Ancients for company, then to bée bold to examine their workes or writings. Which ſcrupuloſity hath ingendred infinite errors in all kindes of knowledge, & in all civill adminiſtration, & ſo in every kinde of Art. But theſe Learned men did not mean any other thing by this queſtion, then to finde ſuch numbers as ſhould beare the ſame proportion together, as thoſe numbers in the queſtion propoſed did bear one to another: which thing you ſhall perceive more plainly by another queſtion of theirs, that is this.

A man lying upon his Death-bed, bequeathed his goods (which were worth 3600 Crownes) in this ſort : becauſe his Wife was great with childe, and he yet uncertaine whether the Childe were Male or Female, hee made his bequeſt conditionally, that if the Wife bare a Daughter, then ſhould the Wife have halfe his goods, and the Daughter ⅓; but if ſhe were delivered of a Sonne, then that Sonne ſhould A queſtion of a Teſtament.

have

have $\frac{1}{2}$ of the goods, and his Wife but $\frac{1}{3}$: Now it chanced her to bring forth both a Sonne and a Daughter ; the question is, How shall they part the goods agreeable to the Testatour his Will ?

S If some cunning Lawyers had this matter in scanning, they would determine this Testament to be quite voide, & so the Man to die intestate, because the Testament was made in sufficient, sith this condition was not expressed in it, and also it might have chanced that shee should have brought forth neither Sonne nor Daughter, as often hath béene sée..e:so is the Will insufficient in that point also.

Maister. Such Scanners should séem too cunning and yet not so cunning as cruell: for the minde of the Testator is to be taken favorably for the aid of the Legataries, when there ariseth such doubt. But let us try this work, not by force of Law, but by proportion Geometricall, séeing the Testator did minde to provide for each sort of them.

Scholar. If the Sonne shall have $\frac{1}{2}$ by force of the Testament so must the Mother have $\frac{1}{3}$. Again, because she hath a Daughter also, therefore ought she to have $\frac{1}{2}$ and the Daughter $\frac{1}{3}$, that is both wayes $\frac{1}{2}$ & $\frac{1}{3}$, which commeth to the wholegods, & $\frac{2}{3}$ more

Wherefore it séemeth also impossible.

Maister. In this matter the minde of th testator is to be understood that such proportio should be betwéen the portion of the VVife, an the Sonne, as is betwéen $\frac{1}{2}$, and $\frac{1}{3}$, that is, th Sonne must have $\frac{3}{6}$, for $\frac{2}{6}$ to his Mother, so shall have 3 to 2, that is as much as his Mother, an halfe as much more ; and the Mother must ha the like rate in comparison to her Daughte

Th

Then muſt I finde out 3 numbers in ſuch proportiſ
on, that the firſt may have as much as the ſecond,
and halfe as much moje (that is) in proportion
ſeſquialtera, and the ſecond to the third in that
ſame proportion : ſuch numbers, be 9, 6, 4.

Scholar. I pjay you Sir, how ſhall I finde out
theſe numbers?

Maſter. That will I gladly tell you.

VVhatſoever the proportion be of any three To finde
three nun—
bers in any
proportiſ
on.
numbers, multiply the Termes of that proportion
together, and the number that amounteth ſhall
be the middle number of the three : then multiply
that middle number by the leſſer term, and divide
that totall by the greater, and the leaſt number of
the three will amount. So if you multiply that
middle number by the greater extreame, and divide
the totall by the leſſer extream, then will the greateſt
number of that Progreſſion amount.

Scholar. Then in this example to finde the To finde
the pro—
pṛrtion
between 2
numbers.
proportion of $\frac{1}{2}$ to $\frac{1}{3}$, I muſt divide (as you taught
me in Diviſion) $\frac{1}{2}$ by $\frac{1}{3}$, and the Quotient will be $\frac{3}{2}$,
that is, $1\frac{1}{2}$. whereby I perceive that the proportion
in this queſtion is as thjée to 2. Therefoje as
you taught me even now, I multiply 3, by 2, and
the ſumme is 6, which muſt be the middle number:
then I multiply the middle number 6 by 2, which is
the leaſt terme, and the ſumme is 12, that I doe
divide by 3, béeing the greater terme, and the
Quotient is 4: ſo is 4 the leaſt number of the thjée.
Then I multiply 6 by 3, whereof cometh 18, and
that I divide by 2, and ſo have I 9 which is the
greateſt number of the thjée.

Maſter. Another way yet may you finde the
third

third number in any Progreſſion, if you have two of them : for if the middle number be one of them which you have, then multiply it by it ſelf (as in this example, 6 by 6 maketh 36) and that totall divide by the other number which you have, and the third number will be the Quotient.

Scholar.   Then I divide 36 (which commeth of 6 multiplied by it ſelf) by 4 the Quotient will be 9 : and if I divide 36 by 9, the Quotient will be 4.   But what if I know the firſt number, and the third, and would have the middle number ?

Maſter. Multiply the 2 numbers together, and in their totall you muſt ſeek the root of that number, and it ſhall be the middle number : but becauſe as yet you have not learned to extract Roots, therefore uſe the firſt forme which I have taught you, till I teach you to extract Roots. And now go forwards with the anſwer to the ſame queſtion.

Scholar. I perceive then, that the Sonne muſt not have $\frac{1}{2}$ of the goods, neither the Mother $\frac{1}{3}$, nor yet the Daughter $\frac{1}{4}$, but yet muſt the goods be divided into ſuch proportion, that the Sonne ſhall have 9 Crownes for 6 to his Mother, and the Mother ſhall have 6 crownes for every 4 to her Daughter. Then I apply it to the Golden Rule in three examples, as followeth.

Where the firſt number is the Addition of thoſe three numbers 9, 6, 4 : and the   19 Z 3600
third is one of them ſeverally :   9
the ſecond is the totall of the   19 Z 3600
goods in that Teſtament : and   9
then by the worke of the Golden   19 Z 3600
Rule, I finde out the fourth   4

number

number in every wozke : that
is foz the Sonne 1705 $\frac{4}{19}$, foz
the Mother 1136 $\frac{16}{19}$, and foz
the Daughter 757 $\frac{17}{19}$: the
which ſummes added together,
doe make the ſumme of the
whole goods as may be ſéen by
this Example.

$$1705 \tfrac{5}{19}$$
$$1136 \tfrac{16}{19}$$
$$757 \tfrac{17}{19}$$
$$3600$$

And this (me thinketh) I doe percefve, becauſe
in this Caſe there is a neceſſary remedy devised
againſt an urgent inconvenience : therefoze thoſe
learned men thought they might uſe the like
liberty in that other queſtion.

Maſter. Pour gueſſe is good, but they had ſo good
reaſon foz them in the one, as they have in the
other : As in another example of theirs, it may better
appeare, as in this.

A man left unto his three Sonnes 7851 *crowns to*
be parted in ſuch ſort, that the firſt Sonne ſhould have
$\frac{1}{2}$, *the ſecond Sonne* $\frac{1}{3}$, *and the third Sonne* $\frac{1}{4}$, *which is*
not poſſible : for $\frac{1}{2}$ *and* $\frac{1}{3}$ *and* $\frac{1}{4}$ *doe make* $\frac{26}{24}$, *or* $\frac{11}{12}$, *that*
is, 1 $\frac{1}{12}$, *ſo it is more then the whole, but reduce theſe*
Fractions *into one denomination, the leaſt that they*
will come to, and they will be $\frac{6}{12}, \frac{4}{12}, \frac{3}{12}$ *and ſo may you*
part the goods into ſuch proportion as theſe 3 *Nume-*
rators beare together, that is, the firſt to have 6
for every 4 *to the ſecond, and the ſecond to have* 4 *as*
often as the third hath 3 : *and ſo their portions will*
be for the firſt, 3623 $\frac{7}{13}$, *for the ſecond*
2415 $\frac{8}{13}$, *& for the third* 1811 $\frac{10}{13}$: *& theſe*
3 *ſhares added together, will make the to-*
tall ſumme of the whole goods, as you may
eaſily ſee in this example.

$$3623 \tfrac{7}{13}$$
$$2415 \tfrac{8}{13}$$
$$1811 \tfrac{10}{13}$$
$$7851$$

Another
queſtion
of a Teſta-
ment.

Another

Another queſtion is there propoſed thus.

Another like queſtion.

There are 450 crownes to bee divided betweene three men, ſo that the firſt man muſt have $\frac{1}{2}$ and $\frac{1}{3}$, the ſecond man $\frac{1}{3}$ and $\frac{1}{4}$; the third man ſhall have $\frac{1}{4}$ and $\frac{1}{5}$.

Scholar. **I marvell that any man ſhould be ſo overſéen, to propoſe that queſtion as a thing poſſible, ſith $\frac{11}{23}$, $\frac{11}{34}$, $\frac{11}{45}$, doe make $1\frac{12}{13}$, that is almoſt double the whole ſumme.**

But I perceive it might be thus propoſed : that as often as the firſt man did receive 50 crownes, ſo often the ſecond man ſhould receive 35, and the third man 27 : for $\frac{1}{2}$ and $\frac{1}{3}$ is equall to $\frac{5}{6}$: and ſo is $\frac{1}{3}$ and $\frac{1}{4}$ equall to $\frac{11}{6}$, and $\frac{1}{4}$ and $\frac{1}{5}$ is $\frac{27}{6}$ and ſo working the queſtion the three figures will appear in this forme : whereby the firſt mans portion is found to be $200\frac{50}{56}$: the ſecond mans part is $140\frac{55}{56}$

112	Z	450
50		$200\frac{50}{56}$
112	Z	450
35		$140\frac{15}{56}$
112	Z	450
27		$108\frac{27}{59}$

the third mans ſhare $108\frac{27}{56}$: **which in the whole both make 450 crownes to be divided betwéen them.**

Maſter. **And thus you are (I thinke) ſufficiently inſtructed in the Rule of Fellowſhip.**

The Rule of Alligation·

NOw *will I goe in hand with the* Rule of Alligation *; which hath his name, for that by it there are divers parcells of fundry prices, and fundry quantities alligate, bound or mixed together : whereby alfo it might bee well called the* Rule of Mixture *; and it hath great ufe in compofition of Medicines, and alfo in mixtures of Metalls, and fome ufe it hath in mixtures of Wines : but I wifh it were leffe ufed therein then it is now a dayes. The order of this rule is this.*

When any fummes are proposed to bee mixed, set them in order one over another, and the common number (whereunto you will reduce them) set on the left hand ; then marks what summes bée lesser then that common number, and which bée greater, and with a draught of your penne evermore linke two numbers together, so that one bée lesser then the common number, and the other greater (for two greater or two fmaller cannot well be linked together) and the reason is this, that one greater and one fmaller, may be fo mixed, that they will make the meane or common number very well : but two leffe can never make fo many as the common number, being taken orderly : no more can two fummes greater then the mean, ever make the mean in due order, as it fhall appeare better to you hereafter. And as it is of neceffity to linke every fmaller (once at the leaft) with one greater, & every greater with one fmaller, fo it is at liberty

to

to linke them oftner then once, and so may there
be to one question many solutions. When you
have so linked them, then marke how much each
of the lesser numbers, is smaller then the mean or
common number, and that difference set against the
greater numbers, which be linked with those smaller,
each with his match still on the right hand, and
likewise the excesse of the greater numbers above the
mean, you shall set before the lesser numbers
which be combined with them.

☞   Then shall you (by Addition) bring all these
differences into one summe, which shall be the
first number in the Golden Rule, and the second
number shall be the whole masse that you will have
of all those particulars : the third summe shall be
each difference by it selfe, and then by them shall be
found the fourth number, declaring the just portion
of every particular in that mixture : As now by
these Examples I will make it plaine.

A question *There are foure sorts of Wine, of severall prices*
of mixing *one of 6 pence a gallon, another of 8 pence, the third*
of Wines. *of 11 pence, and the fourth of 15 pence the gallon,*
Of all these Wines would I have a mixture made to
the summe of fifty gallons, and so the price of each
gallon may be 9 pence.   Now demand I, how much
must be taken of every sort of Wine ?

Scholar. If it shall please you to worke the first
example, that I may marke the applying of it to
the Rule : then I trust I shall bee able not onely
to doe the like, but also to see the reason in the
order of the worke.

Master.

Maſter. Marke then this forme, and the placing of every kinde of number in it.

The Prices
feverall.

The diff-
rences:

The common price

6
8
9
11
15

6 A 12 Z 50
2 B   6   25
       C

1 C 21 Z 50
3 D   1   4 ⅙

12

12 Z 50
 2   8 ⅓
       D

12 Z 50
 3   12 ⅓

Here (you ſee) I have ſet downe the ſeverall prices, which be 6, 8, 11, 15, and have linked toge-ther 6 with 15, and 8, with 11. The common price 9, I have ſet on the left ſide, and the difference between it, and every particular price I have ſet on the right hand, not againſt the ſumme (whoſe difference it is) but againſt the ſumme that it is linked withall, ſo the difference of 15 above 9, is 6, which I have ſet, not againſt 15 but 6, that is linked with 15, & the difference between 6 & 9 (that is 3) I have ſet againſt 15. So likewiſe the difference between 8 and 9 is but 1, that I have ſet againſt 11 ; and the difference of 11 above 9 (which is 2) I have ſet againſt 8. Then adde I all thoſe foure differences, and they make 12, which I ſet for the firſt number in the Golden Rule: the ſecond number I make 50, which is the ſumme
of

of Gallons that I ſhould haue, and the third ſumme
to euery particular difference.   Now if you worke
by the Golden Rule, you ſhall finde the number of
Gallons that ſhall be takn of each ſort of wine : For
the better diſtinction whereof I haue ſet theſe
Letters, A, B, C, D, both againſt the Numbers for
whichthe workes doe ſerue, and ouer the worke
alſo, which ſeuerally ſerue for each of them. And
now (if you liſt to examine the truth of theſe
workes) adde theſe foure ſummes together, and
they will make 50,   that is the totall which I
would haue, as by this example you may eaſily
perceiue.   And (for to proue how the prices
doe agree) doe this: Multiply the totall   25
ſumme 50, by the common price 9, and it   8⅓
will make 450 : then keep that ſumme by   4⅙
it ſelfe, and afterward multiply euery ſeue-   12½
rall ſumme of Gallons, by the price belonging
to the ſame Gallons, and if that ſumme doe agree   50
with this, which you haue kept firſt, then is
your worke well done.   As here 25 is the number of Gallons of 6 pence price, multiply then
25 by ſix, and it maketh 150, which you ſhall ſet
downe, then multiply 8⅓ by 8, which is the
price for the number of Gallons, and it   150
will make 66⁴⁄₆ : ſo againe 4⅙ multiplied   66⁴⁄₆
by 11, doth make 45⁶⁄₆. And laſt of   45⁶⁄₆
all, 12½ multiplied by 15, maketh   187½
187½ : and theſe added together doe   450
make 450,as in the example annexed you may ſee,
wherefore ſeing it doth agree with the former
ſumme of 50, multiplyed by 9, I may juſtly
affirme this worke to be good, and well done.

And

And now to prove how you can doe the like: I pro- The varia-
pound the same Question, onely willing you to use tion of this
some other forme of combining or linking the summes. question.

Scholar. **That shall I prove with your favour,
and therefore I combine 8 with 15, and 6 with 11,
and then the form will be as followeth.**

$$
\begin{cases} 6 \\ 8 \\ 9 \quad 11 \\ 51 \end{cases}
\quad
\begin{array}{c|c}
2 & A \\
6 & B \\
3 & C \\
1 & D
\end{array}
$$

A
$$12\,Z\,50 \\ 2 \quad 8\tfrac{2}{6}$$

B
$$12\,Z\,50 \\ 6 \quad 25$$

C
$$12\,Z\,50 \\ 3 \quad 12\tfrac{1}{6}$$

D
$$12\,Z\,50 \\ 1 \quad 4\tfrac{1}{8}$$

12

**Whereby amounteth the same summe in totall
of the** differences as **did before : and yet now the**
differences be altered as the **combination is changed, whereof I understand the reason by your former worke. And therefore here appeareth no**
strange thing, but that now I have $8\tfrac{2}{6}$ gal-   50
lons of 6 pence, and 25 gallons of 8 pence,   200
and 12 gallons and $\tfrac{2}{6}$ of 11 pence, and so $137\tfrac{2}{6}$
consequently 4 gallons and $\tfrac{1}{4}$ of 15 pence : $62\tfrac{1}{4}$
so that multiplying $8\tfrac{2}{6}$ by 6, it maketh   ———
50, and then 25 multiplyed by 8, maketh   450
200: likewise $12\tfrac{2}{6}$ multiplyed by 11 yeeld $137\tfrac{2}{6}$,
and $4\tfrac{1}{4}$ multiplyed by 15, maketh $62\tfrac{2}{6}$, which 4
summes added into one, will yield in the totall
450, which agréeth with the multiplication of
50 (being the totall summe of Gallons) by 9 the
common or mean price.

Master. **Seeing you conceive this worke so**
X   well

well, I will propound another example unto you of moje variety in the Alligations oj combinings, as thus.

Aqueſtion of Spices. *A Merchant being minded to make a bargain for ſpices, in a mixt maſſe (that is to ſay) of Cloves, Nutmegs, Saffron, Pepper, Ginger, and Almonds: the Cloves being at 6 ſhillings, Saffron at 10 ſhillings, Pepper at 3 ſhillings, Ginger at 2 ſhillings, and Almonds at 1 ſhilling.*

Now would he have of each ſojt ſome, to the value of 300 pound in the whole, and each pound one with another to bear in pjice five ſhillings: How much ſhall he have of each ſojt?

Scholar. That will I try thus.

Firſt I ſet downe thoſe 6 ſeverall prices, and at the left hand I ſet the common price five ſhillings. Then I linke them thus, one with 10, two with 6, and 3 with 8: as in the example following.

Maſter. I had minded to have combined them in moje variety: but I am content to ſee your own

own work first, and then moze varieties in combination may follow anon.

Scholar. Then to continue as I began I seek the difference betwéen 1 and 5, (which is 4) and that I set againste 10, then against 1 I set 5, which is the excesse of 10 above 5, so I gather the difference betwéen 2 and 5, which is 3, and that I set againste 6, because it is combined with 2, and likewise the difference of 6 above 5 (which is 1) I set againste 2. Then take I the difference of 3 from 5, which is 2, and that I set againste 8 : and befoze that 3, I set the difference of 8 above 5, which is thzee. Then gather I all these differences by Addition, and they make 18, which I set foz my first Number in the Golden Rule; and so appeareth by those workes, that of Almonds I must take $83\frac{1}{3}$ pound, of Ginger $16\frac{2}{3}$ pound, Pepper 50 pounds, of Cloves 50 pound, of Nutmegs $33\frac{1}{3}$ pound, and of Saffron $66\frac{2}{3}$ pounds.

Then foz triall hereof, I multiply every parcell by his severall price, as $83\frac{1}{3}$ which is the summe of Almonds, I multiply by one which is their pzice.

Also $16\frac{2}{3}$ the summe of Ginger, I multiply by two, which is the pzice of it : and so each other in his kinde, as this Table annexed doth repzesent, and then adding them all together I finde the totall to be 1500, which also will amount by the multiplication of the grosse masse of 300, by the common pzice 5, wherefoze it appeareth well wzought.

$$83\frac{1}{3}$$
$$33\frac{1}{3}$$
$$150$$
$$300$$
$$266\frac{2}{3}$$
$$666\frac{2}{3}$$

$$\overline{}$$

$$1500$$

Ææ 2   Master.

Master. Now I will make the alligation to
prove your cunning somewhat better : but becaufe
you fhall not thinke your felfe preffed too much, I
will alfo note the differences, as by this Example
you may fée, where I have alligated 1 with 6

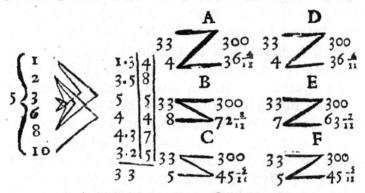

and 8, and therefore have I fet againft 1 both
their differences, that is 1 and 3 : Likewife, becaufe 2 is combined with 8 and 10, I fet before
him their differences, 3 and 5. Againft 3 I have
fet onely 5, which is the difference of 10 with
whom 3 is combined onely. Likewife 6 is onely
alligate to 1, and therefore is the difference of 1
from 5, which is 4, onely fet againft it : 8 is linked
with 1 and 2, and therefore hath fet againft
him, both their differences, 4 and 3 : and 10 is
joyned with 2 and 3, therefore hath he their differences, 3 and 2. And becaufe of eafe for you, in
another columne I have fet the differences reduced
into one number, for every feverall fort, and have
alfo added them together whereby appeareth that
they make 33, & fo confequently you fée the works
of the Golden Rule fet forth. For the fix Drugges

I

I have added the letters A. B. C. &c, as before.

But I would not wish you to cleave still to Note these Elementary aides, but accustome Memory to trust her self: so shall occasion of negligence best be avoyded. And as for the proof try it at more leisure, because the time now is short, and you sufficiently instructed in that proof. And there resteth divers things behinde yet, of which I would gladly give you some taste, before your departure.

Scholar. But if it may please you to let mee see all the variations of this question, before you goe from it, for methinketh I could vary it two or three wayes more yet.

Master. I am content to see you make two or three variations: but I would be loth to stay to see all the variations: for it may be varied above 300 wayes, although many of them would not well serve to this purpose.

Scholar. I thought it impossible to make so many variations.

Master. Marvell not thereat, for some questions Note. of this Rule, may be varied above 1000 wayes; but I would have you forget such fantasies till a time of more leisure. And now goe forward with some variation of this question.

Scholar. For the first variation, I linke the first number 1 with 1 and 10, and 2 I combine with 9 and 10: then joyne I 3 with 6, 8, and 10, as in this forme.

X 3 A

A
$$43 \, Z \, 300$$
$$8 \, Z \, 55\tfrac{35}{43}$$

D
$$43 \, Z \, 300$$
$$5 \, Z \, 34\tfrac{31}{43}$$

B
$$43 \, Z \, 300$$
$$6 \, Z \, 41\tfrac{37}{35}$$

E
$$43 \, Z \, 300$$
$$6 \, Z \, 41\tfrac{37}{43}$$

C
$$43 \, Z \, 300$$
$$9 \, Z \, 62\tfrac{34}{45}$$

F
$$43 \, Z \, 300$$
$$9 \, Z \, 62\tfrac{34}{43}$$

$$5 \left\{ \begin{matrix} 1 \\ 2 \\ 3 \\ 6 \\ 8 \\ 10 \end{matrix} \right.$$

3.5	8
1.5	6
1.35	9	43
3.2	5
4.2	6
4.32	9	43
	43

And so doth there appeare the proportion of weight for every kinde of Drugge in this mixture. Now for the tryall.

Master. Nay stay there: you shall not néed to make tryall in one example so often, or if you list to doe it by your selfe, I am content. But now set forth (for declaration that you conceive the Rule) two or three examples of severall Combinations, and then will we passe to some other example, and so end this Rule.

Scholar. As it pleaseth you, so will I doe. And these be the varieties: in which as the combinations are severall, so doth it plainly appeare, that the differences by which the proportion of each severall kinde is taken are also severall. And yet I sée in the three first of these five varieties, and in the one

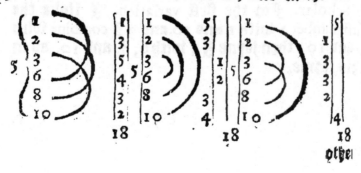

other

$$
5 \begin{cases} 1 \\ 2 \\ 3 \\ 6 \\ 8 \\ 10 \end{cases}
\qquad
\begin{array}{c|c}
1.3.5 & 9 \\
5.3 & 8 \\
5 & 5 \\
4 & 4 \\
4.3 & 7 \\
4.3 & 9 \\ \hline
& 42
\end{array}
\qquad
5 \begin{cases} 1 \\ 2 \\ 3 \\ 6 \\ 8 \\ 10 \end{cases}
\qquad
\begin{array}{c|c}
1.3 & 4 \\
4.5 & 9 \\
3 & 3 \\
4.3 & 7 \\
4.5 & 9 \\
3 & 3 \\ \hline
& 35
\end{array}
$$

other before, the totall summe of the differences to be one, that is to say, 18, whereby I perceive that the variety of their mixture doth depend on the variety of their differences severall, and not of the variety of their totall summe.

Master. So is it. And seeing you conceive it so well, I will make an end of this Rule, onely exhibiting unto you one Question or two of the mixture of Metals, that by it you may devise others like, and exercise your selfe therein also, because the use of it serveth often in businesse of charge, not so much for Goldsmiths, as of coynage in Mints. First, I demand of you this question: *If a Mint-master have Gold of* 22 *Karects, and some of* 23 *Karects, some of* 24: *Againe, some* 15, *some* 16, *and some of* 18 *Karects, and would mix them, so that he might have* 100 *Ounces of* 20 *Karects: How much must he take of each sort?*

Scholar. To know that, I answer in order thus:

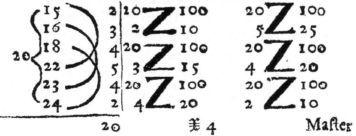

                    20   ℈ 4   Master

Master. You have wrought the question well, but how chanced you made no doubt of that new name Karect.

Scholar. Because I thought it out of time to demand such questions now seeing you make so much hast to end : and againe in this case the proportion of the number is sufficient for my purpose in this worke, trusting that another time you will instruct me as well of this , as of sundry other things, which as I have heard you talke of, so I have a great desire to them.

Mast. Your answer is reasonable, and your request and trust (with Gods helpe) I intend to satisfie, & now to go forward with this matter, let me see your examination of this lastwork.

```
10
15
20
25
```

Scholar. First for the one part I adde together all the particular sums, as they appeare in the worke, and they make 100, as here by their Addition doth appeare.

```
20
10
100
```

And so it seemeth that the sums are well gathered : but for the further triall of them I multiply first 20 which is the common or meane sun of the Karects by 100, which is the sum of th whole Masse, which I would have, and t maketh 2000. Then I multiply every parti cular sum by the Karects, that it doth contai as 10 by 15, and that maketh 150.

```
150
240
360
550
460
240
```

Likewise I multiply 15 by 16 , and yieldeth 240: so 20 by 18, maketh 360. A 25 by 22, yieldeth 550: likewise 20 by 23, bringe forth 460: and last of all, 10 multiplyed by 2 yieldeth 240 : which summes all joyned togeth make 2000, that doth agree with the like sum

```
2000
```

befo

before, wherefore I may well say, that the worke is good. And now (if it please you) I would set forth some varieties of this question) to prove my wit.

Master. Goe to, let me see.

Scholar. Here be foure varieties,

```
        ⎧ 15 ⎫  3.4 | 7             ⎧ 15 ⎫  2.3 | 5
        ⎪ 16 ⎪  3   | 3             ⎪ 16 ⎪  3.4 | 7
   20   ⎨ 18 ⎬  2   | 2        20   ⎨ 18 ⎬  4   | 4
        ⎪ 22 ⎪  2   | 2             ⎪ 22 ⎪  5   | 5
        ⎪ 23 ⎪  5.4 | 9             ⎪ 23 ⎪  5.4 | 9
        ⎩ 24 ⎭  5   | 5             ⎩ 24 ⎭  4.2 | 6
              28                          36
```

```
        ⎧ 15 ⎫  2.3.4 | 9           ⎧ 15 ⎫  4     | 4
        ⎪ 16 ⎪  4     | 3           ⎪ 16 ⎪  4     | 4
   20   ⎨ 18 ⎬  3     | 3      20   ⎨ 18 ⎬  2.3.4 | 9
        ⎪ 22 ⎪  5     | 5           ⎪ 22 ⎪  2     | 2
        ⎪ 23 ⎪  5.2   | 7           ⎪ 23 ⎪  2     | 2
        ⎩ 24 ⎭  5 4   | 9           ⎩ 24 ⎭  5.4.2 | 11
              36                          32
```

And more yet could I make, but not like to the number that you speak of in the variation of the other question.

Master. That will I teach you at more leisure, seeing it is a thing rather of pleasure then of any necessity.

But now for your exercise in this Rule, one other question I will propose. A Mint-master hath six Ingots of silver, of sundry finenesse, some of four Ounces fine, and some of five Ounces, some of six, and other of eight, some of 11, and other of 12, and his desire is to mixe 500 pounds weight, so that in the whole masse

A question of mixing of silver.

every

every pound weight should beare 9 Ounces of fine silver : How much shall he take say you of every sort of silver ?

Scholar. To finde out that, I set the numbers thus in order.

And gathering the differences it will appeare, that, of the first sort there must be 43 $\frac{11}{23}$ of the second like much: of the third sort 65 $\frac{5}{23}$: and of the fourth sort as much : of the fifth sort 195 $\frac{15}{23}$ and of the sixth sort 86 $\frac{22}{23}$, which in the whole will make 500 pound weight, and in ounces after 9 ounces fine 4500, that is of the first sort 173 $\frac{21}{23}$, and of the second sort 217 $\frac{2}{23}$, of the third sort 391 $\frac{7}{23}$, of the fourth sort 521 $\frac{17}{23}$, of the fifth sort, 2152 $\frac{4}{23}$, and of the sixth sort 1043 $\frac{11}{23}$, which all together doe make 4500 ounces, agréeable to the multiplication of 9 by 500.

Master. This is well done of you, therefore now make three or foure varieties, and so an end of this Rule.

Scholar. These four varieties I set for examples.

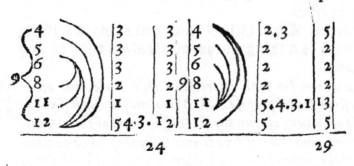

Master.

$$
9 \begin{cases} 4 \\ 5 \\ 6 \\ 8 \\ 11 \\ 12 \end{cases}
\quad
\begin{matrix} 23 \\ 3 \\ 2 \\ 2 \\ 5.3.1 \\ 5.4 \end{matrix}
\quad
\begin{matrix} 5 \\ 3 \\ 2 \\ 2 \\ 9 \\ 9 \end{matrix}
\qquad
9 \begin{cases} 4 \\ 5 \\ 6 \\ 8 \\ 11 \\ 12 \end{cases}
\quad
\begin{matrix} 3 \\ 3 \\ 2.3 \\ 2.3 \\ 3.1 \\ 5.4.3.1 \end{matrix}
\quad
\begin{matrix} 3 \\ 3 \\ 5 \\ 5 \\ 4 \\ 13 \end{matrix}
$$

30   33

Maſter. And by theſe it appeareth, that you can finde out moze, with which I will not now meddle, ſave onely (foz to ſhew you an eaſie help dzawing the lines of Combination) I will ſet fozth 2 varieties here.

$$
9 \begin{cases} 4 \\ 5 \\ 6 \\ 8 \\ 11 \\ 12 \end{cases}
\quad
\begin{matrix} 2 \\ 2.3 \\ 3.2 \\ 3 \\ 5.4.3 \\ 4.3 \end{matrix}
\quad
\begin{matrix} 2 \\ 5 \\ 5 \\ 3 \\ 12 \\ 8 \end{matrix}
\qquad
9 \begin{cases} 4 \\ 5 \\ 6 \\ 8 \\ 11 \\ 12 \end{cases}
\quad
\begin{matrix} 3 \\ 2.3 \\ 2.3 \\ 2.2 \\ 4.3 \\ 5.4.2.1 \end{matrix}
\quad
\begin{matrix} 3 \\ 5 \\ 5 \\ 5 \\ 8 \\ 12 \end{matrix}
$$

35   38

And this ſhall ſuffice now foz the Rule of Alligation or mixture : foz by theſe examples may you eaſily conjecture ſuch other as do appertain to it, as well foz the due wozking, as foz variety of dzawing the lines of Combination.

Scholar. Sir, albeit it pleaſed you erewhile to put me from my muſing at the many varieties that may fall in theſe Combinations, and termed them phantaſies, yet my phantaſie giveth me, that the conſideration of this ſhould in many other examples and caſes of impoztance be very nedfull, and the knowledge of it moſt pzofitable : Therefoze ye may well think, that at another time convenient

I

I will request you to aid me herein.

Master. Truth it is, that this consideration may fall in practice as well Politick as Philosophicall, and sundry waies in them be applied: Therefore when time shall fall fit, for the discussing of this consideration, you shall not want my helping hand.

The Rule of Falshood.

The occasion of the name.

Now will *I briefly also teach you the* Rule of Falsehood, *which beareth his name, not for that it teacheth any fraud or falshood, but for that by false numbers taken at all adventures, it teacheth how to finde those* true numbers *you seek for.*

Scholar. So might any other Rule be called the Rule of Falshood, for they worke by wrong numbers, and by them finde out the right numbers: so doth the Rule of Alligation, the Rule of Fellowship, and the Golden Rule, partly.

Mast. In the Golden Rule, the Rule of Fellowship, & the Rule of Alligation, although the numbers that you worke by, be not the true numbers that you seek for, yet are they numbers in just proportion, and are found by orderly worke, whereas in this Rule the numbers are not taken in any proportion, nor found by orderly worke, but taken at all adventures.

And therefore I sometimes being merry with my friends, and talking of such questions, do call unto them such Children or idiots, as hapned to be in the place, and so take their answer, declaring that

that I would make them solve those questions,
that seemed so doubtfull.

And indeed I did answer to the question, and
worke the Triall thereof also by those answers
which they happened at all adventures to make :
which numbers seeing they bee taken as manifest
false, therefore is this Rule called the Rule of false
Positions, and for brefnesse, The Rule of Falshood :
which Rule for readinesse of remembrance, I have
comprised in the few verses following, in form of
an obscure Riddle.

> Ghesse at this worke as hap doth lead,
> By chance to truth you may proceed,
> And first worke by the question,
> Although no truth therein be done.
> Such falshood is so good a ground,
> That truth by it will soon be found.
> From many bate too many moe,
> From too few take too few also :
> With too much joyn too few again :
> To too few adde too many plain :
> In crosse wise multiply contrary kinde,
> And all truth by falshood for to finde.

The sense of these Verses, and the summe of this
Rule is this.

When any *question is proposed* appertaining to this
Rule, first imagine any number that you list, which
you shall name the *first position*, and put it in stead of
the true number, and then worke with it as the *question* importeth : and if you have missed, then is the
last number of that worke either too great or too
little : that shall you note as here after shall be

The expo
sition of
the Rule.

taught

taught you, and you shall call it the *first errour*.

Then begin againe, and take another number, which shall be called the *second position*, and worke by the *question* : if you have missed againe, note the excesse or default as it is, and call that the *second errour*. Then multiply crosse-wise the *first position* by the *second errour*, and againe, the *second position* by the *first errour*, and note their totalls severally by the names of totalls : Then marke whether the two errours were both alike, that is to say, both too much, or both too little : or whether they be unlike, that is, the one too much, and the other too little : for if they be like, then shall you *subtract* the one totall from the other (I mean the lesser from the greater) and the remainer shall be your *dividend* : so must you abate the lesser errour out of the greater, and the residue shall be the *divisor*. Now divide the *dividend* by that *divisor*, and the *quotient* will shew you the true number that you seek for. But and if the errours be unlike, then must you adde both those totalls (which you noted) together, and take that whole number for the *dividend*, so shall you adde both errours together, and that whole number shall be the *divisor*, and the *quotient* of that division shall give you the true number that the *question* seeketh for, and this is the whole Rule.

Scholar. This Rule seemeth so unlike any other, that without some example I shall not easily understand it.

Master. With a good will : propose halfe a score sundry questions and examples of variety for the better understanding of the worke hereof : and for the first take this example. *A maso*

wa

was bound to build a wall in 40 dayes , and it
was covenanted ſo with him, that every day of Maſon-
that he wrought, he ſhould have for his wages 2 ry the firſt
ſhillings 1 peny, and every day that he wrought not, example.
he ſhould be amerced 2 *ſhillings* 6 *pence, ſo that*
when the wall was made and the reckoning taken of the
dayes that he wrought, and of the other that he
wrought not, the Maſon had clearly but five ſhil-
lings five *pence for the worke. Now do I demand how*
many dayes he did work of thoſe 40, *and how many he*
did not worke.

Scholar. I pray you expreſſe the order of the
worke, that I may partly by imitation, and partly by comparing it with the Rule, be able again to
doe the like.

Maſter. This order ſhall you kéep in the worke
of this Rule : firſt take ſome number (as you liſt)
at adventure ; as for example, I ſay he played 1 2
dayes, and wrought 28 dayes. Now caſt you the
wages of every day and ſée whether it will agrée
with the ſumme of 5 ſhillings 5 pence.

Scholar. The 28 dayes that he wrought after
25 pence the day, yieldeth 700 pence : Then 1 2
dayes that he wrought not, at 30 pence each day,
doth amount to 360 pence, which if I abate out of
700 pence, there reſteth 340 : but you ſay he had
not ſo much.

Maſter. He had but 65 pence, and by this ſuppoſition he ſhould have had 340 : therefore is this
ſumme too much by 275, which ſumme I   1 2
muſt ſet down after this ſort as you ſée   X
here, where firſt I have made a croſſe
(commonly called S. Andrews croſſe) and   275†
at

at the oter co2ner on the left hand I haue fet the
firft pofition 12: and at the other co2ner under it
I haue fet 275 which is the firft errour, with this
figure †, which betokeneth too much, as this line
——————— plaine without a croffe line betokeneth to
little.

On the right hand of the croffe I haue left two
like roomes fo2 the fecond pofition and his errour.
Therefo2e to p2ofecute the wo2ke, I fuppofe he
played 16 dayes, and wrought 24.

Scholar. I was a while in doubt why you named the dayes of his working, feeing they be not
fet in the figure: and I doubted how you knew
them, o2 elfe whether you did fuppofe them at all
adventures, as you did the dayes that he played:
but now I gather that feeing 40 dayes is the whole
time limited, then the dayes that he played being
fuppofed, the reft of 40 muft needs be the dayes
that he wrought, and therefo2e 28 followed 12 of
neceffity, and 24 followeth 16 alfo of neceffity,
but yet I fcarce perceive why you fet not in the figures as well 28 as 12.

Mafter. It fo2ceth not which of them I take, fo
that in the fecond pofition I take the numbers of
the fame nature that is here both of working dayes,
o2 both of idle, but now examine you this fecond
pofition.

Scholar. If he played 16 dayes, then abating 16
times 30 pence, the fum will be 480 pence, and fo2
24 dayes that he w2ought, every day pielding 25
pence, the totall is 600 pence: fo that abating 480
out of 600, there refteth 120, and as you fay, t
fhould be but 65: therefo2e it is to much by 55
tha

that muſt be ſet on the right hand of the figure, at the neather part, and over it on the ſame ſide 16, which is the ſecond poſition, thus.

And as I gather by your words, it were all one if I did ſet 28 in ſtead of 12, and 24 in ſtead of 16.

Maſter. So were it. But this ſhall you marke, that, of what nature ſoever the two poſitions be, of the ſame nature is the quotient. Therefore when the poſitions in this queſtion are 12 and 16, which both being numbers of the playing dayes, the quotient ſhall declare the true number of playing dayes : whereas if the poſitions had béen 28 and 24, which are ſuppoſed to be the working dayes, then would the quotient declare the true number of the working dayes, and not of playing dayes, as it will doe now. And therefore to continue the work of this queſtion, and to finde the true number of playing dayes, I muſt multiply croſſe-wiſe the firſt poſition by 55 that is the ſecond errour, and the totall will be 660. Then I multiply 275 and 16, and it yeeldeth 4400.   Now becauſe the errours are alike, that is to ſay, both too much, I muſt ſubtract 660 out of 4400, and ſo remaineth 3740, which is the dividend. Againe I muſt ſubtract the leſſer errour 55 out of 275, that is the greater errour, and there will remaine 220 which will be the diviſor : then dividing 3740 by 220 the quotient will be 17. Wherefore I ſay now conſtantly, that 17 is the true number of dayes that the Maſon played : and then it followeth that he wrought 23 dayes : and ſo is the queſtion anſwered.

Now for the order of triall of this work, there need- The proof
eth none other triall but only this, to worke with this of this rule

P   *number*

number according to the question, and if it agree, then appeareth the number to be that you would have.

And here now seeing he wrought 23 dayes, and must have for every day 25 pence, the whole summe commeth to 575. Then again, seeing he played 17 dayes and must abate 30 pence for every day, the whole sum of the abatement will be 510. Therefore I subtract 510 out of 575, and there will remaine 65, which maketh 5 shilling 5 pence, the cleere wages of the Mason for his worke according to the question.

Scholar. Now I trust I understand the worke and the Rule so well (and the better by this proof) that I can be able to doe the like : And for a proof, I take the same question, all save the last number, where I will suppose that he had 10 shillings for his wages cleere. And now to guesse of the number of the dayes he wrought, I suppose first that he wrought 20 dayes : then say I if he wrought 20 dayes, his wages must be 500 d. then did he play other 20 dayes, for which must be abated 600 d, and then he loseth 100 d. And so am I at a stay, for it is not like to your former worke.

Master. You should have required of me some question, and not have taken a question of your owne phantasying, untill you were more expert in this Art, for so might you as well happen to an impossible question, as on a possible : but now to go forward, consider that this number is too little by 220, seeing he should gaine by your supposition 12 pence, and in this position he loseth 100, that both make 220, which you shall set downe for the first errour, with this signe —, betokening too little

as here in this forme following doth ap-   20
peare.

   And now for the rest goe forward your
selfe once againe.   ———220

   Scholar. As my errour hath uttered my folly, so
it hath procured me better understanding.

   Now therefore considering this position not to
solve the question, I take another, supposing that he
wrought 30 dayes.   Then for his wages he must
be allowed 750 pence, and for the 10 dayes which
he wrought not, he must abate 300 pence, and so
remaineth cleer 450 pence, but it should be onely
120 pence, therefore it is too much by 330, which
I set downe in the figure with the former position
and his errour, and the figure appeareth thus :

   Now first, I multiply in crosse   20   30
wayes 20 by 330, and it will be
6600.

   Then again I multiply 30 by ——220- 330
220, and it will be also 6600. Wherefore if I
shall subtract the one out of the other, there will
remain nothing to be the Dividend.

   Master. In this you forget your selfe again;
for in as much as the signes in the errours be
unlike, therefore must you worke by Addition, adding together those two totalls to make the Dividend, and also adding the two errours to make the
Divisor.   And because you shall no more forget
this part of that Rule, take this briefe remembrance.

      Vnlike require Addition ;
         And like desire Subtraction.

   Scholar. You mean, that if the errours have like
signes, then must the Dividend, and the Divisor

bée made by Subtraction, as is taught before: And if those signes bee unlike (as in this last example they bee) then must I by Addition gather the dividend and the divisor. Therefore must I adde 6600 to 6600, and it will bée 13200, which will be the dividend. Then againe I adde 220 to 330 and it will be 550, which must be the divisor: wherefore dividing 13200 by 550. the quotient will be 24, whereby I know that the Mason wrought 24 dayes, and then it followed, that he played 16 dayes.

Master. Examine your worke, whether it be agréeable to the question or no.

Scholar. For 24 dayes worke the wages must be 600 pence, and for 16 dayes which the Mason wrought not, there must be abated 480 pence, and then remaineth cleare to the Mason 120, as the question importeth: wherefore it is evident that 24 is the true number of dayes that he wrought.

Master. Although you séem now to understand this worke, yet to acquaint your minde the better with the new Trade of this Rule, I think it good to propose to you 5 or 6 examples more before I make and end of it.

Scholar. Sir, I thanke you that you doe so consider my commodity and profit in knowledge, for undoubtedly it is practice and exercise that maketh men prompt & expert in every kinde of knowledge.

Master. You say well, so that they follow some certain precepts to governe and rule their practice by, else may practice procure custome of errour, and a repugnance to exactnesse of knowledge: namely as long as the errour is not plainly known to the vulgar sort, But return to your work.

There

There is a servant that hath bought of Velvet and *Damask for his Master* 40 *yards, the Velvet at* 20 *shillings a yard, and the Damask at* 12 *shillings, and* *when he commeth home, his Master demandeth of him,* *how much he hath bought of each sort : I cannot tell (saith he) exactly : but this I know that I paid for Damask* 48 *shillings more then I paid for Velvet: now must you guesse how many yards there is of each sort.*

Scholar. Although the guesse seemeth difficult, yet I wil probe what I can doe : for I remember your saying, that it forceth not how fond or false the guesse bee, so it be somewhat to the question, and not an answer of a contrary matter.

Therefore first I imagine that he bought 20 yards of damask, for which hee should pay after the former price 240 shillings : then must he needs have of Velvet other 20 yards, (to make up the 40 yards) and that would cost 400 shillings. So that the totall of the price of the Damask is lesse then the summe paid for Velvet 160 shillings, and should bee more by 48. Therefore the first errour is 208 too little. Then begin I againe, and suppose he bought of Damask 30 yards, that cost 360 shillings, then had he but 10 yards of Velvet, which cost 200 shillings : and now the price of the Damask is greater then the price of the Velvet by 160 shillings, and should be but 48, therefore is the second errour 112 too much, which I set in forme of figure, as here doth appeare. Then do I multiply in crosse wayes 208 by 30, and the summe will be 6240. Also I multiply 112 by 20 , and there wil amount 2240. And in as

$$20 \qquad 30$$

$$208 — \qquad 112 \dagger$$

much

much as the signes of the errours be unlike, I know I must worke by Addition, therefore adde I these two totalls together, and they make 8480, which is the Dividend : then adde I also the two errours together, 208, and 112, and they make 320, which is the Divisor : wherefore dividing 8480 by 320, the quotient will be 26 ½, which is the true summe of yards of Damask that he bought, and in Velvet 13 yards ½, and that appeareth by examination, thus : 26 ½ yards of Damask at 12 shillings the yard maketh 318 shillings : then in Velvet he had but 13 yards and ½ and cost 270 shillings, at 20 shillings the yard.   Now subtract 270 out of 318, and there will remaine 48, which is the number of shillings that the Damask did cost more then the Velvet.

Master.   Now shall you have a question of another kinde.

A question
of debt,
the third
example.
There are three men that do owe money to me, and I have forgotten what the totall summe is, and what the particulars be.

Scholar. Why, then it is impossible to know the debt.

Master. Peace, you are to hasty, there is more help in it then yet you see, I have three severall notes, whereby it appeareth that I did conferre their debts together, and found the debt of the first & the second to amount to 47 pound. the debt of the first man and the third man did make 71 pound, and the second man his debt with the third, did rise to 88 pound.   Now can you tell what every man did owe, and what was the whole summe ?

Scholar. Nay, in good faith, but as I perceive that it must be found by conjecture, so will I guesse
at

at it supposing that the first man did owe 20 pound, and the second man 30, and the third.

Maſter. Nay ſtay there, you are too farre gone already : you may not ſuppoſe a ſeverall ſumme for every man, for it is enough to ſuppoſe one ſumme for the firſt man, and let the other riſe as the queſtion importeth. Therefore ſeeing you ſet the firſt man his debt to bee 20 pound, the ſecond man cannot owe 30 pound, for the declaration is, that their debts added together did make 47 pound, ſo muſt the ſecond man his debt be but 27 pound. Now the ſecond debt with the third muſt make 88 : therefore ſubtract 27 out of 88, and there will remain 61, as the third man his debt. Then ſaith the declaration, that the firſt and third mans debts do make 71 : but by this ſuppoſition they make 81, that is 10 too much, which I muſt ſet for the firſt errour. Now worke you the ſecond poſition.

Scholar. I ſuppoſe the firſt mans debt to be 24 pound : then muſt the ſecond mans debt (by your declaration) be but 23 pound, ſeeing both they make but 47 pound. And the ſecond man his debt with the third, doe make 88 pound, and the ſecond man oweth but 23 : therefore the third man muſt owe 65 pound. Now the third mans debt with the firſt, ſhould make by the declaration 71 pound, and they doe make 89 pound, that is 18 pound too much, and that is the ſecond errour, which I ſet down with the firſt, and their poſitions in this forme, and then I doe multiply in croſſe wayes 20 by 18, and it is 360. And 10 by 24 maketh 240. Alſo becauſe the ſignes of the errours be like,

20   24

10†   18†

like,

like, I muſt worke by ſubtraction : therefore I ſubtract 240 out of 360, and there reſteth 120, which is the Dividend : then do I ſubtract 10 out of 18 by the ſame reaſon, ſo is the Diviſor 8, which is found 15 times in 120 : therefore I ſay that the firſt man did owe 15 li. and then the ſecond man muſt owe 32 li. for thoſe two doe make 47 li. and the third mans debt is 56 : for ſo much remaineth if I abate 15 out of 71, or if I take 32 out of 88.

The fourth ex. ample.

Maſter. *For the fourth example, take this eaſie queſtion for the variety in work. Two men having ſeverall ſummes, which I know not, doe thus talke together : the firſt ſaith to the ſecond, if you give me 2 ſhillings of your money, then ſhall I have three times ſo much money as you. The ſecond man anſwereth, It were more reaſon that our ſummes were made equall, and ſo will it be if you give mee 3 ſhillings of your money. Now gueſſe what each of them had.*

Note.

Scholar. I imagine that the firſt had 9 s.

Maſter. Conſider evermore in your imagination that you take a likely ſumme, as in this queſtion, take ſuch a ſumme, that having 2 added unto it may be divided into three parts even.

Scholar. Why ? I remember you ſaid before it forceth not how fondly ſoever I gueſſed.

Maſter. As for the poſſibilitie of the ſolution, it truth : but for eaſineſſe in worke, the apteſt numbers are moſt convenient.

Scholar. I thought no leſſe, and therefore tok 9 as an apt number to be parted into three but I perceive I ſhould have conſidered the apneſſe of that partition after the addition of th

uil

unto it, and then 7 had been moze méet.

Master. That is truth, and then should the second man his summe be 5 : foz although he have now but the third part of 9, that is 3, yet you must remember that he lent the first man 2, & so had be 5.

Scholar. Then to go fozward : if the second man had thzee of the first man, then should he have 8 and the first man but 4 ; so hath he double to the first man, yet he said in the question they should have equall : wherefoze it appeareth that he hath 4 too much.

Therefoze I note that errour with his supposition, and guesse again that he hath 10 shillings : whereunto I adde 2 shilling; bozrowed of the second man, and then he hath 12 shilling, so the second man hath remaining but 4, whereunto if I adde the 2 that he lent to the first man so had be but 6 shillings at the beginning.

Then take 3 shillings from the first man, and give to the second, and then hath the first man but 7, & the second hath 9, which are not equall, but there are 2 too many, wherefore I set down both the positions with their errours, as befoze you see, and multiply acrosse, so commeth there 40 and 14 : and because the signes be like, I take 14 out of 40, and so resteth 26 to be divided : then likewise I take 2, out of 4, and there resteth 2, by which I divide 26, and the quotient will be 13, which is the summe that the first man had. And so appeareth that 2 being added thereto; the summe will be 15, so hath the second man but 5, and befoze he had 7 : then take 3 from the first, and put to this 7, and so have each

each of them 10, and that is equall as the question would.

Master. *For the fifth example take this question.* *One man said to another, I think you had this yeare* *two thousand Lambs: so had I said the other; but* *what with paying the tythe of them, and then the seve-* *rall losses, they are much abated: for at one time I lost* *halfe as many as I have now left, and at another time* *the third part of so many, and the third time $\frac{1}{4}$ so ma-* *ny. Now guesse you how many are left.*

Scholar. Because there is mention made of cer-taine parts, I must take a number that may have all these parts, that is to say, $\frac{1}{2}$ $\frac{1}{3}$ and $\frac{1}{4}$ which will be 24, howbeit 12 hath the same parts. Therefore I take first 12 to be the number that doth remaine, so hath he lost 6, 4, and 3 that is 13, and the whole 25. but it should be 2000.

Master. You are deceived yet still, you have forgotten the 10 part, which must be defalked, that is 200, so there remaineth but 1800, and now go on againe.

Scholar. Then to finde the errour, I take 25 out of 1800, and there remaineth 1775 to few, which I set for the errour. Then for the second position I take 24, whose halfe is 12, the third part 8, and the quarter 6, whereby riseth 50, which is too little by 1750 therefore I set downe both the positions, with their errours thus:

And multiply in crosse wayes 1775 by 24, where-of commeth 42600. Also I multiply 1750 by 12, and there ariseth 21000. And because the signes are like

12   24

——1775   1750——

like, I doe subtract the one from the other, and so remaineth the Dividend 216000. Then doe I subtract 1750 out of 1775, and there resteth 25, by which I divide 21600, and the quotient is 864, whereof the halfe 432, and the third part is 288, the quarter is 216; which all being added together, will make 1800: And if you adde thereto the tenth which was abated before, then will the whole summe be 2000.

$$
\begin{array}{r}
864 \\
432 \\
288 \\
216 \\
\hline
1800
\end{array}
$$

And now doth there come a question to my memory which was demanded of me, but I was not able to answer to it: And now me thinketh I could solve it.

Master. Propose your question.

Scholar. *There is supposed a Law made, that (for furthering of tillage) every man that doth keepe sheep, shall for every ten sheep ear and sow one Acre of ground: and for his allowance in sheep pasture, there is appointed for every foure sheep one Acre of pasture. Now is there a rich Sheep-master which hath 7000 Acres of ground and would gladly keep as many as he might: by that Statute I demand how many sheep he shall keep.*

A question of sheep and tillage: the sixth example.

Master. Answer to the question your self.

Scholar. First, I suppose hee may keep 500 sheep, and for them hee shall have in Pasture after the rate of four sheep to an Acre, 125 Acres, and in Arable ground 50 Acres, that is, 175 in all: but this errour is too little by 6825. Therefore I guesse again that he may keep 1000 sheep, that is, in Pasture 250 Acres, and in tillage 100 Acres, which make 350, that is too little by 6650. Both these errours with their positions, I set down as

you

you fée, and multiply them croſſe 6825 by 1000, and it maketh 6825000, alſo I multiply 6650 by 500, and there commeth 3325000, which ſumme I ſubtract out of the former, and there remaineth 3500000 for the Dividend: likewiſe I ſubtract the leſſer errour out of the greater, and there reſteth 175, by which I divide 3500000 (the Dividend aforeſaid) and the quotient will be 20000, ſo that by this rate he that hath 7000 Acres of ground, may kéep 20000 ſheep.

$$500 \qquad 1000$$

$$6825 - 6650$$

Maſter. You have done well, notwithſtanding both this laſt queſtion, and the next before might bée wrought without the ſecond poſition by the Rule of proportion as this: When in this queſtion you found in the firſt errour that for 500 ſheep there muſt be 175 Acres, then might you reduce it to the Golden Rule, thus:

$$175 \quad Z \quad 500$$

$$7000 \qquad 20000$$

If 175 Acres will admit in allowance 500 ſheep, then 7000 will have 2000. And ſo by one poſition, with the help of the Golden Rule may you anſwer that queſtion.

Likewiſe for the queſtion of Lambs, when you had found that 12 came of 25, you might have ſet the figure as followeth, and have ſaid:

$$25 \quad Z \quad 12$$

$$1800 \qquad 864$$

If 25 do leave but 12, what ſhall 1800 leave? and it would appear to be 864.

Scholar. Sir, I thanke you for this aid, for it doth much ſhorten the worke of this Rule.

Maſter. Yet againe I will ſhew you another way to anſwer to this laſt queſtion without the Rule

of

Another way of working.

Another way yet.

of falſe poſition, and that by the Rule of Fellowſhip, foz it appeareth in the propoſing of the queſtion, that ten ſheep muſt have in paſture 2 Acres and ½, and foz them muſt there be eared but one Acre; ſo it followeth, that foz 2 Acres eared, there muſt bee 5 ſet to paſture and if you put them both into one ſumme, they will make 7. Therefoze look what propoztion 7 being this totall, both bear to 5 and to 2, ſuch propoztion ſhall any totall in this queſtion beare to the paſture ground, and the eared ground.

Scholar. This ſerveth wondzous aptly. Therefoze to pzove it, I demand this by the fo1mer ſuppoſition: If a man have 300 Acres, how much ſhall he leave in paſture, and how much ſhall he turne to tillage? You ſay that as 7 is to 5, ſo ſhall 300 be to the Acres of paſture: and as 7 is to 2, ſo is 300 to the Acres of tillage, whereof foz both I have ſet examples here following, whereby appeareth that of paſture, there ſhall bee 214 ²⁄₇ Acres, and of Tillage 85 ⁵⁄₇, which both ſummes added together, doe make 300.

$$7\diagdown\begin{matrix}5\\214\tfrac{2}{7}\end{matrix}\quad 300$$

$$7\diagdown\begin{matrix}2\\85\tfrac{5}{7}\end{matrix}\quad 300$$

Maſter. *Now take another Example: A man hath* *three ſilver Cups with one Cover, the Cover weigheth 18 ounces, the ſecond Cup weigheth even halfe the weight of the firſt and the third. Now if the Cover be put to the firſt Cup, they weigh juſt as much as all the three cups do weigh: & if the Cover be joyned with the ſecond Cup, they weigh as much as the ſecond twice, & the third: & if the Cover be put to the third cup they will make twice as much as the firſt & ſecond cup. Now try you what was the juſt weight of every cup.*

Another queſtion: the ſeventh example.

Scholar.

Scholar. I do set the weight of the first Cup to be nine Ounces, then in as much as these two (that is to say, the cover and the first Cup) doe weigh the weight of the three Cups, I see that the three Cuppes must weigh 27 ounces. for so much is 18 and 9, Also because the first and the third doe weigh double so much as the second, therefore it is the third part of that weight, that is 9, and then would it follow, that the third Cup also should weigh 9 ounces; but then the question saith that the Cover being joyned to the second Cup, they weigh as much as the second twice, and the third once, that should be 27, and so it doth; that being joyned with the third Cup, they should weigh twice as much as the first and the second, that should be 36, and they weigh but 27, so is that errour 9 to little. Then beginne I again, and say, that the first Cup doth weigh twelve ounces which I joyne with the Cover, and they make thirty ounces : then seeing the second is of that weight, it must needs weigh ten ounces, and the third must weigh 8 ounces, seeing the first and the third must weigh 20 ounces. Now put I the Cover to the second Cup, and they weigh 28 ounces, which should be even so : then joyne I the cover with the third Cup, & so should it weigh twice the first and the second, that is 44 ounces, and they weigh but 26, that is 18   9   12 to little : those errours with their positions I set down, and multi-ply in crosse wayes 9 by 12,   9—— 18—— whereof commeth 108 : Also 9 by 18, and that yeldeth 162: and in as much as the signes be like I abate the lesser out of the greater, and there doth remaine

remaine 54. Then doe I alſo abate the leſſer errour from the greater, & ſo remainth 9, by which I diuide 54, and the quotient is 6, which I take for the true weight of the firſt Cup, which being ioyned with the Couer, muſt weigh as much as the three Cups, ſo doe they weigh but 24 ounces. Then ſeeing the ſecond Cup is the third part of that weight, for the other two Cups (you ſay) muſt weigh double his weight, the weight of the ſecond Cup is 8 ounces, and ſo the weight of the third Cup muſt be 10 ounces. Now put the Couer to the ſecond Cup, and it will make 26 ounces : that muſt be the weight of the ſecond twice, and the third once, that is, twice 8, and once 10, and ſo is it. Again put the Couer to the third Cup of 10 ounces, and they muſt weigh, twice as much as the firſt and the ſecond, that is 28, and ſo is all agreeable.

Maſter. Then anſwer to this Queſtion.

A queſtion of water : the eight example.

There is a Ciſterne with four Cocks, containing 72 *barrels of water : and if the greateſt Cocke be opened, the water will avoid clean in ſix houres ; at the ſecond Cocke it will aske eight houres : at the third Cock it will avoid in no leſſe then nine houres : and at the ſmalleſt it will require twelve houres. Now I demand in what ſpace will it avoid, all the Cocks being ſet open ?*

Scholar. Firſt I imagine it will avoid in two houres.

Maſter. Then muſt there avoid by the firſt Cock ⅓ of the water, that is 24 Barrels, and by the ſecond Cock ¼, that is 18, and by the third Cock ²⁄₉, that is 16 Barrels, and by the ſmalleſt Cock ⅙,

that

that is 12 Barrels, all which summes put together, do make 70, as by their Addition it doth appeare, but it should be 72, therefore the errour is too few.

Scholar. Then will I begin again by your favour, because I think I underſtand the work, and put three houres for the due time: ſo ſhall there run out at the greateſt Cock $\frac{1}{2}$, that is, 36 Barrels, and at the ſecond hole $\frac{1}{8}$, that is 27, and at the third Cock $\frac{1}{3}$, that is 24, and at the ſmalleſt hole $\frac{1}{4}$, that is 18 Barrels, which all together doe make 105, and ſhould be but 72, ſo is it too much by 33: therefore do I ſet the errours in order of the figure with their poſitions, and worke by multiplication, in croſſe, ſaying, 2 times 3 is 6, & 2 times 33 mak 66, & becauſe the ſignes are unlike, I muſt adde theſe 2 totalls together, which make 72: alſo I adde the two errours, and they make 35, by which I divide 72, and the Quotient riſeth $2\frac{2}{35}$, whereby I ſee that all the Cocks being ſet open, the water will avoid in two houres, and $\frac{2}{35}$ of an hour.

```
24
18
16
12
—
70
```

```
2        3
   ✕
2 —— 33 †
```

Maſter. This exerciſe maketh you to grow expert in the Rule. Therefore I will enure you ſomewhat more with a queſtion or two.

A queſtion of partners:

The ninth example.

There were two men that had been partners, and had in account between them 300 Duckets; whereof the one ſhould have for his part 180, & the other 120: but in the parting of them, they fell at variance; ſo that each of them catched as many as he could: yet afterward being reconciled they agreed that he which hath gotten moſt part of them, ſhould lay down $\frac{1}{4}$ of them again, and he that had gotten leaſt, ſhould lay down $\frac{1}{3}$ of thoſe which

which he had taken, and then parting them into two
equall parts, each man to have halfe thereof, and so
had they their just portions as they ought to have: now
I demand of you what each of them had gotten by the
scrambling.

Scholar. I suppose he that had least, got 108
Duckets, then the other had 192: wherefore in
laying down again of the 192, there was put
down ¼, that is, 144, and so had he left but 48.
Also of the 108, there was laid down 36, that is ⅓,
& so he had left 72. Then I put together 144 & 36,
and it maketh 180, which I part into two parts
even, and so commeth 90 to be given to each of
them: which summe put to 72 maketh 162; and
joyned to 148, it maketh 238: and now I doubt
how I shall go forward.

Master. You need not to take but one of them, Note
which you list, the greater or the smaller, for all
commeth to one purpose: and so may you compare
it that you take to any of the other summes, re-
membring that you make comparison to the same
in the second works: as for example of the first
part. If you compare 138 with the lesser summe
one, that is, 120, so is it 18 too much, and if you
compare it with the greater summe, then is it 42
too little. Again, if you compare 162 to the
greater summe, the errour will be 18, as it was
in the other: but it will have a contrary signe: and
if you compare it with the lesser summe, it will be
42 too much: so that the errour both wayes is
either 18 or 42: & as for the signes it little forceth,
for in them is nothing considered here, but like-
nesse and unlikenesse, which in this case doth neither

Z　　　　　further

further oz hinder : But now go on with the worke.

Scholar. If it be so, then am I out of my greatest doubt. Then I joyne that 90 (which I found as the half of the latter partition) unto 48, which is left with the one man and so hath he 138, which (I may say) is 18 too many, foz the least should be but 120 that errour do I note, and then make a new position, supposing the one man to have 204, and the other to have 96, wherefoze of the 204, there must be laid down 153, and so remaineth with him 51. Also of the 96, there must bee laid down ⅓, that is 32, and so resteth with that man 64: Now of the 153 and 32, I make one summe, as 185, which I must divide into two equall parts, and so each man shall have 92½, whereunto if I adde their former pozitions reserved, then the one shall have 156½, and the other hath 143½ Wherefoze take the lesser summe now again, as I did befoze, that is, 143½ and find that he hath to many by 23½, foz he should have but 120, and so have I foz my two positions two errors, which I set down as here may be seen, each error under his position, and then by the Rule I doe multiply in crosse wayes 108 by 23½, and there riseth 2538, which I note, then again I multiply 96 by 18, and thereof amounteth 1728.

108   96

X

18 †   23½

Now because the signes are both like, that is, both too many, I must worke by Subtraction, and so abating 1728, out of 2538, there will rest foz the Dividend 810 : then foz the Divisoz I subtract 18 out of 23½, and there remaineth

eth 5 ½, by which I divide 810, and the quotient will be 147 $\frac{3}{11}$, which is the juſt portion of him that had the leaſt ſumme. And if I doe ſubtract it out of 300, being the totall ſumme, then will there remain 152 $\frac{8}{11}$, as the poꝛtion that the other did get.

Maſter. Foꝛ the pꝛoofe of this woꝛke, you may chuſe whether you will examine thoſe numbers according to the foꝛme of the queſtions oꝛ elſe woꝛke by other two poſitions, foꝛ to finde the ſecond number : and if thoſe poſitions bꝛing the ſame numbers that did amount by the firſt two poſitions, then doth each woꝛk confirm other.

Scholar. By your patience, I will pꝛove both wayes, not onely to ſæke their agræement but alſo to accuſtome my minde to thoſe woꝛkes : foꝛ I perceive it is exerciſe that muſt be the chiefe engraver of theſe Rules in my memory.

Maſter. You conſider it well : then go to.

Scholar. Firſt, I will by two other poſitions try to finde the portion of him which had moſt.

Maſter. Although you may doe it with any poſitions, yet to ſee the agræement of your woꝛke the better, take the ſame poſitions that you did befoꝛe, comparing them now to the greater, as you did befoꝛe unto the leſſer.

Scholar. Then I ſuppoſe that he that had moſt, had 192, ſo had the other 108. Now if I take ¼ out of 192, that will be 144 and there will reſt to that man but 48. And from the ſecond which had 108, if I take ⅓, that is 36, there will remain to him 72 : then joyning 144 with 36, it will make 180, the halfe whereof being 90. If I adde to

each

each of thofe two mens portions remaining with
them, the one fhall haue 138, and the other 162,
of which two I take the greater (that is 162) and
fée it to be 18 too few; for it fhould be 180, that
errour I note under this pofition. Then for the
fecond pofition I take (as I did before) 204 for
the one, and fo refteth 96 for the other: then take
I $\frac{1}{4}$ of 204 and it will be 153, and there refteth to
him 51. Alfo of the 96 I take $\frac{1}{3}$ that is, 32, and
there remaineth to him 64; now put I that 32 to
153, and it yéeldeth 185: which being parted in
equall values, maketh 92$\frac{1}{2}$ to be added to each
mans remainder, and fo the one hath 143$\frac{1}{2}$, and the
other 156$\frac{1}{2}$: wherefore I take the greateſt fumme,
and it is 23$\frac{1}{2}$ too little, that do I note alfo, and
fet both thefe errours under their pofitions, as in
this Example following doth appeare.

And then multiplying 192 by 23 $\frac{1}{2}$, there doth
arife, 4512.

Again, I multiply 204 by 　　　192　　　204
18, and it maketh 3672,
which I doe fubtract out of
4512, becaufe the fignes be　—18　　—23$\frac{1}{2}$
like, and there refteth 840 for the dividend, then
fubtracting 18 out of 28$\frac{1}{2}$ there will remain 5$\frac{1}{2}$,
which I muft take for the Divifor. And fo divi-
ding 840 by 5$\frac{1}{2}$, the quotient will be 152$\frac{8}{11}$, whereby
I haue found an agréeable fumme to that which I
found by the former pofitions, for him that had moft,
which I doe fubtract out of 300, that is the totall,
there will reft 147$\frac{3}{11}$, which was the portion of
him that had the leaft part.

Maſter. So by divers pofitions, you fée, that one
　　　　　　　　　　　　　　　　　　　doth

doth confirme the worke of the other. Now examine thoſe two numbers by the forme of the queſtion, and ſo ſhall you proue your worke good alſo.

Scholar. If that he which gat moſt had 152 $\frac{6}{11}$,
then muſt he lay down $\frac{3}{4}$ of this ſumme, that is
114 $\frac{6}{11}$, and ſo ſhall remain with him but onely
38 $\frac{1}{10}$ The other which had leaſt, that is 147 $\frac{1}{11}$,
muſt put down of his ſumme $\frac{1}{3}$, that is 49 $\frac{1}{11}$, and ſo
doth there remain with him yet 98 $\frac{3}{11}$. Then do
I adde together 114 $\frac{6}{11}$, and 49 $\frac{1}{11}$, and it will make
163 $\frac{7}{11}$ which I muſt part into equall parts, and
that will be 81 $\frac{9}{11}$, to be giuen to each of them :
putting 81 $\frac{9}{11}$ vnto 38 $\frac{1}{11}$, there doth amount 120
iuſt, which is the true Portion of him that ſhould
haue the leſſer ſumme : and adding 81 $\frac{9}{11}$, 98 $\frac{3}{11}$, the
totall will be 180, the true portion of the other.
And ſo is the work by this proofe alſo tryed to be
good. And this I marke by the way, that in their
ſcrambling, he got moſt (as it chanceth often) that
ought to haue had leaſt by juſt partition.

Maſter. Let your ſtudy be to learn truth and
juſt Art of proportion, and to diſtribute and part
according thereunto, as often as occaſion ſhall be
miniſtred. And here would I make an end of this
Rule, ſaue that I remember one pleaſant queſtion
which I cannot ouerpaſſe, which I will declare
ſomewhat largely, becauſe you ſhall as well vnder-
ſtand ſome reaſon in the pleaſant inuention, as
apt proceeding in the witty working thereof.

Hiero *King of the Syracuſans in Sicilia had cauſed* The tenth
to be made a Crown of Gold of a wonderfull weight, example
to be offered for his good ſucceſſe in wars : in making of Gold
whereof the Goldſmith fraudulently took out a certain and Silver.

portion

portion of Gold, and put in Silver for it, so that there was nothing abated of the full weight, although there was much of the value diminished.

Which thing at length being uttered (as no evill can alwayes lie hid) the King was sore moved: and being desirous to know the truth without breaking of the Crown, proposed the doubt to Archimedes, unto whose wit nothing seemed unpossible, which although presently he could not answer unto, yet he had good hope to devise some policy for that invention, and so musing thereon, as he chanced to enter into a Bain full of water to wash him, he observed, that as his body entred into the Bain, the water did runne over the Tub, whereby his ready wit of such small effects conjecturing greater works, conceived by and by a reason of solution to the Kings question, and therefore rejoycing exceedingly, more then if he had gotten the Crown it selfe, forgat that he was naked, and so ranne home, crying, as hee ranne, ἔυρηκα, ἔυρηκα, I have found, I have found. And thereupon caused two massie pieces, one of Gold, and another of Silver, to be prepared, of the same weight that the said Crown was of : and considering that Gold is heavier of nature then Silver, and therefore Gold of like weight with Silver, must needs occupie lesse roome, by reason it is more compact and sound in substance, he was assured that putting the masse of Gold into a vessell brimfull of water, there would not so much water runne out, as when he should put in the silver masse of the like weight. Wherefore he tried both, and noted not onely the quantities of the water at each time, but

but alſo the difference or exceſſe of the one above the other, whereby he learned what proportion in quantity is between Gold and Silver of equall weight.   And then putting the Crown it ſelf into the veſſell of water brim-full (as before) marked how much water did run out then, and comparing it with the water that ranne out when the Gold was put in, noted how much it did exceed that: and likewiſe comparing it to the water that ranne out of the Silver, marking how much it was leſſe then that, and by thoſe proportions found out the juſt quantity of Gold that was taken out of the Crown, and how much Silver was put inſtead of it: but ſeeing Vitruvius which writeth this Hiſtory, doth not declare the particular worke of this tryall, it ſhall be no inconvenience to ſuppoſe an example for declaration ſake, wherein although the true juſt proportion be not expreſſed, yet the forms of tryall ſhall be truly ſet forth.   And for an example, I ſuppoſe the weight of the Crown to be 8 pound, and ſo of each the other two Maſſes. And when the Maſſe of Gold was put into the water, I imagine that there ranne out two pound of water: and when the maſſe of Silver was put in, I ſuppoſe there ran out 3 pound $\frac{1}{2}$. Againe, when the Crowne was put in, there ran out two pound $\frac{1}{4}$: Now to know what quantity of Silver was in the Crown, work by the Rule of falſe poſition, and imagine that there was two pound of Silver, then muſt there be ſix pound of Gold, then ſay thus by the Rule of Proportion.   If eight pound of Gold do expell two pound of water, what ſhall ſix pound expell? and it will be 1 pound $\frac{1}{2}$.   Againe, for the Silver ; if eight pound of Silver expell three pound $\frac{1}{2}$ of water, what shall two pound of Silver put out? it will be $\frac{7}{8}$, now adde those two weights of water together, and they will make two pound $\frac{1}{8}$, and it should be by the supposition two pound $\frac{3}{4}$, so is it too much by $\frac{1}{8}$.

Scholar. Now do I understand the worke as I thinke, therefore I pray you let me worke the rest of the question. And because this first supposition did erre, I note that position & his errour, & take a new position, esteeming the Silver to be but one pound, so must there be in gold 7 pound. Then say I, if eight pound of Gold do yeeld two pound of water, what shall seven pound yeeld? and it will be 1 pound $\frac{3}{4}$. Againe, if 8 l. of Silver expell 3 pound $\frac{1}{2}$ of water, what shall 1 pound expell? and it will be $\frac{7}{16}$. Now must I adde those two summes together, and they make two pound $\frac{3}{16}$, and they should make 2 pound $\frac{3}{4}$, so is it too litile by $\frac{9}{16}$. Therefore I set the positions with their errors in order as here followeth: And then I multiply in crosse wayes 2 by $\frac{9}{16}$ and it maketh $\frac{9}{8}$: Likewise 1 multiplyed by $\frac{1}{8}$ — maketh $\frac{1}{8}$. And because the signes be unlike, I must adde these two summes which make $\frac{1}{4}$: and that is the dividend.

2　　1

X

$\frac{1}{8}$ †

Againe, I must adde $\frac{1}{8}$ to $\frac{9}{16}$, and it will be $\frac{11}{16}$ that is the Divisor. Now I shall divide $\frac{1}{4}$, by $\frac{11}{16}$ and the quotient will bee $\frac{16}{12}$, that is, 1 $\frac{1}{3}$: whereby I know that there was but 1 pound and $\frac{1}{3}$ of Silver into the Crowne, and so much Gold taken out for it.

Master. Prove it now by examination, according to the question.

Scholar. If there were 1 pound $\frac{1}{3}$ of Silver, then was there of Gold 6 pound ⅔. Now ſay I by the Rule of proportion: if 8 pound of Gold expell two pound of water, what ſhall 6 pound ⅔ expell, it will be 1 pound ⅔.

$$8 \diagdown 2 \qquad 6\tfrac{2}{3} \diagdown 1\tfrac{2}{3}$$

$8 \diagdown 3\tfrac{1}{2}$ Againe, if 8 pound of Silver expell $1\tfrac{1}{3} \diagdown \tfrac{-7}{11}$ three pound ½ of water, what ſhall 1 ⅓ expell? It will bee $\tfrac{-7}{12}$. Now muſt I adde together 1 pound ⅔ and $\tfrac{-7}{12}$, and they will make 2 pound $\tfrac{-2}{36}$, that is, 2 pound ¼, according to the ſuppoſition of the queſtion: whereby I perceive the work to be well done. And I cannot but much rejoyce of this excellent invention, ſo my deſire is kindled vehemently to be perfectly inſtructed in every part thereof, & namely in this point, whether the proportion between water and Gold be ſuch that for 8 pound of Gold put into a veſſell full of water, there ſhall runne out two pound of water, and for as much Silver, whether 3 pound ½ of water would abide.

Maſter. I perceive your meaning, and conjecture your imagination to be thus, that if you knew the exact proportion between Gold and Silver, and water, both in their weight and quantities, then could you eaſily finde out the mixtures of them, which thing I have reſerved for another worke that intreateth of ſuch matters eſpecially. And at this time you muſt conſider that you learn Arithmeticke, which intreateth of the manner to ſolve doubtfull queſtions touching number, without regard what matter is ſignified by that number: elſe were it neceſſary in Arithmeticke, to teach all Arts, ſeeing in it may be moved queſtions of all Arts.

But ſeeing you are ſo deſirous to know theſe things,

A queſtion
of the pro-
portion of
gold, ſilver,
and quick-
ſilver, un-
to water.

things, I will tell you in such a sort, that you shall practise your Art in finding it, and propose it in forme of a question. Gold beareth a greater proportion to water then Silver doth, and their two proportions be in proportion together, as 48 to 25. But to help you somewhat in this Riddle, you shall note that the proportion of Quick-silver unto water, is the just middle number proportionall in progression Geometricall between the proportion of Gold and Silver unto water.

And this proportion is $\frac{2\,2\,\circ}{2\,1}$. Now if you will know the juſt numbers of theſe 3 proportions, then muſt you finde out 3 numbers in Progreſſion Geometricall, whereof the middlemoſt muſt be $\frac{2\,2\,\circ}{2\,1}$, & the firſt muſt be unto the laſt, as 25 to 48. And thus I will leave you to finde thoſe numbers when you be at leiſure.

Scholar. Yet Sir, I thank you heartily for thus much, for now I ſee the poſſibility to finde them out. Howbeit, becauſe this queſtion ſeemeth ſtrange, if it might pleaſe you to inſtruct me ſomewhat in the order of working for it, I ſhould the more eaſily finde the true working.

Maſter. You deſire too much if you will ſtudy for nothing: Therefore to occaſion you to ſtudy the better, I will leave this doubt wholy to your own ſearch: But as touching the generality of the Rule, Archimedes needed not to take two Maſſes of gold and Silver equall in weight with the Crown, for the proportion might as well be found in any other weight, yea, although the Maſſe of Gold were of one weight, and the Maſſe of Silver of another. As for example: if the Crown were of 8 pound weight as I did ſuppoſe, and I have not

ſo

so much other fine Gold, but onely one pound, and trying that by water, and finding that it doth expell but $\frac{1}{4}$ of an ounce of water; yet then by it I may inferre, that 8 pound of Gold would expell 6 ounces of water. And likewise of Silver, whereof if I had but two pound, and finde that it doth expell thrée ounces of water, then might I affirme, that 8 pound would expell 12 ounces, that is, one pound weight : and so is it good as if the thrée Masses were all of one weight. And thus for this time I will make an end of this other part of Arithmeticke.

Scholar. Although I cannot sufficiently thank you for this, yet your promise made me to looke for the Art of Extraction of Roots, whereof hitherto I have learned nothing.

Master. I will not breake my promise, but intend (God willing) to performe it within this thrée or foure moneths, if I perceive this my paines to be well taken in the mean season. And you shall not repent the tarrying for it : for it shall be increased by the tarrying. And in the mean time you shall take this Addition, not for the second part of Arithmeticke which I promised, but for an augmentation of the first part, unto which I would have annexed the Extraction of Rootes square and cubie, namely for Examples of the Statute of Assise of Wood, but that in the second part I must write of divers other Roots, and thought it best to reserve those Rules also with their Examples unto the same second part.

Scholar.

Scholar. Sir, although I cannot recompence your goodnesse, yet I shall alwayes doe mine endeavour to occasion you not to repent your benefit on me thus imployed.

Master. That recompence is sufficient for your part.

FINIS.

The third Part,

OR,

Addition to this BOOKE,

Entreateth of briefe Rules, called
Rules of Practife, of Rare, Pleafant,
and Commodious effects abridged into
a briefer method then hitherto hath
been publifhed.

With divers other neceffary
Rules, Tables, and Queftions, not
onely profitable for Merchants, but
alfo for Gentlemen, and all other Occupiers whatfoever, as by the Con-
tents of this Booke may
appear.

Set forth by JOHN MEILIS,
School-mafter.

The first Chapter of this Addition entreateth of brief Rules, called Rules of Practife, whith divers neceffary queftions, profitable not onely for *Merchants, but alfo for all other Occupiers whatfoever.*

THE working of Multiplication in practife, is no other thing then a certain manner of **Multiplying** of one kinde by another: whereupon is broght forth, the Product of the propofed number, which is accomplifhed by the meanes of Divifion in taking the *halfe*, the *third*, the *forth*, the *fifth*, or fuch other parts of the fumme, which is to be multiplied.

And for the better underftanding of fuch converfions, you fhall underftand that in the manner and ufe of thefe Rules of Practife, you ought firft to know the even or aliquot parts of a fhilling, which in this Table following doth appear.

$$Item \begin{cases} 6 \\ 4 \\ 3 \\ 2 \\ 1 \end{cases} pence\ is\ the \begin{cases} \frac{1}{2} \\ \frac{1}{3} \\ \frac{1}{4} \\ \frac{1}{6} \\ \frac{1}{12} \end{cases} of\ a\ fhilling.$$

Wherein as you fee according to the order of thefe rules of Practife: at 6 pence the yard of any thing, you muft take $\frac{1}{2}$ of your number which is to be multiplied, and the product that commeth thereof fhall

be

be ſhillings, if any unite doe remain it is 6 pence.

For 4 d. take the $\frac{1}{3}$ of the number that is to be multiplied, and the product alſo produceth ſhillings, if any unites doe remain, each one ſhall be worth in value 4 pence. The like is to be underſtood of the other 3, &c.

I *Example.*

At 6 d the yard, what   379 yards ?
                          ‾‾‾‾‾‾‾‾‾‾‾‾‾‾
                          189 ſ ————— 6 d

II

At 4 d the yard, what   1 0 4 yards ?
                         ‾‾‾‾‾‾‾‾‾‾‾‾‾‾‾‾
                          3 4 ſ ———— 8 d

III

At 3 d the yard, what   5 0 1 4 yards ?
                          ‾‾‾‾‾‾‾‾‾‾‾‾‾‾‾‾
                          1 2 5 3 ſ ——— 6 d

IV

At 2 d the yard, what   5 3 2 yards ?
                          ‾‾‾‾‾‾‾‾‾‾‾‾‾‾
                          8 8 ſ ———— 8 d

V

At 1 d the yard, what   4 0 9 ?
                          ‾‾‾‾‾‾‾‾‾‾‾‾
                          3 4 ſ ———— 1 d

Here you may ſee in the firſt example, that 379 yards at 6 d. the yard, are worth 189 ſ. 6 d in taking the $\frac{1}{2}$ of 379. And in the ſecond example the 104 yards at 4 pence the yard, are worth 34 ſ, 8 d, in taking the $\frac{1}{3}$ of 104. Likewiſe in the third example, 5014 yards at 3 d the yard, bringing forth 1253 s. 6 d in taking the $\frac{1}{4}$ of 5014. Alſo in the fourth Example at 2 d the yard, maketh 88 s. 8 d.

And laſtly, in the fifth Example : 409 yards at 1 d the yard amounteth to 34 s. 1 d. in taking the $\frac{1}{12}$ of
                                                    409. and

409. and so is to be done also of all other questions the like, when the number of the pence is any of the even or aliquot parts of 12 d.

Item, to bring the Products of these shillings, and all other the like into pounds is very easie in dividing of it in your minde by 20. for it is to be understood that as often as 20 is found in that Product, so many pounds doth it contain : which with facility to performe, alwayes strike off the figure towards your right hand, with a right down dash of your pen, for the 0 that appertaineth to the 20. And then begin at the left hand, in taking the half off the rest. And if that at the last any unite doe remain, the same shall be joyned with the figure that is cut off which shall represent the odde shillings contained in that work.

As for example, in your third question at 3 d the yard, which amounteth to 1253 s. 6 d. the Product whereof maketh, 62 li. 13 s. 6 d. as here you may see is easily performed by this Example.

$$\begin{array}{c} 1 \\ 12\,5\,|\,3 \\ \hline 62\text{--}13\text{——}6 \end{array}$$

Also for the working of one peny the yard, it is something harsh and hard to take the $\frac{1}{12}$ of some Products : therefore to ease that hard worke, you shall first bring your delivered summe into groats by taking $\frac{1}{4}$ part of the product, and if any unites remain of that $\frac{1}{4}$ part as sometimes there may they are pence : and must be signified with a line from the groats with their title of *pence* ; and because that 60 groats maketh a *Pound* or twenty shillings, *strike* off the first figure toward your right hand, for the 0 that appertaineth to 60 (as you did even now for the 0 that belongeth to 20 :) Then in taking the $\frac{1}{3}$ of that product, if there doe remain any unites, the same shall

A a   you

you joyne with the figure that you cut off, efteem-
ing them as *groats*, which keep in your minde, and by
taking the ½ part of them, you fhall turn them into
fhillings, and fo have you done : *As for Example*, by
a Queftion or two hereafter propofed, fhall more
plainely by the worke appear.

At 1 d, the yard, what ‖ 5 4 3 6 8 yards ?

1 3 5 9 2 *groats.*

½ li———— 2 2 6-10 s.-8 d.

Here in taking the ½ part of 1 3 5 9, in coming to
the laft worke, the ⅔ part of 39 being taken, the *remainder* is 3, which joyned with the two that was
cut off, maketh 32 *groats*, which converted into *fhillings*, by taking the ⅓ part, maketh as appeareth 10
fhillings 8 d. Many other wayes there are, but none
more apt for a *young learner* to underftand then this:
wherefore this one way well impreffed in memory is
better then 20 wayes doubtfully underftood.

At 1 peny the yard, what ‖ 4 5 3 3 yards ?

1 1 3 | 3 *groats*-1 d

½ li. 1 8 ————17-9 d

At 1 peny the yard, what ‖ 6 4 7 6 8 yards ?

1 6 1 9 | 2 *groats*

li 2 6 9 -- 1 7-4 d

*N*Ow *followeth alfo to be underftood that if the
number of pence be not an aliquot part of 12,
you muft reduce them into fome aliquot part of 12 :
and after the aforefaid manner, you fhall make of them
two or three Products, as need fhall require, and adde
them together into one fumme. And here for thy
furtherance appeareth a note of the order of their
parts,*

parts, as they are to be taken.

$$\text{for pence} \begin{cases} 5 \\ 7 \\ 8 \\ 9 \\ 10 \\ 11 \end{cases} take \begin{cases} 3 \\ 4 \\ 4 \\ 6 \\ 6 \\ 6 \end{cases} and \begin{cases} 2 \\ 3 \\ 4 \\ 3 \\ 4 \\ 4 \& 1 \end{cases} or \begin{cases} 4 \\ 6 \\ 6 \\ 4.4 \\ 4.4 \\ 4.4 \end{cases} \& \begin{cases} 1 \\ 1 \\ 2 \\ 1 \\ and 2 \\ 3 \end{cases}$$

Here in the firft note of this *Table* at 5 d. you fhall firft take for 3 d the $\frac{1}{4}$ of the number that is to be multiplyed : and likewife for 2 d. the $\frac{1}{6}$ of the fame number, adding together both the Products : But if you will worke by 4 and 1, you muft for 4 d, firft take $\frac{1}{3}$ of the number that is to be multiplied : and for 1 d, take the $\frac{1}{12}$ of the whole fumme, or rather, which is better, for 1 peny you may take the $\frac{1}{4}$ of the product which did come of the 4 pence : becaufe that 1 d, is the $\frac{1}{4}$ of 4 pence. The totall fummes of thefe two numbers fhall be the folution to the *Queftion.* And in like manner it is to be done of all others, as by thefe Examples following fhall appear.

I

At 5 d the yard, what	758 yards ?
3 d	189 ——— 6 d
2 d	126 ——— 4 d
fhillings,	325 ——— 10 d

Otherwife,

At 5 d the yard what	758 yards ?
4 d	252 ——— 8 d
1 d	63 ——— 2 d
fhillings,	315 ——— 10 d

A a 2   II

I I

At 7 pence the Ell, what   563   Elles ?
4 d   187————— 8 d
3 d   140——————9 d
fhillings,   328————— 5 d

I I I

At 8 d the pound, what   112 pound ?
4 d   37———————4 d
4 d   37———————4 d
fhillings,   74————————8 d

Otherwife,

At 8 d the pound, what   112 pound ?
6 d   56——————— 0
2 d   18————— 8
fhillings,   74—————— 8 d

I I I I.

At 9 pence the Elle, what   356 Elles?
6 d   178———— 0
3 d   89———— 0
fhillings   267——— 0 d

V.

At 10 pence the peece, what   795 peeces ?
6 d   397———— 6
4 d   265————— 0
fhillings   662————— 6

V I.

VI.

At 11 pence the pound, what	7576 pounds?
6 d	3788 ——— 0
4 d	2525 ——— 4
1 d	631 ——— 4
	6944 ——— 8 d
	Pounds 347 — 4 s — 8 d

1 Here in this first example, where it is demanded (at 5 d the yard) what will 758 cost? First, for 3 d, I take the $\frac{1}{4}$ of 758: and thereof commeth 189 s. 6 d: Then for 2 d I take the $\frac{1}{6}$ of the same 758, which amounteth to 126 s. 4 d, these two summes added together, doe make 315 *shillings* 10 *pence*: and so much are the 758 *yards* worth at 5 d the *yard*.

Item also for the same again: First for 4 d, I take the $\frac{1}{3}$ of 758; and thereof commeth 252 s. 8 d: then for 1 peny I take the $\frac{1}{4}$ of the same 758, that is to say, of 252 s, 8 d, and it yeeldeth me 63 s. 2 d: which both added together maketh 315 s — 10 d, as before.

2 *Item, for* 7 d *there is taken the* $\frac{1}{3}$ *and the* $\frac{1}{4}$ of the whole summe which is to be multiplyed, and adde them together, that is to say, first, for 4 pence there is taken $\frac{1}{3}$ of 563: which comes to 187 s. - 8 d, as appeareth by the worke, and for 3 d there is taken the $\frac{1}{4}$ of the whole summe, which amounteth to 140 s — 9 d. Both which products added together, doe make 328 s. — 5 d. and so much comes 563 *Elles* to at 7 d the *Elle*.

3 *Item, for the first* 8 d. *there is taken for* 4 d. the $\frac{1}{4}$ of the whole summe, and another $\frac{1}{4}$ for the other 4 d. which added together, as in the example doth

 evidently

evidently appear, amounteth to 74 s.-8 d.

Again, for the fecond work of 112 li. there is taken firft the ½ of the whole fumme for 6 d. which comes to 56 s. then for that 2 d. you have to take ⅓ of the whole fumme, or if you will, the ⅓ of the product that came of 6 d. either of which maketh 18 s.-8 d. thefe two fummes being added together doe make 74 s. 8 d. as in the third example appeareth.

4 Item, for 9 d. there is taken for 6 pence the ½ of the whole fumme, and the ¼ of the whole fumme for 3 d. or otherwife for the 3 d. you may take the ½ of the product that came of 6 d. becaufe 3 pence is the ½ of 6 d. which added together as plainly appeareth in the fourth example, amounteth to 267 s. 0 d.

Item for 10 d. firft there is taken for 6 d. the ½ of the whole fumme, which amounteth to 397 s.-6 d. Then for 4 d. there is found 265 s. both which added together, make 662 fhillings, 6 d. as appeareth in the fifth example : It may alfo be wrought, as appeareth by the fecond note in the Table, by 4 d. twice taken, and the ½ of the product of 4 d. or elfe by the ⅔ of the whole fumme, &c.

Item, for 11 d. there is firft taken the ½ for 6 d. then the ⅓ of the whole fumme for 4 d. Laftly, the ¼ of the laft product for 1 d. All which 3 fummes added together, maketh in fhillings 6944 s. 8 d. and in pounds 347-4 s-8 d.

Item, likewife by the fame reafon, when you will multiply (by fhillings) any number that is under 20. you fhall have in the Product pounds, if you know the even or aliquot parts of 20. which are here in this little Table fet down to fight.

Item,

Item, s. $\left\{\begin{matrix}10\\5\\4\\2\\1\end{matrix}\right\}$ is the $\left\{\begin{matrix}\frac{1}{2}\\ \frac{1}{4}\\ \frac{1}{5}\\ \frac{1}{10}\\ \frac{1}{20}\end{matrix}\right\}$ of one pound.

So that for 10 s. *which is the* $\frac{1}{2}$ *of a pound*, you may take the $\frac{1}{2}$ of the number which is to be multiplyed, and you shall have in your product pounds, if an unite doe remaiñ, it shall be worth ten *shillings.*

Likewise for 5 *shillings you must take the* $\frac{1}{4}$ of the number which is to be multiplied, and if there do remain any *unites*, they shall be fourth parts of a *Pound*, every unite being in value five *shillings.*

For 4 *shillings take the* $\frac{1}{5}$ *of the number* which is to be multiplied : and if there doe remain any *unites*, they shall be fift parts of a *pound*, each unite being in value 4 *shillings.*

For 2 *shillings you must take the* $\frac{1}{10}$ *of the* number to be multiplied, wherefore to take the $\frac{1}{10}$ of any number, you must cut off the last figure of the same number (which is nearest your right hand) from all the other figures with a small right down line or dash with a Pen, and so have you done : for all the other figures which doe remain toward your left hand from the same figure that you do separate, shall be *pounds* : and that figure so separated towards your right hand, shall be so many pieces of 2 s. the which figure you must double to make thereof the true number of shillings, as by the Example shall appear.

Finally, for 1 *shilling needeth small work.*, for it is so many shillings as be proposed in the summe, which to bring into pounds, hath been already taught in the *first rule.*

A a 4   *Example.*

Example.

At 10 ſ. the piece, What	6543 pieces?
½ li.	3271 10 ſ.
At 5 ſ. the Elle, What	4373 Elles?
¼ li.	1093 5 ſ.
At 4 ſ. the yard, What	7839. yards?
⅕ li.	1567. 16 ſ.
At 2 ſ. the pound weight, What	752│7 pound
1/10 li.	752 — 14 ſ.
At 1 ſ the piece, What	775│3 pieces?
1/20 li.	387 13 ſ.

Next followeth in order to be underſtood, that if the number of ſhillings be not ſome even or aliquot part of 20, you muſt then convert the ſame number of ſhillings into the aliquot parts of twenty, and thereof make two or three products as need ſhall require: which done, adde them together, and bring them into Pounds. And here for thy furtherance, I have ſet down a note of the order of their parts, as they are to be taken.

ſ					ſ	
3	2 ¢ 1			13	10. 2 ¢ 1
6	4 ¢ 2	5 ¢ 1		14	10. ¢ 4
7	5 ¢ 2			15	10. ¢ 5
8 } of	4 ¢ 4 } 0?	5.2.1	16 } of	10. 5.1
9	5 ¢ 4	4.4.1		17	10. 5.2
11	10 ¢ 1			18	10. 4.4
12	10 ¢ 2			19	10. 5. 4

For 3 s. according to the tenure that you ſee is expreſſed in the Table, you muſt firſt take for 2 ſhilling
th

the $\frac{1}{10}$ of the number that is to be multiplied. Then for one *fhilling* you muft take the $\frac{1}{2}$ of the product which did come of the fame $\frac{1}{10}$ part : which two fummes added together produceth the effect defired.

Item, for 6 fhillings according to the note fet forth in the Table, firft for 4 f. I take the $\frac{1}{5}$ of the number that is to be multiplied : Then for 2 s. the $\frac{1}{2}$ of the product that came of 4 s. and adde them together.

Or elfe as appeareth alfo in the Table, for 5 fhillings you may take the $\frac{1}{4}$ and the $\frac{1}{2}$ part of the product that came of 5 *fhillings*, and adde them together.

Item, for 7 s. firft for 5 s. take $\frac{1}{4}$ of the product that is to be multiplied, then for 2 s. take the $\frac{1}{10}$ of the number that is to be multiplied, and adde them together, *&c.*

Item, for 8 s. according to reafon, and the intent of the *Table*, for the firft 4 s. take the $\frac{1}{5}$ of the product, and the fame number again for the other 4 s. and adde them together.

Item, for 9 fhillings : firft for 5 fhillings, take *the* $\frac{1}{4}$, then for 4 fhillings take the $\frac{1}{5}$: and adde them together.

Otherwife, as you fee by the intent of the Table, work twice for 9 fhillings as was taught even now for 8 : and then take the $\frac{1}{4}$ of the laft product for the 1 fhilling : but 5 and 4 is the fhorter.

Item, for 11 s. firft difpatch 10 s. for which you muft take the $\frac{1}{2}$ of the product : then laftly, for 1 fhilling take the $\frac{1}{10}$ part of the fumme produced of the $\frac{1}{2}$ of the product, and adde them together.

Item, for 12 fhillings, where I will end with the firft part of my Table. For take the $\frac{1}{2}$ for 10 fhillings.

And

And then for 2 *fhillings*, take the $\frac{1}{5}$ of the fumme tha came of 1 **o** *fhillings*, take and adde them together, or elfe if you pleafe for 2 *fhillings* you may take the $\frac{1}{10}$ of the whole given number.

To write more of the manner of taking the true parts, I omit. The defirous practitioners will (no doubt) conceive it. Alfo the *Table* is fome aid to help the unperfect, whereupon by and by I will fet down three or four of thefe notes in *Examples*, and the reft I will leave to thine own induftry and practice, to labour upon.

This is the order moft commonly ufed in practice, when the number of fhillings is not an *aliquot* part of a pound. But (*loving Reader*) after I have touched the even or *aliquot* parts of a pound that falleth out in pence and fhillings, I will deliver two new Rules that fhall drown this common order quite and clean : wherein fhall be comprehended in one line, or working both of even and odde parts of fhillings under 20 without regard whether it be an *aliquot*, or not an *aliquot* part ; which two Rules (when they come in place) I commit to thy friendly judgement in working.

Now follow the examples upon the notes afore faid.

At 6 fhillings the yard, what	3 2 1 5 yards
4 fhillings	643
2 fhillings .	3 2 1 ———— 1
li.	9 6 4 ———— 10

Other

Otherwife by Multiplication of 6.

		3215	
6 fhillings		1929|0	
li———964—	10 fhillings.
At 7 fhillings the Ell, what		4563 Ells ?
5 fhillings		1140——15
2 fhillings		456——6
li.	1597———1 fhilling.

Otherwife by Multiplication of 7.

	4563	
7 s	3194|1	
	1597———1
At 8 s. the piece, what	7563 pieces ?
4 s	1512——12
4 s	1512——12
pounds	3025——4 s.

Otherwife by Multiplication of 8.

	7563	
8 s	6050|4	
pounds   3025———————4 fhillings.
At 13 s. the piece, what	401 pieces ?
10 s	200———10
2 s	40——2
1 s	20——1
pounds	260——13

Other-

Otherwife by Multiplication.

$$
\begin{array}{r}
4 0 1 \\
\hline
1 2 0 3 \\
4 0 1 \\
\hline
5 2 1 3
\end{array}
$$

13 s.

pounds   260——13 s.

Thefe and fuch like queftions of compound number, which I have here in this fourth rule for orders
fake fet down, for that it hath been heretofore a common courfe of work, I account but fuperfluous. For
in the eight and nineth Rules of this my fimple Addition fhall appear, that the given price or any even or
odde number of fhillings, either under or above 20
fhall be wrought at one or two workings at the moft,
how difficult foever the queftion be.

3 Rule.

To reduce pence into pounds at one opera-tion.

Item, *there refteth yet a kinde of practife, how to
bring pence into pounds at the firft working, whereupon you muft underftand that 240 pence maketh one
pound, or 20 s. In confideration whereof I cut off the
laft figure or 0, and there remaineth but 24 (of which
24) 8 d. is the ⅓ part thereof, 6 d. is the ¼ part, 4 d
is the ⅙ part, and 2 pence is the ¹⁄₁₂ part thereof.*

Whereupon if it were demanded what 1486 yard
or pounds of any thing commeth to, at 8 pen
the yard; in pricking or cutting off the firft figu
towards your right hand, for the 0 that appertaine
240. There is remaining of the faid fumme 14
whereout I taking the ⅓ part, and it commeth to 4
li. and there refteth 1, which 1 I put to the 6. that
prick or cut off, and it maketh 16 pieces of 8 pen
which I double to make into groats, & they make 3
where

whereof the ½ part maketh 10 s. and there remaineth
s. which is 8 d, whereby it followeth, that the 1486
yards at 8 pence the yard maketh 49 li. 10 s. 8 d. as
by the example shall appear.

Item, for 6 d. take the ¼ part of the number from the
prickt figure; and if any unites remain, they are so
many six pences, whereof taking the ½ they are shil-
ings, if there doth remain yet one, it is in value six
ence.

Item, for 4 d. take the ½ part of the number from
the prickt figure; if any unites doe remain, they are
so many groats, which to convert into shillings, take
the ¼ part. And if any yet remain, they are thirds of
shillings, each one in value being worth 4 pence.

Item, for 3 pence take the ¼ part from the prickt
figure, if any unites remain, they are so many pieces of
pence, whereof in taking the ¼ part, maketh shillings:
any thing yet remain, they are the fourth parts of
shillings, each one being in value 3 pence,

Item, for 2 pence: as appeareth also by the Table,
take the ⅙ part of the number from the prickt figure:
any thing remain, they are so many pieces of 2
ence, which by taking the ⅙ part, you shall turne into
shillings, and if any unites remain, they are so many 6
arts of shillings, or pieces of two pence, whether you
will.

If one cost 8 pence, what   1486 ?
                                     _____
maketh pounds   49 — 10 — 8 d.

If one cost 6 pence, what   7865 ?
                                     _____
maketh pounds   196 — 12 - 6 d.
                                                  At

At 4 pence the yard, what   8736 yards?

maketh pounds.   145--12--0 d.

If one cost 3 pence, what   9874 worth?

maketh pounds   123--8--6d.

At 2 d. the Ell, what   7894 Ells to?

maketh pounds   65--15--8d.

BVt if your number of pence be not an aliquot or even part of 24, then must you bring them into the aliquot parts of 24. and make thereof diverse products which must be added together, as by the question hereafter following shall appear.

Item, for 5 d. first take for 3 d. then for 2 d. and adde them together, according to the instruction of the *second Rule* : or else first take for 4 d, then 1 d.

Item, for 7 d. first take for 4 d. then for 3 d. and adde them together.

Item, for 9 d. first take for 6 d. then for 3 d. and adde them together.

Item, for 10 d. first take for 6 d. then for 4 d. and adde them together.

Item, for 11 d. first take for 8 d. then for 3 d. and adde them together : as by these Examples.

Examples.

1 If one yard cost 5 d. what   759|6

           4 pence   126—1?

             1   31—1?

maketh pounds   158———?

Other

Otherwiſe.

I ——— 5 ———	759|6	
3 pence	94 ———	19
2 pence	63 ———	6
maketh pounds	158 ———	5 s.

2 If one coſt 7 d. what　9 8|7 worth?

4 pence	16 ——	9	
3 pence	12 — 6 ——	9	
maketh pounds	28 — 15 - 9d.		

Otherwiſe.

I ——— 7 —	9 8|7		
6 pence	24 — 13 ——	6	
1 peny	4 —— 2 -	3	
maketh pounds	28 — 15 — 9d.		

3 If one coſt 9 d, what　9 8|7　worth?

6 pence	24 — 13 -- 6		
3 pence	12 — 6 — 9		
maketh pounds	37 — 0 - 3d.		

Otherwiſe,

I ——— 9 —— 9 8|7 ?			
6 pence	24 — 13 —— 6		
3 pence	12 —— 6 —— 9		
maketh pounds	37 —— 00 —— 3		

4 If one coſt 10 pence, what　9 8|7?

6 pence	24 — 13 — 6		
4 pence	16 — 9 — 0		
maketh pounds	41 — 2 — 6		

5 If

5 If one coſt 11 pence, what   98|7 ?
        8 pence   ——————
        3 pence   32——18--0
maketh pounds   12——-6--9
                45——-6--9

But if you haue any ſhillings and pence to be multiplied together, then are you to take for the ſhillings
according to the inſtruction of the third Rule : and for
the pence according to the firſt Rule before mentioned : unleſſe you can ſpie the advantage thereof,
and thereby helpe your ſelfe ; as appeareth in this
ſecond example, where firſt I worke for 6 d. which is
to be rebated out of the given number, and I have
719 li. 11 s. my deſire.

At 19 s. 6d. the yard, what 738 yards ?

         738   *Otherwiſe by*
10 s.   369——- 0   *Rebating.*
 5 s.   184——10
                  738
 4 s.   147——12   6 d.   18——9 s.
 6 d.   18——9   li——7——19——11s.
pounds 719——————11 s.

The like again is done by Rebating, as by theſe
two examples appeareth.

At 18 s. the Ell, what   418 Ells ?

2 s.   ——————
                 41-————16
pounds   ——————
                376————4s.
At 16 s. the Ell, what   517 Ells ?

4 s.   ——————
                103————8
pounds   ——————
                413————12s.
                    And

And now I will touch a little the even part of a pound, that falleth out in pence and fhillings, whereof for thofe parts you fhall take fuch like parts out of the given number that is to be multiplied, as the price of that given number beareth in proportion to a pound, which alfo for their better aid is here fet down.

$$
\begin{array}{cc}
1 \text{ s.} & 8 \text{ d.} \\
2 & 6 \\
3 & 4 \\
6 & 8
\end{array}
\Big\}
\text{ is the }
\left\{
\begin{array}{c}
\frac{1}{12} \\
\frac{1}{8} \\
\frac{1}{6} \\
\frac{1}{3}
\end{array}
\right\}
\text{ part of a pound.}
$$

Item, firft for 1 fhilling 8 pence take the $\frac{1}{12}$ part of the given number, and if any thing do remain they are twelve parts of a pound, each one being in value 1 fhilling 8 pence.

Item, for 2 fhillings 6 pence, take the $\frac{1}{8}$ part of the number that is to be multiplied ; and if any thing doe remaine, they are eight parts of 1 pound, each one being in value 2 fhillings fix pence.

Item, for 3 fhillings 4 pence, as appeareth by the Table, you muft take the $\frac{1}{6}$ part of the given number, and if any thing do remain, they are 6 parts of a pound, each one being in value 3 fhillings 4 pence.

Item, for 6 fhillings 8 pence take the $\frac{1}{3}$ part of the number that is to be multiplied : And if any unites doe remaine, they are thirds of a pound, every one being worth 6 fhillings 8 pence.

Other infinite numbers there are, that may be reduced by abbreviation into the proportionate parts of a *pound*, as 16 fhillings 8 pence maketh $\frac{5}{6}$: which 16 fhillings 8 pence is eafily reduced into groats, by multiplying 16 by 3, and thereto adde 2, which maketh 50 groats.

B b   Then

Then set 60 the groats of a pound under
50 : cutting off the the two Cyphers as is
here performed.   16—8

And then have you brought 16 *shillings*   3
8 pence into the knowne parts of a *pound*,   5|0
which maketh   6|0

But yet gentle Reader, for thy further instruction,
I have hereunto annexed in a *Table*, how pence and
shillings bear proportion to a pound, which I commit to thy friendly benevolence; it will be some aid
unto the ungrounded Practitioner : but I count him
the best workman that can presently reduce his given
price into the known and proportionate parts of a
pound.

A *Table* of the *Aliquot parts of a pound or* 20 *fhillings.*

s.	d.	l.		s.	d.	l.
0	2	$\frac{1}{120}$		8	4	$\frac{5}{12}$
0	3	$\frac{1}{80}$		8	9	$\frac{7}{16}$
0	4	$\frac{1}{60}$		9	0	$\frac{9}{20}$
0	6	$\frac{1}{40}$		10	0	$\frac{1}{2}$
0	8	$\frac{1}{30}$		11	0	$\frac{11}{20}$
1	0	$\frac{1}{20}$		11	3	$\frac{9}{16}$
1	3	$\frac{1}{16}$		12	0	$\frac{3}{5}$
1	8	$\frac{1}{12}$		12	6	$\frac{5}{8}$
2	0	$\frac{1}{10}$		13	0	$\frac{13}{20}$
2	6	$\frac{1}{8}$		13	4	$\frac{2}{3}$
3	0	$\frac{3}{20}$		13	9	$\frac{11}{16}$
3	4	$\frac{1}{6}$		14	0	$\frac{7}{10}$
3	9	$\frac{3}{16}$		15	0	$\frac{3}{4}$
4	0	$\frac{1}{5}$		16	0	$\frac{4}{5}$
5	0	$\frac{1}{4}$		16	8	$\frac{5}{6}$
6	0	$\frac{3}{10}$		17	0	$\frac{17}{20}$
6	6	$\frac{13}{40}$		17	6	$\frac{7}{8}$
6	8	$\frac{1}{3}$		18	0	$\frac{9}{10}$
7	0	$\frac{7}{20}$		18	4	$\frac{11}{12}$
7	6	$\frac{3}{8}$		18	9	$\frac{15}{16}$
8	0	$\frac{2}{5}$		19	0	$\frac{19}{20}$

Here follow four examples upon the
four Notes delivered.

At 1 s. 8 d. the yard, what   3884   yards?

maketh pounds   323 —— 13 —4d.

     At 2 s. 6 d. the yard, what   4563 yards?

maketh pounds   570 —— 7 —— 6d.

     At 6 s. 8 d. the Ell, what   7562 Ells?

maketh pounds   2520 —— 13 —4d.

     Now by custome you are able to work by all sorts of
summes being delivered in shillings and pence, as one
shillings one peny, two shillings two pence three
shillings three pence, and so of all other: wishing
you to have some consideration of your questions,
when they are set down, for there are many subtile
abbreuiations, and great advantages to be gotten, and
easily to be perceived.

     As of 3 s. ——— ———8 d. of 2 s. and 1 s. 8 d.

     Of 4 s, ——— ———2d. of 3 s. ——— ———4d.&
10 d. which 10 d. is $\frac{1}{4}$ of 3 s. ——— . ———4d.

     Of 5 s. ——— — 8 d. of 4 s. 1 s. ———4d.

Of 5 s. 10 d. of 5 s. and 10 d. which 10 d. is $\frac{1}{6}$ of 5 s.

     And by this mean when you have taken one pro-
duct, you may oftentimes upon the same take another more briefly then upon the summe which is to
be multiplied &c.

2 Rule.

     *N*Ow (*Gentle Reader*) *that you have seen the ver-*
tue of the even or aliquot parts of a pound in
shillings alone, and also in the aliquot parts of shil-
lings and pence: according to my promise hereafter
                                     followeth

followeth a briefe and easier method for any even num-
ber of shillings, either under or above 20, then ever
yet hath been published; Notwithstanding M. Hum-
phrey Baker, whose travell is worthy commendation,
and whom for knowledge sake I reverence, hath in
some part touched this first part, though not in this
method. The work of the Rule both pleasant, ready,
and brief, as by the variety of the examples delivered
thereupon shall appear. And first I will set forth a
question, thereby the better to expresse or teach you
the order thereof : which is this.

If one cost 6 s. what 8574?

    1   6 s.   8574

maketh pounds   2572 ——— 4 s.

To the understanding of this example, after you
have set down your given number in form of the Rule
of 3, with a line drawn under it, you shall presently
set a prick under your first figure 4 toward your
right hand drawing from the prick, as heretofore hath
been practised, a little short line, thereto set down the
shillings anon, which done, multiply the first figure
4 by 6, the value of your price; (which here you see
standeth in sight above the line) it maketh 24, which
is one pound foure shillings. The one pound keep to
carry to the next place, and the foure shillings set
down at the end of the prescribed line towards your
right hand. Thus have you done now with 6 above the
line, and also with 4 in the first place (for the prick under 4 doth signifie that 4 hath done his office.) Then
secondarily for a generall Rule take but the ½ of
the given price, which here is the number that
shall now continue the rest of the multiplication, and end the work, whereupon I multiply

M. John
Mellis his
first Rule.

Note a ge-
nerall rule

               B b 3   3 into

3 into 7, ftanding in the fecond place it maketh 21, and with the 1 pound I kept in 22 ; fet down 2 and keep 2 in mind, working according to the Rule of multiplication, delivering the tenths in minde in their due place, which done, the product from the prick to your left hand reprefenteth the pounds, and the other at the end of the fhillings, as appeareth by the examples.

If one yard coft 2 s. what	7536 ?	
1 2 s.	7536	
maketh pounds	753——12 s.	
If one yard coft 4 s. what	8792 ?	
1 4 s.	8792	
maketh pounds	1758——8 s.	
If one piece coft 6 s. what	9537 ?	
1 6 s.	9537	
maketh pounds	2861——2 s.	
If one piece coft 8 s. what	7509 ?	
1 8 s.	7509	
maketh pounds	3003——12 s.	
If one coft 12 s. what	5794 ?	
1 12 s.	5794	
maketh pounds	3476——8 s.	
If one coft 14 s. what	3705 ?	
1 14 s.	3705	
maketh pounds	2593——10 s.	
If one coft 18 s. what	5703 ?	
1 18 s.	5703	
maketh pounds	5132——14 s.	

If

If one cost **22** s. what   953 ?

    1   22 s.   953

maketh pounds   1048------6 s.

Let these suffice (gentle Reader) for an entrance into even numbers. And now I will shew the like rule for any odde or uneven part of a pound.

TO help you to the understanding of these other questions that hereafter follow : where in my first Example the given number is 6487 at 3 s. the yard : I multiply 3 above the line into 7, it maketh 21. The one shilling is set down, and the 1 pound I keep. Now am I to take the ½ of three, which because it is an odde number I cannot.

Therefore I shall keep and continue my multiplication by three still, and work by the ½ of the rest of the given figures or number, to wit, 648. And first the ½ of 8 which is 4 multiplied into 3, maketh 12, thereto joyn the 1 li. in minde, it maketh 13. set down 3, keep one. Then again multiply by two the ½ of four it maketh six, and with one in minde it maketh 7. Then lastly, take the ½ of six, which is 3. saying, 3 times 3 is 9, which 9 set down, and so is the question answered as appeareth by the practice, and examples following.

(margin: ... ... the second Rule.)

At 3 s. the yard, what   6487 ?

    1   3 s.   6487

maketh pounds   973 ------

If one yard cost 5 s. what   4269 ?

    1   5 s.   4269

maketh pounds   1067 ------ 5 s.

At 7 s. the Ell, what	6489 ?	
1 7 s.	6489	
maketh pounds	2271	3s.
If one Ell coft 9 s. what	2807 ?	
1 9 s.	2807	
maketh pounds.	1263	3s.
At 11 s. the Piftolet, what	8263 ?	
1 11 s.	8263	
maketh pounds	4544———13 s.	
If one piece coft 13 s. what	4629 ?	
1 13 s.	4629	
maketh pounds	3008———17 s.	

But now note (gentle Reader) when the given price falleth upon an odde number. as 3, 5, 7, 11, 13, &c. then it is to be prefuppofed that the given fumme to be multiplyed, muft be a fum made of even numbers, 2, 4 6, 8, 10, &c. elfe cannot the queftion be wrought at one line or working.

Providing alwayes that it may bear an odd figure in the firft place towards your right hand, as appeareth in thefe fix examples, which laft were wrought, and fuch like, &c. which may beare an odde number for the price, and be done at one line or working very well.

But if the given price be an odde number, and the fum to be multiplyed odde numbers alfo : then can it not be done at one working, but requireth the aid o two workings, for odde with odde will not agree which notwithftanding to bring to paffe, take thi for a generall Rule. Firft, worke for the even num ber, contained in that queftion, or given price accoi din ding as you have learned, and then afterwards for the one odde shilling, take the $\frac{1}{2}$ of the summe given to be multiplied, omitting the first prickt place, as was taught for the working of one shilling in my first Rule of Practice, and adde those two together, and you shall have your desire.

Examples.

At 3 s. the yeard, what	7539 yards ?	
2s.	753	18
1s.	376	19
maketh pounds.	1130	17 s.

At 7 s. the Ell, what	7539 ?	
5 s.	7539	
	2261	14
2 s.	376	19
maketh pounds	6238	13

At 13 s. the yard, what	7534 ?	
10 s.	3767.	0
2 s.	753.	8
1 s.	376.	14
maketh pounds	4897	2

And thus have I abridged into these two rules how to bring any nūber of shillings whatsoever they be, into pounds, with a briefer Method, then ever yet hath been published which I commend unto thy friendly censure and judgement in the use and practice thereof.

Note this well..

If

If one coft 6s. 5d. what   1231 ?

       6s.   369 ——— 6

       4d.   20 — 10 — 4

       1d.   5 —— 2 — 7

maketh pounds   394 —— 18 ⋯ 11

At 14s. 2d. what   2825 ?

      14s.   1977 —— 10

      2d.   23 — 10 — 10

maketh pounds   2001 ⋯ 0 —— 10

At 16s. 4d. what   2531 ?

      16s.   2024 —— 16

      4d.   24 —— 3 ⋯ 8d.

maketh pounds   2066 —— 19 — 8

At 3s. the Piftolet, what   8325 ?

maketh pounds   1248 ———⋯ 15s

At 7s. the Crown, what   6529 ?

                 2284 ——— 3s.

At 9s. the piece, what   6567 ?

maketh pounds.   2955 ———⋯ 3s.

These three laft queftions may feeme fomething harder, yet they are eafie enough, if you mark them well, If I fhould explain them, then are they too eafie. Therefore I leave them to whet the minds of the defirous.

10 Rule.   ITem *when any one of the fummes, which is to be multiplied, is compofed of many Denominations, and the given number but of one figure alone ; then fhall you multiply all the Denominations of the other*

                                      fumme

fumme by the fame one figure, beginning firft with that fumme which is leaft in value toward your right hand, and bring the product of thofe pence into fhillings, and the product of the fhillings into pounds, as by this example appeareth.

At 3 li. 7s. 4d. a yard, what are 9 worth ?
maketh pounds. 30--6s-0d.

BVt *if any of the fummes that are to be multiplied, there be a broken number ; Firft worke for the* 11 Rule *whole according to the inftructions that you have learned, and then take fuch part of the given price, as that broken number beareth in proportion to the price, as in the examples following. After you have wrought for 3s. & for 6d. then are you to take the $\frac{1}{2}$ of 3s.6d. for the $\frac{1}{2}$ yard, and adde that to the fumme : So adding all the 3 products together, which make 43 li. 2s. 9d. the juft price of 245 $\frac{1}{2}$ Ells, and thus muft you doe of all other.*

At 3s. 6d. the Ell, what   245 $\frac{1}{2}$?

3 s.	36 —— 18	
6d.	6 —— 3	
$\frac{1}{2}$		1 -- 9
maketh	43 —— 9 -- 5 $\frac{3}{4}$	

At 16s. 4d. the piece, what   14 $\frac{3}{4}$?

16s.	11 ——	—— 4	
4d.	0 ——	4 —— 8	
$\frac{3}{4}$		12 —— 3	
maketh pounds	12 —— 0 —— 11		

If

If one peice cost 4 li. 3 s. 6½ d. what 12 pieces ?

4 li.	48———	
3 s.	1	16

6 d.   6———6
½

maketh pounds   50———2———6

The proofe.

If 12 pieces cost 50 li. 2 s. 6d. what one piece ?
maketh pounds   4 ——— 3 ———6½

14 Rule. Item, *touching the manner how to understand the order of this question, and others the like, first seek how many times 12 is contained in 50, which is 4 times, and so resteth 2 Pound, which 2 Pound converted into shillings, and joyned with the other 2 shillings, maketh 42 shillings : wherein is found 12 three times resteth 6 s. which turned into pence, putting thereto the 6 pence in the first place, it maketh 78, wherein 12 is found sixe times, resteth 6 d. which containeth 12, but ½ a time, put that ½ to the 6d. and then the solution is 4 li. 3 s. 6½ d. as appeareth by the Practice thereof.*

15 Rule. Item. *The like is to be done of any thing that is bought or sould after five score to the hundred, or the Quintall. As for example.*

If 100 pound cost 27 li. 13 s. 4 d. what one pound ?

27 li.

27 li. —13 s.— 4 d.   But to work it more
          2   neatly, it is by a little un-
                             derftanding, ended thus.

s. 5 | 53
     | 12

1 | 10

5 | 3
                             27 li. — 13 s. — 4 d.
d. 6 4 | 0   20
       | 10 0. or ⅖
                             s, 5 | 53
Maketh 5 s. 6 d. | 12

I have wrought this at lẽgth d. 6 | 40
for the aid of the young | 100
learner, becaufe he fhould
underftand how all the mul-   Maketh 5 s. 6⅔ d,
plication is fet down.

Item *to the underftanding of this & fuch like quefti-*
ons, the right down line is all the guide, which is pul-
led down clofe by 20. *as you fee in the example, where*
27 li. 13 s. *is reduced all into* s. & *maketh* 553 s.

The 5 *towards the left hand being feparated with the*
hanging or right down line, is the juft number of fhil-
lings, that anfwereth the queftion. Nextly, 53 s. *is mul-*
tiplied by 12, *to reduce them to pence, putting to the* 4 d.
it yeeldeth for the multiplication of the firft figure
two 110. *the one beyond the line towards the left hand,*
is 1 *penny towards the reft of the price : then* 53 *alfo*
multiplied by 1 *yeeldeth* 53 : *but the* 5 *behinde the*
line towards the left hand, is alfo 5 *pence more, to-*
wards the price, which 1 *and* 5 *I adde together under*
the line, it maketh 6 d. *So is there found now as appear-*
eth by the Titles of fhillings and pence, 5 s. 6 *pence.*

Finally. I come now on this fide the line towards
                                              the

the right hand, and under 12 I finde first 10, and then 3, which added together, maketh 40. under which 40, you must put the 100, and it maketh $\frac{40}{100}$ which abbreviated, commeth to $\frac{2}{5}$. So the just price of one pound after 5 score to the hundred, maketh 5 s. 6 $\frac{2}{5}$ d. One example more, and so I will leave this Rule.

If 100 cost 10 $\frac{3}{4}$ d. what   9874 ?

6d	246 ——————17	
4d	164 ——————11 ——————4
$\frac{2}{4}$ d	20 ——————11 ——————5
$\frac{2}{4}$ d	10 ——————5 ——————8 $\frac{1}{2}$

Maketh	li.	4	42	5	parts of a peny, $\frac{5}{4}$
			20		
	s.	8	45		
			12		
			45 $\frac{1}{2}$ 91		
	d.	5	100{ 100		

Also the like may bee done of the usuall weights here in *England*, (which is 112, for every hundred weight) in case you know the aliquot parts of a hundred weight, which are these, 56 li. 28 li. 14 li. and 7 li. For 56 li. is the $\frac{1}{2}$ of 112 li. 28 li. is the $\frac{1}{4}$ of 112 li. 14 li. is the $\frac{1}{8}$ and 7 li. is $\frac{1}{16}$ part.

Therefore for 56 li. take the $\frac{1}{2}$ of the summe of the money that 112 li. weight is worth.

For 28 li. take the $\frac{1}{4}$ of the summe of money that 112 li. weight is worth.

For 14 li. take the $\frac{1}{8}$ of the summe that 112 li. is worth.

And for 7 li. the $\frac{1}{16}$ of the summe of money that 112 li. is worth.   As

As for example ; at 17 li. 19 s. the hundreth pounds weight, that is to say, the 112 li. what shall 3 quarters and 7 pound cost ?

1. C.	17 li.	19 s.	3 q;	7 li.
2 quarterns	8	19	6	
1 quartern	4	-9	9	
7 pounds	1	-2	$5\frac{1}{4}$	
Maketh pounds	14	11	$8\frac{1}{4}$	

The second Chapter treateth of the Reduction of divers measures to others value by Rules of Practice.

Ow will I shew a few examples of practice 18 Rule
in reducing of Measures, as Ells, Yards,
*Braces, Pawnes of Genee, &c. Much
more would I have touched but that I
feare the Booke will rise to too great a
Volume.*

In 864 Ells of *Antwerpe,* how many yards of *London ?*

864	864
432	216
216	648

maketh 648 yards of *London.*

ITem, *in these and such like questions of* Flemmish *measure, to be brought into yards* Engl. sh. *first take the ⅓ of the given number as appeareth in the first ex-*
ample

ample towards your left hand. Then take halfe of that product, or the ¼ of the given number, and adde these two products together, as they shall be yards English; *as by the example you may perceive.*

The second example toward your right hand is yet briefer then the first, whose worke is this; Take the ¼ of the delivered number, and that product subtract out of the given number, and the rest sheweth your desire. Of these two wayes use which you thinke best.

The Proof.

In 648 yards *London,*
How many Ells of *Antwerpe?*

$$648$$
$$216$$

maketh   864   Ells of *Antwerpe.*

15 Rule.

ITem, *for the understanding of this worke, first take the ⅓ part of the yards of* London, *which found adde that ⅓ part and the yards together, as appeareth by the Practice, and the product sheweth the Ells of* Antwerpe.

Item, in 20 yards of *London,*
How many Ells of *Antwerpe?*

maketh   426⅔ Ells,
320 yards
106⅔
426⅔ Ells   ⅔

Proof,
426⅔ Ells,
106⅔
320 yards,

16 Rule.

Other Reductions,

ITem, *you shall understand, that forasmuch as six braces of* Millain, *make five Ells of* Antwerpe, *whereupon according to the* Rules of Practice, *you may reduce the one into the other, by the like reason aforesaid,*

in

in taking the $\frac{1}{5}$ part & then subtract the same, to make Ells of Antwerpe. And again by the contrary taking the $\frac{1}{5}$ part with adding the given number, to turn the Ells to Braces. As for example.

In 876 Braces, how many Ells of *Antwerpe* ?

<pre>
      876   The contrary.
      146   730 Ells Flemmiſh.
Ells 730 Antwerpe.   146 Braces.
        Ells 730 Antwerpe.
             182½
   Yards   547½ Engliſh.
</pre>

Thus appeareth that 876 Braces by Practice, make 730 Ells *Flemmiſh*, which Ells *Flemmiſh* reduce into *Engliſh* yards.

So again upon the same firſt queſtion of Braces, I would know how many yards *Engliſh* they make.

After the rate that 100 Braces are

<pre>
       worth 62 ½ yards.
       876 Braces.
       438
       109½
</pre>

I anſwer, 547½ yards.

ITem, *To the underſtanding of this work*, *and such like*, firſt take the $\frac{1}{2}$ of the given Braces, and after take the $\frac{1}{4}$ of that halfe, or the $\frac{1}{8}$ of the given number, and adde them together, and the Products are alſo yards Engliſh.

ITem, *three Ells of* Rochell *make* 5 *Ells at* Lisbone. *So likewiſe three Ells at* Lions *make* 5 *Ells at* Antwerpe.

To worke thefe & fuch like, double the Ells of Lions, *and the Ells of* Rochell, *and from their Product fubtract the* $\frac{1}{21}$ *and the reft fhall be the Ells of* Antwerpe, *or the Ells of* Lisbone.

I. *Example.*

In 63 Ells of *Lions,* how many Ells of *Antwerpe?*

$$\begin{array}{r} 63 \\ 63 \\ \hline 126 \end{array}$$

$\frac{1}{2}$　21

Anf. 105 Ells *Ant.*

In 100 Ells of *Rochel,* how many Ells of *Lisbone?*

$$\begin{array}{r} 110 \\ \hline 200 \end{array}$$

$\frac{1}{2}$23$\frac{2}{3}$

Anf. 166$\frac{2}{3}$ Ells of *Lisb.*

Touching the proofe or return of thefe and fuch like queftions for a generall Rule, you fhall firft take the $\frac{1}{3}$ of the given number : and adde that $\frac{1}{3}$ and the given number together, and the $\frac{1}{2}$ of that product fhall be your defire.

Example.

In 105 Ells of *Antwerpe,* how many Ells of *Lions?*

$$\begin{array}{r} 105 \\ 21 \\ \hline 126 \end{array}$$

$\frac{1}{3}$)
$\frac{1}{2}$)

Anf. 63 Ells of *Lions.*

In 166$\frac{2}{3}$ Ells of *Lisbone,* how many Ells of *Rochell?*

$$\begin{array}{r} 166\frac{2}{3} \\ 33\frac{1}{3} \\ \hline 200 \end{array}$$

$\frac{1}{3}$)
$\frac{1}{2}$)

Anf. 100 of *Roch.*

Queftions

Queſtions of Factoridge and Intereſt, briefe and truly reſolved by the Rule of Practice without Time.

1 Queſtion.

AT 5 ſhillings *per Centum*, what comes 8860 li. 15 s. 4d. unto ?

Anſwer. Note that 5s. is the fourth of 20. s. I take the $\frac{1}{4}$ part of 8860 li. 15s. 4d, which makes 2215 li. 3s. 10d. Now the Root is 100, which you ſhould divide by, ſo cutting the 2 laſt figures away of the pounds, you háve 22 li. then mul-tiply 15 li. by 23s, ſo adde the 3 unto, you ſhall have 303s. cut away the two laſt figures, there reſt-eth 03s. Laſtly, there remains 3s. which I multiply by 12 to bring into pence, and ſo I finde od. and $\frac{42}{100}$ remaining, which being abbreviated make $\frac{23}{50}$ parts of a peny, ſo I find that there is gained 22 li. 3s. od $\frac{21}{50}$ parts of a peny.

li. 22. 15. 03. 10

	20
s. 3	30
	12
o d.	0
	4
	4 23
	100|50

2. Queſt. At 10 s *per centum*, what comes.

Anſwer. Note that 10 s. is the $\frac{1}{2}$ of 20s. I take the $\frac{1}{2}$ 1448 li. 16s. 8d. which makes 724 li. 8s. 4d. cut off the two laſt figures, and there reſteth 7 li. then multiply the 24 li. by 20s. and adde the 8s. and it maketh 488s. cut

1448 li. 16s.8d. unto ?

li. 7. 24. 8. 4.

	20.
s. 4 88	
1 2	
1 80	
8 8	
d. 10 6 0 5	
10 0 3	

C c 2   thi

the two laſt figures off, and there reſteth 4s. then
multiply 88s. by 12d. and take in 4d. and there
reſtech 1060d. cut off the two laſt figu es, and there
reſteth 10d. and $\frac{60}{100}$ which is $\frac{3}{5}$ of a peny : ſo the
whole ſumme is 7 li. 4s. 10$\frac{3}{5}$ which is the anſwer to
the queſtion.

3. *Queſt.* At 15 ſhillings
per centum, what comes,

Anſwer. Note 15. that
is $\frac{1}{2}$ and $\frac{1}{4}$ of 20. take the $\frac{1}{2}$ of
1008 lib. 12 ſhillings, there
reſteth 504 li. 6s. then take
the $\frac{1}{4}$ adde them together, the
totall will be 756 li. 9s. cut off
the 2 laſt figures, reſteth 7 li.
then multiply by 20 s. & take
in your 9s. it maketh 1129 s.
cut off the two laſt figures,
there reſteth 11s. then multi-
ply by 12 d. there commeth 348d. cut off the laſt
two figures, there reſteth 3d. and $\frac{48}{100}$ which being abbreviated maketh $\frac{12}{25}$ parts of a peny, ſo ſhall you find
7 li. 11s. 3d. $\frac{12}{25}$. which is the anſwer to the queſtion.

```
1008 l.12s. od. unto?
 504.6 ———— 0
 252.3 ———— 0
7|56.9 ———— 0
  |20
11|29
  |12
   58
   29
3|48|24|12
100|50|25
```

4. *Queſt.* At 1 li. *per cen-*
tum, what comes

Anſwer. Cut away the two
laſt figures, and multiply by
20, and 12, and take in your
ſhillings and pence. And you
ſhall finde 8 li. 13 s. 8 pence
$\frac{4}{5}$ as doth appeare by this
work.

```
li. 8|68 li. 13s.4d.
              unto?
     |20
 s. 13|73
     |12
    1 50
    7 3
 d. 8 8 |o |4|
      10|0 |5|
           5 Queſt
```

5. *Queſt.* At 2 li. *per cen-* 5608 li. 6s. 8 d. unto?
tum, what comes

Anſwer. Multiply the whole
ſumme by two lib. thus, then
cut off the two laſt figures of
your pounds, as you did before, and you ſhall finde
112 pound, then multiply by
20 and by 12, taking in your
ſhillings and pence, and you
ſhall finde 112 li. 3s. 4 d. which is either for Factor
or Broker, &c.

```
                2
 112|16.|13.4
        |20
     3|33
      |12
       70
       33
      4|00
```

6. *Queſtion.* At 3 pound,
per centum, what comes

Anſwer. Multiply the ſum
by 3 pound, thus ; then cut
off the two laſt figures, and
you ſhall finde 24 pound, then
multiply by 20, and by 12.
taking your ſhillings and
d. and you ſhall finde os .6 d.
$\frac{27}{50}$ parts of a peny, which is
ſomething above a half-peny.

which is either for Factor
800 li. 18s. 2d, unto?

```
                3
 2402, 14. 6.
       |20
   0|54
   |12
    1 4
    54
    6|54 27
     100 50
```

7. *Queſtion.* At foure *per
centum*, what comes

Anſwer. Multiply by 4 li.
thus, cut off the two laſt figures, Multiply by 20, and
by 12. taking in your ſhillings
and pence, and you ſhall finde
11 li. 19 s. od. $\frac{25}{2}$ parts of a
peny, which is ſomething
above a farthing.

298 li. 15s. 9d. unto?

```
                4
 11|95.3.0.
    |20
 19|03
    |12
    06
 03
 036 18 9
 100 50.25
```

Cc 3   8 *Queſt*

8 *Queſt.* At 5 li. $\frac{3}{4}$ per *centum*, what comes

Anſwer, Multiply by 5 li. thus, then take the $\frac{1}{2}$ of the whole ſumme and place the figures even, then take the $\frac{1}{2}$ of that $\frac{1}{2}$ & adde all three ſumms together, cut off the two laſt figures, then multiply by 20. and by 12. taking in your ſhillings, and pence, and you ſhall find 210 l. 7 s. 7 d. $\frac{-9}{10}$ parts of a peny, which is the anſwer to the queſtion.

9 *Queſt.* At 6 l. $\frac{1}{2}$ per *centum*, what comes

Anſwer, Multiply by 6 li. and then take $\frac{1}{2}$ of the whole ſumme, adde them both together, then multiply by 20, and by 12, taking in your odd ſhillings and pence, and you ſhall find 369 li. 10s. 0 d. $\frac{-2}{20}$ parts of a peny, which is the anſwer to your queſtion.

10 *Queſt.* At 7 li. $\frac{1}{2}$ per *centum*, what comes

Anſwer, Multiply by 7 li. then take the $\frac{1}{2}$ adde them together, cut off the two laſt figures, then multiply by 20, you ſhall finde 290 li. 3s. The anſwer to the queſtion.

3658 li. 16s. 8d.
       unto 5

18294	03	4
1829	08	4
914	14	2
210|38	05	10

   |20

7\|65
 |12

1 30
6 6
7\|9 0
1 0\|0

5684 li. 12s. 6d.
     unto 6

34|07	15	0
2842	06	3
369|50	01	3

  |20

05
01
15 \|2
100 \|20

3868 li. 13s. 4d.
      unto   7

27080. 13. 4
1934. 06. 8
290|50. 0

  |20

3 \|00

      11 *Queſt.*

11 *Queſt.* At 8 li. *per* 2560 l. 17 s. 9 d unto?
Centum, what comes

Anſwer, Multiply 8 li. cut
off the two laſt figures, multiply by 20, and by 12, and
you ſhall finde 204 li. 17 s.
5 d. $\frac{1}{25}$ parts of a peny.

```
2560l.17s.9d unto?
              8
    _____
204 87.02. 0
    | 20
17  | 42
    | 12
    ___
     84
    42
    _____
     5 04 | 2 | 1
    _____
    100  50 25
```

Queſtions of Intereſt with Time, wrought by Practice.

1 Queſtion.

AT 6 *per Centum* what
comes unto for 1 month

Anſwer, Multiply by 6 lib.
there commeth 2813 li. 00 s.
0 d. then take for 1 moneth
the $\frac{1}{12}$ of the Totall, and you
ſhall finde 234 li. 8. 4 d. of
the two laſt figures of the li.
Multiply by 20 & by 12, taking in your odde money, and
you ſhall finde 2 li. 6. 10 d. $\frac{2}{5}$
parts of a peny, which is the
anſwer to the queſtion.

```
468 li.  16 s. 8 d?
2813.   00. 0
_____
li. 2 34. 08. 4
      20
s.  6 88
      12
    _____
    180
     88
_____
d.1 0 6 | 0 | 3
_____
     10 0  15
```

2. *Queſt.*At 7 pound ½ *per centum,* what comes unto for 2 months.

Anſwer. Multiply by 7 pound, then take ½, adde them two together, then for your two months take the ½ of the Totall, multiply by 20 and 12, taking in your odde ſhillings and pence, and you ſhall finde 47 pounds 10 ſhillings 1 peny ₁₄⁄₁₀ parts of a peny, which is the anſwer to the queſtion.

```
3800 li. 12s. 8d.
              7
26604. 08 8
 1900. 06 4
28504. 15 0
 47|50 15 10
   |20
 10|15
   |12
   30
  1 6
 1|9|0|
 10|0|
```

3. *Queſtion,* At 8 pound *per centum,* what comes unto for 3 months.

Anſwer, Multiply by 8 pound, then take for your 3 monthes ½ of the Totall, multiply by 20, and by 12, adding in your odde ſhillings and pence, and you ſhall find 197 pound 5 ſhillings 11 pence, ₂¾ parts of a peny, your demand.

```
9864 li. 16s. 4d.
              8
78918.  10 8
197|29. 12 8
   |20
 5 |92
   |12
   192
   93
 11|12 |6|3
 10|0|50|25
```

4 *Queſt.*

4 *Quest.* At 6 pound ½ *per centum,* what comes unto for 4 months.

Answer, Multiply by 6 li. Then take ½, adde both together, then for your 4 months take ⅓ part of the whole, cut away your two last figures, multiply by 20, and by 12, adde in your odde shillings, and pence, and you shall finde 131 pounds, 14 shillings 11 pence $\frac{12}{50}$ parts of a peny, your demand.

6080 li. 13s. 0d.

       6

36483	18 0
3040	06 6
39524	04 6

131|7½ 14. 10
    |20
14|94
 |12
 188
 94
11|38 |19
  |100|50

5 *Question,* At 8 *per centum,* what comes unto for 5 months.

Answer, Multiply by 8 li. then for 5 months take ¼ and ⅙ of the Totall, cut off the 2 last figures of your pounds, Multiply by 20 and by 12, adde in your odde shillings and pence, and you shall find 100 pound 13 shillings 4 pence, your demand.

3020 li. 00s. 00d.

       8

24160 00 00
 6040 00 00
 4026 13 04
10,|66 13 04
  |20
13|33
 |12
 70
 33
 4|00

6 *Quest.*

6 *Queſtions*, At 8 *per Centum*, what comes unto for 6 months.

Anſwer, Multiply by 8 li. then for your 6 moneths take the ½ of the Totall, cut off the two laſt figures of your pounds. Multiply by 12. taking in your odde ſhillings and pence, and you ſhall finde 322 lib. 8s. 5d. 12/25 parts of a peny, your deſire.

7 *Queſt.* At 8 li. *per Centum*, what comes unto for 7 moneths.

Anſwer, Multiply by 8 li. then for your 7 months take ⅓ and ¼ of the Totall, cut off the the two laſt figures of your pounds, then multiply by 20 and 12, taking in your odde mony, and you ſhall finde 275 li. 2s. 11 d. ⅘ your deſire.

```
8060 li. 12s. od.
            8
64484.    16s.
322|42. 08. 0
   |20
 8|48
   |12
   |96
   |48
|76|28|9
5|100|5|25


5896. 00 od.
          8
47168 0 0
15722.0. 0
11792.13.4
275|14.13.4.
   |20
  2|93
   |12
   190
   93
11|2|0|1
10|0|5
```

8 Queſt.

8 *Queſtion*, At 8 per *centum*, what comes unto for 8 months.

Anſwer, Multiply by 8 li. then for 8 months take $\frac{2}{3}$ of the totall, cut off the two laſt figures of your pounds, then multiply by 20, and by 12 adde in your odde money, and you ſhall finde 116 li. 5 ſhillings, 9 pence, $\frac{3}{25}$ your deſire.

$$3680 \text{ li. } 08\text{s. } 0\text{d.}$$
$$8$$

29443.	04	0
9814.	08	0
9814.	08	0

116 | 28　16　0
20
5 | 76
12
152
76

9 | 12 | 6 | 3 |
100 | 50 | 25 |

9 *Queſt.* At 8 li, *per centum*, what comes unto for 9 months.

Anſwer, Multiply by 8 pound, then for your nine moneths take $\frac{1}{2}$ and $\frac{1}{4}$ of the whole ſumme, cut off the two laſt figures of the pounds, then multiply by 29, and by 12. taking in your odd ſhillings and pence, and you ſhall find 221 pounds, 1 ſhilling 11 pence, $\frac{7}{25}$ which is ſomething above a farthing.

$$3684 \text{ li. } 19\text{s. } 0\text{d.}$$
$$8$$

29479,	12	0
14739.	16.	0
7369.	18.	0
2209.	14	0

20
94
12
188
94

11 | 28 | 14 | 7
100 | 50 | 25

10 *Queſt.*

10 *Queſt.* At 6 ¼ *per centum*, what comes unto for 10 months.

Anſwer, Multiply by 6 pound, then take the ½ and ¼ of 100 pound, adde all three ſummes together, then for the 10 months take ⅔ & ⅓ of the Totall, add them together, cut off the two laſt figures of the pounds, multiply by 20, and 12, adding in your ſhillings and pence, cutting off the laſt figures of your ſhillings and pence, you ſhall finde 5 pound 12 ſhillings, 6 pence, your deſire.

	li.	os.	od.
100.			
			6
600.	0	0	
50.	0	0	
25.	0	0	
675.	0	0	
337.	10	0	
225.	00	0	
5 62.	10.	0.	
20.			
12 50			
12			
100			
50			
6 00			

11 *Queſt.* At 8 pound *per centum*, what comes unto for 11 months.

Anſwer, Multiply by 8 pound, then for 11 months take ⅔ and ¼, from the Totall, adde all three ſummes together; cut off the two laſt figures of your pounds. Multiply by 20 and by 12, adding in of your ſhillings and pence, cutting off the two laſt figures of your ſhillings and pence, and you ſhall finde 65 pound 0 ſhillings 7 d. 12 ½ ⅓ parts of a peny, your deſire.

	li.	16s.	od.
886			
			8
7094	08	0	
2364	16	0	
2364	16	0	
1773	12	0	
65 03	04	0	
20			
0 64			
12			
128			
64			
7 68 34 7			
100 50 25			

12 *Queſt.*

12 *Quest.* At 8 pound *per centum*, what comes unto for 12 months.

Answer, Multiply by 8 pound, cut off the two laft figures of the pounds, Multiply by 20, and by 12, adding in your fhillings and pence, cut off the two laft figures of your fhillings, and the two laft of your pence, and you fhall finde 726l. 8s. 11d. $\frac{12}{25}$ parts of a peny your defire.

9080 li. 12s. 2d.
          8

726|44. 17 4
     |20
  8|97
    |12

11|68 34|17
  1005c|25

The third Chapter teacheth of the Order and work of the Rule of *Three* in broken numbers after the Trade of *Merchants*, digreffing fomething from Mafter *Record*, which is comprehended in three Rules.

NOw, that I have fomewhat intreated of the Rules of Practife, *I will give a few inftructions, after my fimple order, for the working of the* Rule of Three *in broken numbers, wherein I fhall need to fay the leffe, becaufe I hope the ftudious Learner, that hath travelled any thing in the* Grounds of Arts, *is not unfurnifhed of knowledge capable to underftand me.*

But before I deliver any inftructions for broken Numbers, I will propofe a queftion which fhall be wrought three fundry wayes, thereby to fhew as it were, three Degrees of Comparifon : how farre the Rule of Three in broken, for more fpeed of worke, differeth from the whole, which

I

I rather fet downe for a view, that the ftudious herein
may be more defirous to attaine broken, leaving an
more to difcourfe in Dialogue forme, but onely to
give inftructions where need is : and in the reft to
put forth the queftions, with their anfwers.

My firft queftion is thus.

If one yard coft 6s. 8d. what are 789 worth at
that rate ?

The firft   1 ———— 6s. ———— 8d. ———— 789
way.   1 2   80
        ————————————————   ————————————
                  80   63 120 d.

Here the Product of the fumme are pence, according to the nature of the middle number.

            11
          1370 1
          63120 (5260 (263
          12221 1220
             111

        I anfwer ———— 263 li.

The fe-   1 ——— 6⅔ s ———— 789
cond way.
        ————————————————————————— 20
          3   20 ———————————

                              15780 s.

Here the Product of the fummes are fhillings according to the nature of the middle number.

            1
          15780 (5260 (263
          3333   1220
                 li.

The third   1 ——— ⅓ ———— 789
way   3   1   1

                  789   Here

Here the Product is pounds, according to the title of the second number.

$$\substack{1 \\ 789(263 \\ 333}$$

I answer, 263 li.

Now that you have seen the three former vertues of the Rule of three; whose Products have first brought forth pence, next shillings, and lastly pounds, I will deliver three notes in order following : and with them a dozen questions, that shall shew the work of the Rule of three in broken Numbers or Fractions.

1 The first foure shall be sundry questions of a *Note these* Fraction comming in the second place. *three well*

2 The second foure shall be of two Fractions comming in the second or third place.

3 The third foure of Fractions in all three places.

Note, upon the first Rule for a Fraction comming in the second place.

My first Question is this.

If one yard cost me 3 *shillings* 4 *pence, what are* 1 Rule. 756 *worth at that price?*

In setting down the question to perform the work, The first I turn foure pence into the part of a shilling, which is variety. $\frac{1}{3}$ and then the question standeth thus :

$$1 \text{————} 3\frac{1}{3} \text{———} 756.$$

To the ready working of this question, and all such other like, my first note is this, which take for a generall Rule ; That when any one Fraction shall come,

come, either in the second or third place, that the Denominator of that Fraction or Fractions, must alwayes be brought unto the Number, or Numerator of the first place ; and thereby multiply the one into the other.

And this benefit is always gotten by the vertue of bringing the Denominator of the second Numbers Fractions unto the first place : For the Fraction in the middle number is now released : and the Product that cometh of the Multiplication, is of the nature and like the denomination of the whole number in the second place, which here are shillings.

Note this.

Whereupon now to worke the Question, I bring 3, the Denominator of the Fraction in the second place, unto the first Number 1, with a line set under thus $\frac{1}{}$ and the third under it thus, $\frac{1}{3}$ saying, once 3, is 3 my Divisor : that done, reduce $3\frac{1}{3}$ saying, 3 times 3 is 9, and the other 1 over 3 make 10, my second number in the *Rule of Three*, by which 10 I doe multiply my last number 756, as appeareth by the worke thereof, and it yeeldeth 7560 shillings, my Dividend.

Then dividing 7560 by 3, my Divisor, it yeeldeth in quotient 2520 shillings, which maketh 126 pounds, as appeareth here most plainly, both by the Example, and the worke.

At 3s. 4d, the yard, what 756 yards ?

$$1 \text{———} 3\frac{1}{3} \text{———} 756$$
$$\overline{3} \qquad\qquad \overline{10} \qquad\qquad \overline{10}$$
$$7590s.$$

$$\begin{array}{c|c} 1 & 1 \\ 7560 & 2520 (126. \\ 333\ 3 & 2220 \end{array}$$

I answer 126 li.

Yet

Yet otherwise upon the same question, altering the price now into the proportion it beareth to a pound, for the 3s. 4d. is $\frac{1}{6}$ part of a pound : which Example first standeth thus as appeareth on the left hand, and afterwards wrought as appeareth on the right hand.

$$1 \longrightarrow \tfrac{1}{6} \longrightarrow 756$$

$$1 \underline{} \tfrac{1}{6} \underline{} 756 \quad 6$$

$$\frac{\quad 1 \quad\quad\quad 1 \quad}{126 \text{ pounds.}}$$

The second veriety. v2

As soon as I have carried 6 the Denominator of my middle number unto my first place, as before hath been taught, I pull down 1, the numerator of 6, with a line under 6, thus. $\frac{6}{1}$, and that one in custome I pull down in sight ; being the figure that I will multiply my third or last number by, according to the tenour of the *Rule of Three*. And because one can neither multiply nor yet Divide (though here it is set down in form of multiplication, the rather for your understanding) the Product of the multiplication according to the declaration of this my first Rule or Note, is converted into the Title of my second number, which here are pounds. Now followeth the division performed in my Divisor 6, to make an end of that question.

$$\frac{1\frac{?}{8}}{756} (126. \textit{ which maketh } 126 \text{ li. } \textit{as before.}$$
$$666$$

And thus much for the variety in working that question.

And now followeth another.

D d   Mj

My ſecond Queſtion.

If one yard of Cotten coſt 8 ¼ d. what 859 ?

1	8¼	859
4	33	33

2577
2577
28347

r

r|32r　　　)|1　　　li.　　　s.　　　d.

28347 (7086 (590 　(29————10————6¼

4444　12zz　220

This Queſtion was alſo wrought like the firſt, and bringeth forth 29 li. 10 s. 6¼d. the price of 859 yards.

My third Queſtion.

If ſeven pounds of any thing coſt 3 li.——10s. what comes 987 pounds to ?

li.

7	3½	987
2	7	7
14		6909

00
14r
234|7
6909 (493 —2|1
　　　　　　14|2
1444
xx

I anſwer, 483 li.——10s.

Notes

Notes upon my second Rule for two Fractions coming in the second and third place.

My first Question is this.

IF one Ell coſt 13 s.——— 4 d. what halfe a quar-ter or ⅛ of an Ell ?

Anſwer, *Firſt bring* 13 s.——— 4 d. *into the parts of a pound, which is* ⅔, *and then will the queſtion ſtand thus.*

$$1 \text{———} \tfrac{2}{3} \text{ li.} \text{———} \tfrac{1}{8}$$

ITem, for the performance of this work, doe as before was taught in the firſt Rule : firſt *bring* 3 *the Denominator of the ſecond Fraction unto your firſt number* 1, ſetting a line under it thus : ⅓ Saying once 3 is 3, that done, bring 8 the *Denominator of the third Fraction*, ſetting it under 3, and multiply them toge-ther, ſaying, 3 times 8 maketh 24 which 24 is your *Diviſor.* (Now have you done with the *Denominator* 8) therefore you ſhall put a line under thus, ⅓. And the like line alſo under ⅛. ſetting or pulling down under them their own *numerators*, that is, 2 under 3, and alſo 1 under 8, as appeareth in the Example, which *numerators* for a generall rule evermore to be pulled downe of cuſtome in ſight, to multiply the one by the other, according to the tenour of the *Rule of Three.* Then I multiply the one by the other, ſaying, once 2 is 2, which ſignifieth 2 pound, being of the nature and like denomination of the middle number, which 2 pound is to be reduced into ſhillings, otherwiſe it cannot be divided by my firſt number 24.

Then dividing 40, by 24, the quotient bringeth

forth

forth $1\frac{2}{3}$. So much is $\frac{2}{3}$ of an Ell worth after that rate. Otherwise, although 2 pound could not be divided by 24, yet it might have been abbreviated to $\frac{1}{12}$ of a pound: which is worth 1s. 8d. as before.

li.

$$1 \text{———} \frac{2}{3} \text{———} \frac{1}{3} (1$$

3	2	1 z (6
8	20	4⊘ (1$\frac{2}{3}$s.
24	40	24

Second Question.

IF one pound of any weight cost 13 shillings 4 pence, what are $\frac{2}{3}$ of the pound worth after that rate?

Answer, Reduce the 13 shillings 4 pence into the parts of a pound: which is $\frac{2}{3}$. and then will the question stand thus.

li.

$$1 \text{———} \frac{2}{3} \text{———} \frac{7}{8}$$

*I*Tem, *for the understanding of this, if you mark well the last Example, this and the rest lieth open, and needs small instruction. For as you did last, so again, bring the Denominator of the second and third Fraction, unto the first figure* 1, *multiplying the one into the other; which maketh also* 24, *your Divisor.*

Then making a line under 3 thus, $\underline{3}$ and a line under 8 thus, $\underline{8}$ and pulling downe their *Numerators* under each figure, that is 2 under 3, and 7 under 8, which as I said before for a generall rule, I pull down of custome in sight, to be the two Numbers, that of duty ought to be multiplied together; which done, I bring 2, being the lesser figure under 7,

multi-

multiplying them together, it maketh 14, which are
of the nature of the middle number : that is to wit,
pounds, which 14 cannot aptly be divided among 4.
therefore are reduced into *shillings*, as is plainly to be
seen in the example : then 280 *shillings* parted among
24 yeeldeth for his *quotient* 11s. 8d. your desire and
the just price of $\frac{7}{8}$ of an *Ell*. Otherwise 14. though
it could not be divided by 24. might by *Mediation*
or *Division* in broken *Numbers* have been divided or
abbreviated to $\frac{7}{13}$, which in effect being reduced to
his known parts, maketh 11s. 8d. as before. But
my good will and meaning is to aid young beginners:
therefore have I reduced the 14 *pound* into *shillings*,
which is the easier way.

Now followeth the example.

$$
\begin{array}{cccc}
\text{I} & 2 & 7 & |\text{I} \\
3 & 3 & 8 & x \\
8 & 2 & 7 & 4\ (6 \\
24 & & 3 & 280\ (11\frac{2}{3}\text{s.} \\
& & 14 & 284 \\
& & 20 & x
\end{array}
$$

280 s. I answer, 11$\frac{2}{3}$s.

The third example.

If one yard cost me 2s.——6d. what 345 $\frac{1}{4}$ yards ?
Answer, First put 6d. into the parts of a shilling,
and then the question standeth thus :

I————2$\frac{1}{2}$————345$\frac{1}{4}$.

Item, to the ready understanding of this, and all

D d 3   such

such like, according as before hath been declared,
bring the *Denominators* of the second and third
Fractions unto the first place, multiplying them the
one into the other, all which make 8 for the common
Divisor. Then next reduce your second number:
saying, two times 2 is 4. and 1 is 5. as was taught in
the Example aforesaid. Lastly, reduce your third
number 345 ¼ all into fourths, and they make 1381.
which 1381 is to be multiplied by 5. according to
the tenour of the *Rule of Three* : which done, maketh 6905 shillings, and divided by 8. your *Divisor*
yeeldeth in Quotient 863 ½ shillings, which maketh
in pounds 43 pound, 3 shillings, 1 ½ : and so much
are the 345 ¼ yards worth at that price.

The same question wrought again by two shillings
6 pence, is now converted into the parts of a pound,
and standeth thus :

$$1 \longrightarrow \tfrac{1}{8} \longrightarrow 345\tfrac{1}{4}.$$

Item, After I have brought here my second and
third *denominator* unto my first place, and found 32
to be my *divisor* ; having thus finished my first place
with all things unto him belonging (which is meant
of bringing and multiplying the *denominators* of the
second and third *Fractions* into him) I then goe in
hand to see what is to do in my second place, where
presently of custome I pull down my numerator 1
under 8. being the figure in sight that shall multiply
my third number.

Then lastly, I reduce 345 ¼ all into fourths as
afore was practised. which maketh 1381 the which
1381 I am to multiply by 1 my second number
they are nothing increased, but by the *Metamorphos.*

of my work they are now 1381. pound, being of the
nature of the middle number, as I have often shewed
you, which divided by 32 my divisor yeeldeth 43
pound, and $\frac{1}{32}$, which $\frac{1}{32}$ of a pound reduced into
known numbers, make 3 shillings 1 $\frac{1}{2}$ pence, as before.

Example.

$$1 \underline{\quad\quad \frac{1}{8} \quad 345\tfrac{1}{4}} \quad 1381 \quad (43 \quad \underline{\quad 5 \quad}$$
$$32 \quad\quad 1 \quad\quad 1381 \quad\quad\quad 32$$
$$1381$$

NOw follow foure other questions, which are in
all three places broken numbers; or whole and
broken together.

Item, First, for the finding out of your *Divisor*,
you shall take this for a most certaine, and generall
Rule : That you must multiply the *Numerator* of the
first *number* in the question, by the *Denominator* of
the second : And that *Product* againe, by the *Denominator* of the third : And the totall thereof shall be
your *Divisor*.

Secondly, for a generall rule to find out your *Dividend, multiply* the *Denominator* of the first number
by the *Numerator* of the second, and the whole
thereof by the *Numerator* of the third. And the totall thereof shall evermore be your *Dividend*.

*Now for an example, I propose this question, there-
by to make my meaning more plaine, and to shew you,
as I have done in the rest, the manner and order of the
work.*

If $\frac{2}{3}$ *of any weight or measure cost* $\frac{5}{6}$ *of a pound, or*

2 os, *what are ⅔ of the like weight or measure worth after that rate?*

Example.

$$\frac{2}{3} \text{———} \frac{5}{6} \text{———} \frac{3}{8}$$

ITem, for the more plainer underſtanding hereof, and all other the like in broken Numbers: Firſt, you ſhall pull down two, the *Numerator* of the firſt Number or Fraction, with a line under, thus 3¼: that done, according as you have learned before, bring 6, the *Denominator* of the ſecond Fraction, and ſet it under two, multiplying the one into the other, which maketh 12. Then laſtly, bring 8. the *denominator* of the third *Fraction*, and ſet it under 1 2. multiplying that 1 2 by 8. which amounteth to 96. or elſe for more brief, multiply 6 by 8, ſaying, ſix times 8 makes 48. which 48 ſet under 2. and multiply the one into the other, it maketh 96. as before. And this 96 is the firſt number in the *Rule of Three*. That ſhall alwayes for a moſt generall Rule be your *diviſor.*

Secondly, to work for your *dividend*, you ſhall, (as it hath beene ſufficiently declared before) pull down 5. the *numerator* of your ſecond *Fraction*, and ſet it under 6, with a line under thus __6.

That done (as you know) you are to pull down 3, the *numerator* of the third *Fraction*, and ſet it under 8, with a line under it, thus, __8, multiplying the one into the other, according to the Tenour of the *Rule of Three*; which maketh 15. Then according to my Note, forget not to bring the *denominator* of the firſt *Fraction*, which is three, under 15. and multiply them together, which maketh 45. which 45 is your *dividend*, and are of the nature of

denomi

denomination of the middle number, as I have taught you before : And therefore are 44 li. which aptly cannot be divided by 96. Therefore you shall reduce the 45 li. into s. as you see performed in the Example, which amounteth to 90 s. which divided by 96 your *divisor* it yeeldeth 9 s. and $\frac{1}{20}$ of a shilling, which in lesser termes is $\frac{2}{8}$: which $\frac{1}{4}$ in mony maketh $4\frac{1}{2}$d. and so much will the aforesaid $\frac{1}{8}$ cost, as by the work following shall appear.

The Example.

$$\frac{2}{3} \quad \frac{5}{6} \quad \frac{3}{8}$$

$$\begin{array}{ccc} 2 & 5 & 3 \\ 6 & & 5 \\ \hline 12 & 15 & \end{array}$$

$$\begin{array}{ccc} 12 & 15 & \\ 8 & 3 & \\ 96 & 45 & \\ & 20 & \\ & 900 & \end{array}$$

Otherwise though 45 could not be divided by 96. yet by Division in broken numbers it might have been abbreviated to $\frac{15}{32}$ of a pound, which reduced into known parts, will make 9 s. $4\frac{1}{4}$d. as before.

Now my second example shall be the proof of this question.

If $\frac{1}{8}$ yards cost $\frac{15}{32}$ of a pound, or 20 shillings, what shall $\frac{2}{3}$ cost ?

Answer, Work as was taught you before, and you shall have your desire.

Here

$$\frac{3}{8} \text{———} \frac{15}{32} \text{———} \frac{2}{3}$$

$$
\begin{array}{cc}
3 & 15 & 2 \\
32 & 2 \\
\hline
96. & 30 \\
3 & 8 \\
\hline
288 & \cdot 240
\end{array}
$$

Here as appeareth by the work, the Multiplication being ended, 240 is to be divided by 288, which to some perchance may seem hard, yet notwithstanding is the work good. Therefore abbreviate 240. by 288. as you see here is practised : and the end of your abbreviation shall come to ⅚ your desire, $\frac{240}{288} \frac{2}{3} \frac{5}{6}$.

Otherwise, $\begin{array}{c} 240\,120\,60\,30\,5 \\ 288\,144\,72\,36\,6 \end{array}$

Otherwise, $\begin{array}{c} 340\,40\,5 \\ 288\,48\,6 \end{array}$

The third Question.

If ¾ ells cost 13s. —d. what 156 ½ ells ?

Answer, To work this question the shortest way : reduce 13 shillings 4 pence, into the parts of a pound, which is ⅔.

Then as you did afore, after you have set down the question, the *Numerator* of the first *Fraction* 3 is pulled down under 4. and *Denominators* of the other two *Fractions* multiplyed into him, which maketh 18. your *Divisor*.

Then the *Numerators* of the second *Fraction* is pulled downe, under 3 in custome now in sight, ready to *multiply* my third number, by which is performed as soon as the last numbers 156 ½ is reduced into halfs.
Then

Then laftly, I multiply that product by 4 the *Denominator* of the *Fraction* : it yeeldeth 2504. which I divide by 18. and my quotient is 139. pound, and $\frac{2}{18}$ or $\frac{1}{9}$ of a pound remaining, which is worth 2s——2⅔d. And fo much will 156½ Ells coft, as by the work following doth appear.

3	2	156½	x
4	3	313	7⁴
3	2	2	x762
6		626	2504 139½
18		4	1888
		2504	11

li.

The fourth queftion.

If 2 ½ Ells coft 1 ⅔ pounds, what commeth 29 ¼ Ells to ?

Item, to the workemanfhip of this queftion, firft reduce your fecond *number* in faying, three times 1 is 3, and two is 5, Then bring the *multiplication* of the *Denominators* of the fecond and third *Fractions* which maketh 12. and *multiply* that 12 by 5 your firft *numerator*, and it maketh 60. which is your *divifor*.

Then the *Reduction* of the fecond Number, which is 5, *multiplied* by 117 the Product of the laft *numbers* reduction, make 585, which 585 yet refteth to be *multiplied* by 2, the *Denominator* of the Fraction in the firft place, yeeldeth 1170, which divided by your *Divifor*, 60. yeeldeth 19. pound, 10 fhillings, as appeareth by the work thereof.

Thus having now touched the 12 queftions whereof I firft pretended, which with diligence and oft

practife,

practiſe, I truſt are ſufficient to aid the deſirous unto the working of any broken numbers. I will now treat of divers neceſſary rules incident unto traffick as here-after followeth.

The fourth Chapter teacheth of loſſe, and gaine, in the Trade of Merchandiſe.

IF one yard coſt 6s.——8d. and the ſame is ſold againe for 8s.——6d. the queſtion is what is gained in one hundred pounds laying out on ſuch commodities:

Anſwer, The *Rule of Three* direct, applyed two manner of wayes to doe the ſame : the one is to ſay, If $6\frac{2}{3}$ give $8\frac{1}{2}$, what giveth 100? *Multiply* and Divide, and looke what your quotient bringeth forth, above your laying out, is the neat gaines and ſolution to your queſtion : If you follow the worke, your ſolution will bring forth 127li.——10s. which is 27li.--10. more then your principall, and ſo much is gained in the 100 pounds laying out.

Item, to work in the other way, which I take the neareſt, ſeek the difference betwixt the juſt price and the other price, which is one ſhilling ten pence, then ſay by the *Rule of Three*.

If $6\frac{2}{3}$s. gain $1\frac{5}{6}$s. what ſhall a 100 pound gain?

Multiply and divide, and you ſhall find 27li.-10s. and ſo much is gained in 100 li. laying out.

You may uſe which of theſe two wayes you think good.

The

The proofe.

If a yard of cloth be delivered for 8s. 6d. whereupon was gained after the rate of 27li. 10s. in 100 pounds laying out: The queſtion is, what the yard coſt at the firſt hand?

Anſwer, Put your gain 27li. ———— 10s. to 100 pounds, all maketh 127li. ———— 10s. Then ſay, If 127li. 10s. give but a 100 pounds, what giveth $8\frac{1}{2}$s. Work, and you ſhall finde 6s. 8d. the true ſolution to your queſtion.

Yet another example or proofe upon
the firſt Queſtion.

If one yard coſt 6s. ——— 8d. the queſtion is, at what price the ſame is to be ſold again, for to gain 27li. 10s. in 100 pound laying out?

Anſwer, *Say by the* Rule of Three, *if a* 100li. *gain* 127li. 10s. *what giveth* $6\frac{2}{3}$s? *Multiply and divide, and you ſhall find* 8s. 6d. *your true ſolution.*

If one Ell coſt 7s 8d. and be ſold again for 8s. 6d. *The queſtion is*, What is gained in 20 pound laying out in ſuch commodities.

Anſwer, *Seek the difference betwixt the juſt price, and the other price which is ten pence, and then apply the* Rule of Three, *as before is taught, ſaying, If* $7\frac{2}{3}$s. *give* $\frac{5}{6}$ *ſhilling, what giveth* 20li? *Multiply and divide, and you ſhall finde* 2li. $3\frac{11}{23}$s. *and ſo much is gained in* 20li. *laying out.*

The proofe alſo by an example of Loſſe.

A *Merchant hath bought* Holland *cloth at* 8s. 6d. *the ell, which proveth not to his expectation, whereupon he is content to loſe* 2li. $3\frac{11}{23}$ *in* 20 *pounds*
laying

laying out. The queſtion is, what price ought to be made of the Cloth, abating this loſſe?

Anſwer, Doe as before in Gains hath been taught, putting 2 li. 3$\frac{11}{23}$s. to your 20 pound, all together makeths 22 li. -3$\frac{11}{23}$. Then ſay by the *Rule of Three*. If 22 li 3 $\frac{11}{23}$s. give but 20li. what ſhall come of 8 $\frac{1}{3}$s? work, and you ſhall finde 7s. -8d. the juſt price that the Ell ought to be ſold for after the rate of this loſſe.

Thus it appeareth evidently, as in company the *Rule* is appliable as well to gain as loſs.

If, 20$\frac{1}{4}$ *yards coſt* 36 li. 10s. *how ſhall I ſell the ſame again* $\frac{2}{3}$ *of the Principall, or to make of* 3, 4. *which is all one.*

Anſwer, By the *Rule of Three*, If 3 doe give 4, what will 36$\frac{1}{2}$ give? Multiply and divide, and you ſhall finde 48$\frac{2}{3}$ pounds, Then ſay again, if 20$\frac{1}{4}$ yards do give 48. $\frac{2}{3}$ pounds, as well principall as gain, what will one yard be worth at that price? Multiply and divide, and you ſhall find 2 li. $\frac{25}{243}$.

If one Ell of Cloth coſt me 8s. 8d. *and afterwards I ſell* 10$\frac{1}{4}$ *Ells thereof for* 5 li. 13s. 4d. *I would know, whether I winne or loſe : and how much upon the* 100 *pounds of mony.*

Anſwer, See firſt at 8s. 8d. the Ell, what 10$\frac{1}{2}$ Ells comes to, and you ſhall finde 4li. 11s. and I ſold the ſame for 5li. ——13s. ——4d. ſo that I did gain upon the 10$\frac{1}{2}$ Els 22 ſhillings 4d. Then if you would know how much is gained in 100 pounds, I ſay by the *Rule of Three*, if 4 li. ——11s. did gain 22 ——4d. what will 100 pounds gain? Multiply and divide, and you ſhall finde 24 li. ——10s. ——10d. $\frac{10}{91}$ and ſo much is gained in the 100. pound of mony.

If 12$\frac{1}{2}$ *yards coſt me* 11 *pound five ſhillings, and I ſell*

sell the yard again for 16 *shillings, the question is whether I doe winne or lose, and how much in or upon the pound of mony?*

Answer, Look what the 12 ½/6 yards come to at 16s the yard, and you shall finde ten pound. But they cost 11 pound 5 shillings. So there is lost upon the whole 1 pound 5s. Then to know how much is lost in the pound, say by the *Rule of Three,* if 11 ¼ pound doe lose 1 ¼ pound, what will 1 pound lose? Multiply and divide, and you shall finde 2 s. 2 d ⅔. and so much is lost in the pound of mony.

If I sell the 100 *weight of any commodity for* 4 *pound, whereupon I doe lose after ten pound in the* 100 *pound, I demand how much I shall lose or gain in the* 100 *pound, if in case I had sold the same for* 4 *pound ten shillings.*

Answer, Say, if 90 pound yeeldeth 100. how much will 4 give? Multiply and divide, and you shall find 4 4/9. Then say again, if 4 4/9 give me 4 ½ what will 100 come to? Multiply and divide, and you shall finde 101 pound ¼ which is more then 100 pound by 1 li. 5 shillings: and so much is gained in the 100. pound.

A Merchant hath sold Currants for the summe of 430 *Pound, and he hath gained therein after ten pound, in the* 100 *Pound. The question is, to know how much is gained in all.*

Answer, Say by the *Rule of Three.* If 100 pound doe gain ten pound, what will 430 pound gain? Multiply and divide, and you shall finde 43, and so much hath hee gained in all.

If one yard be worth 28 ½ *s. for how much shall* 10 *yards be sold to gain after* 8 *pound,* 6 *shillings & pence in the* 100 *pound?*

Answer,

Answer, First, adde 8li.——6s,——8d: to 100. Then say, if 100li. do give 108 ⅓s. for principall and gain, what will 28 ½s. principall yeeld ? Multiply and divide, and you shall finde 30 ⅞s. Then say, again, by the *Rule of Three*, if 1 yard do give 30 ⅞ s. (which is as well the principall as the gain) what shall ten yards give ? Multiply and divide, and you shall finde 15 li. 8s. 9d. And for the same price shall the ten yards be sold, for to gain after the rate of 8li.—6s.—8d. upon the 100.

A branch or Proof out of this Question.

A Merchant hath sold clothes for 15 li.—8s.—9d. and he hath gained in the whole the summe of 1 li.— 3s. —— 9d. The question is to know how much he hath gained in the 100. pound ?

Answer, To know this, first rebate the gains from the price, and there will remain 14 li. 5s. 0d. Then say by the *Rule of three* direct. If 14 li. ¼ give me 1 li. 3 ¾, what will 100li. give ? Multiply and divide, and you shall finde 8li. 6s.-8d. the effect desired, the proof is apparent in the question before.

Yet another branch or proofe of the first Question.

If ten yards be delivered for 15 li. 8 s 9d. where-upon was gained after the rate of 8li. 6s. 8d. upon the 100. pound, the question is, what the yard did cost at the first hand ?

Answer, First, say by the *Rule of three*, if ten with principall and gain yeeld 15 li. 8s. ¾ shillings, what shall 1 yeeld. Multiply and divide, and you shall find

30⅔s. Then say again by the *Rule of Three*, if 108 ⅓ principall and gain give but 100 what shall 30⅞ of principall and gain yeeld? Work, and you shall finde 28½s. And so much did the yard cost at the first peny.

If one yard cost 36 s. how much shall 12 yards be sold for, to gain after the rate of 10 li. in the 100?

Answer, First say, If 100. give 110 li. principall and gain, what will 36s. give? Multiply and divide, and you shall finde 39⅗s. Then say again by the *Rule of Three*: If one yard of principall and gain yeeld 39⅗ shillings, what shall 12 yards gain? Multiply and divide, and you shall finde 23 li. 15. ⅕s. which ⅕s. in knowne number, is 2⅖d. And for the same price shall the 12 yards be sold, to gain after the rate of 10 in the 100.

The Proof.

If 12 yards be sold for 23 li. 15s. 2⅖d. whereupon is gained after 10 li. in the 100. the question is, what the yard cost at the first peny?

Answer, First say, If 12 give 23 li. 15⅕s. what one yard? Multiply and divide, and you shall finde 39⅗s. Then say againe by the *Rule of Three*, if 110. pounds give but a 100. what shall 39⅗s. give? Work, and you shall find 36s. the just price of the yard at the first hand.

Item, *When one Merchant selleth wares to another, and he giveth to the buyer 1 li. 6. 8d. upon the score, or 20 li. the question is, How much shall the buyer gain upon the 100 li. after that rate?*

Answer, First, adde 1 li. 6s. 8d. unto 20 li. and they are 21⅓. Then say, If 20 pound give 21⅓, what

shall 100. give? Multiply and divide, and you shall
finde 106 $\frac{2}{3}$. So the buyer getteth after the rate of
6 $\frac{2}{3}$ li. upon the 100 li.

Gentle Reader, other necessary questions appertaining
to Losse and Gain, you shall have in the eighth
Chapter of this Treatise.

The fifth Chapter entreateth of Losse and
Gain upon time, wrought by *the double Rule
of Three, or by the Rule composed* : which is
contained in foure speciall selected branches, or
questions of divers formes, *each one of them spring-
ing from the first Question*, and each one of them
also being a proof to other, *&c.*

F one yard cost me 2s. 8d. *ready money, and
after I sell the same again for* 2s. 10d.
*to be paid for it at the end of three months:
the question is, what I gain upon the* 100
li. *in* 12 *moneths.*

Answer, First say, if 2 $\frac{2}{3}$ gain $\frac{1}{3}$, what shall 100 li.
gain? Multiply and divide, and you shall finde 6 $\frac{1}{4}$
li. Then say again, by the *Rule of Three*, if three
months gain 6 $\frac{1}{4}$ pound, what shall 12. months gain
Work, and you shall finde 25 li. and so much shall
I gain in 12 moneths after that rate.

Item, You may also work it all at one working
by the first part of the *Rule of Three* composed, saying
if 2 $\frac{2}{3}$ d. in three months doe gain $\frac{1}{6}$ of a shilling, (which
is 2d.) what will 100 li. gain in 12 months? Which
for thy further encouragement, the work of this or
example I have here put down, to verifie that I affirm
in the first part of this *Ground of Arts*, that th
Ru

Rule, and so all others, more rejoyceth in *Broken*, then in *Whole*.

s.	months	s.	li.	mo.
$2\frac{2}{3}$ ——	3 ——	$\frac{1}{6}$ ——	100 ——	12
8		1	20	
$\frac{8}{3}$ 2			2000	
24 72000			3	
6 *14444* (500 (25			6300	
144 *144* 220			12	
1			7000	

Where the multiplication and the division being ended, maketh **25** li. your desire.

If a yard be delivered for 2s. *10d. to be payed at* 3 *months, whereupon was gained after the rate of* 25 *li. in the* 100. *for* 12 *months, the question is now, what the yard cost at the first hand?*

Answer, First say, if 12 months gain 25 li. what shall 3 months gain? Work, and you shall finde $6\frac{1}{4}$ li. Then say again the second time, if 10 $\frac{1}{4}$ li. give but 100. what shall $2\frac{2}{3}$ s. give? Work and you shall finde 2s. 8d. which is the just price that the yard cost at the first hand.

If one yard of Cloth cost me 2s. *8d. ready mony, for what term shall I sell the same again for* 2s. *10d. so that I might gain after the rate of* 25 *pound upon the* 100. *pound in* 12. *moneths?*

Answer, First say, if $2\frac{2}{3}$ gain $\frac{1}{6}$, what shall 100 li. gain? Multiply and divide, and you shall find $6\frac{1}{4}$ Then say again for the second work. If 25. pound be come of twelve months, what shall come of $6\frac{1}{4}$? Work, and you shall find 3 moneths, the just term of time that the Cloth ought to be delivered at 2s. 1od. to gain 25li. upon the 100li. in 12 moneths.

If one yard coſt me 2s. 8d. ready money, for what price ſhall I ſell the ſame again to be paid at the end of 3 moneths, ſo that I may gain after the rate of 25 pound in the 100 pound for 12 moneths?

Anſwer, Firſt ſay, if 12 gain 25li. what ſhall 3 moneths gain? Multiply and divide, and you ſhall finde 6¼li.   Then ſay for the ſecond worke, if 100li. give 106¼, what giveth 2⅔s? Work and you ſhall finde 2s. 1od. and for that price muſt the yard be ſold to gain after 25 pound in the 100 pound for 12 moneths.

Many other of theſe queſtions I might here have delivered, but for feare the Booke would riſe to too thick a volume, and ſo to make the price ſo much the dearer, whereby it might not be ſo portable to my Countrie-men, as I wiſh it. But theſe 4 I have of purpoſe framed in this order, having relation one to another, aſſuring you that what queſtion ſoever may be propoſed within the compaſſe of this Rule you ſhall finde by one of theſe 4 to make a ſolution. And moreover, divers others are yet to be delivered where the Creditor giveth divers dayes of payment, which can never be well wrought, nor yet underſtood, unleſſe you can firſt finde by Art the juſt times that all thoſe payments, how different ſoever they be, ought to be paid at once: whereupon firſt I think good here to give ſome inſtructions unto ſuch a Rule; for it is the onely aid for the finiſhing of ſuch queſtions as hereafter ſhall follow.

The

The sixth Chapter entreateth of Rules of Payment, which is a right necessary Rule, and one of the chiefest handmaids that attendeth upon buying and selling, &c.

Example.

A Merchant doth owe a summe of mony, whereof the $\frac{1}{3}$ is to be paid at sixe moneths, and the $\frac{1}{2}$ at eight moneths, and the rest at a year. If he would pay all at one payment, the question is, what time ought to be given him?

Answer, I have omitted the quantity of the summe, for you shall understand the Rule is appliable, and yeeldeth a true solution to what summ soever shall be proposed : But now for order sake in teaching, I do imagine the summe to be 60 pounds, whereupon the manner of this worke is to multiply the proportionate part of the money by the time, as in company. Then 20. being the first payment, and the $\frac{1}{3}$ of 60 which $\frac{1}{3}$ multiplied in broken numbers by 6, his time of payment, maketh $\frac{6}{3}$, which in whole numbers, as appeareth by the Example in the Operation, maketh two moneths : next 30. which is the $\frac{1}{2}$ multiplyed by his terme 8. yeelds 4. moneths, then the rest which is 10 li. must

$$\frac{1}{3} \text{ by } \frac{6}{1} \qquad \text{2 Moneths.}$$
$$\frac{6}{3}$$
$$\frac{1}{2} \text{ by } \frac{8}{1} \qquad \text{4 Moneths.}$$
$$\frac{8}{8}$$
$$\frac{1}{6} \text{ by } \frac{12}{1} \qquad \text{2 Moneths.}$$
$$\frac{12}{8}$$

needs be abbreviated into the proportion it beareth to 60. which is $\frac{1}{6}$, which $\frac{1}{6}$ multiplied by his time 12 moneths, produceth $\frac{12}{6}$, maketh two moneths. All which added together, as appeareth in the Operati-

E e 3 on,

on, maketh 8 moneths, which is the juft time that all thofe payments ought to be paid at once.

A Merchant hath 800 li. *to pay, the* $\frac{1}{6}$ *thereof ready money, the* $\frac{1}{4}$ *at two moneths, the* $\frac{1}{2}$ *at* 4 *moneths, and the reft at a yeare. The queftion is, if he would pay all at one payment, what time ought to be given him?*

Anfwer, The ready money is never multiplied, then $\frac{1}{4}$ multiplied by two moneths, as you did before, maketh $\frac{1}{2}$, then $\frac{1}{2}$ by 4 produceth 2 moneths, as appeareth here in the operation : But now for the reft of the money you cannot multiply it untill you have fought what proportion it beareth to 800 pounds. Therefore you muft fubtract the ready mony, the $\frac{1}{4}$ and $\frac{1}{2}$ out of the principall. The reft will be 66 $\frac{2}{3}$ li. which you muft look what part it beareth to the principall, which you fhall finde to be $\frac{1}{12}$, the fame you muft alfo multiply by his time 12 months and it yeeldeth 1 month, fo all make 3 $\frac{1}{2}$ months, as appeareth in the operation.

A Merchant is to pay 1200 li. *in three terms, that is to wit,* 400 li. *at two weeks: and* 600 li. *at foure months: Laftly,* 200 li. *at five months: The queftion is, in what time they ought to be paid at once.*

Anfwer, Proportionate the parts, and you fhall finde that 400 is $\frac{1}{3}$ part, and for 600. you fhall finde $\frac{1}{2}$, and likewife 200 is the $\frac{1}{6}$ part which multiply by their times as before & you fhall have $\frac{2}{3}$ weeks, more eight weeks : and laftly 3 $\frac{1}{3}$ weeks, which together maketh 12. weeks, or three months, your defire.

A Merchant is to pay 600 *pound in three termes, whereof* 100 *pound is paid prefent, more* 300 *pound at twenty dayes, and the reft at five months, accounting thirty*

thirty dayes to a month. The queſtion is, what time ought theſe payments to be paid at once ?

Anſwer, Worke, and you ſhall finde 2 months.

The ſeventh Chapter treateth of buying and ſelling in the Trade of Merchandize, wherein is taken part ready money, and diverſe dayes of Payment given for the reſt. and what is wonne or loſt in the 100 pound forbearance for twelve months more or leſſe, according to the quantity of money, or proportion of time, *&c.*

A Merchant hath bought ſatins which coſt 8 s. *the yard ready money : and he ſelleth the ſame again to another man for* 10 s. *the yard, but he giveth* 2 *dayes for the payment, that is to ſay,* 3 *months for the one halfe, and* 5 *months for the other halfe. The queſt, is to know how much the ſeller doth gain upon the* 100 li. *in* 12 *moneths after that rate.*

Anſ. Seeke firſt, by the Rules of payment, at what time thoſe 2 payments ought to be paid at once, and you ſhall find 4 moths, at which time the ſecond Merchant ought to have paid the whole entire paymēt & therfore ſay by the firſt part of the *Rule* of 3. compoſed, If 8 s. in 4 months do gain 2 s. what will 100. li. gain in 12 months.

s.	m.	s.	li.	m.
8	4	12	100	12
4			20	
23			2000	
			2	
x			4000	
x6				
48000		3s.		
3*2*2		(5100	(75	
3		320		
		E		

Multiply and divide, and you shall finde 75. pounds, as appeareth in the example, & so much doth the first Merchant gain upon the 100 li. in 12. months.

A Merchant hath sold 50. Clothes, at 9 ½ li. the piece, to be paid the one ½ at foure moneths, the ⅓ at five moneths, and the ⅙ at 7 moneths, and the sellers minde is to take no more but after 8 li. in the 100. for 12 moneths. The question is now, what the first Merchant gaineth in the sale of these Clothes after that rate.

Ans. First, looke what the 40 Clothes come to at that price, and you shall find 475 li. Then secondly according to your direction in the Rules of payment, seeke at what time all the payments are to be performed at once. And you shall finde $4 \frac{2}{6}$ moneths. Then thirdly say, by the first part of the *Rule of 3.* composed. If 100 li. in 12 months gain 8 li. what will 475 li. gain in $4 \frac{2}{6}$ months? Work, and you shall find 15 li. and $\frac{11}{36}$ of a pound, which is the neat gains that the first Merchant hath after the rate aforesaid.

A Merchant hath bought Holland at 7 s. 3 d. the Ell ready money, and he selleth the same again, for 8 s. 4 d. the Ell, to be paid ¼ part in ready money, more ⅓ part at 2 moneths, and the rest at 4 moneths ; The question is now to know how much the first Merchant doth gain upon the 100 li. in 12. months after that rate.

Answer, According to the direction delivered you in the Rule of payment, the ready mony is not to be multiplied. Then working for the other 2 payments find out the true proportion at what time they ought to be paid at once, you shall find for ⅓ at two moneths ⅔ of a month, And the rest of the money which is $\frac{5}{16}$ multiplyed by his term 4 months, yeeldeth 1 months, both which added together make $2 \frac{1}{3}$ months, th

the just time that both the payments ought to be performed at once. And therefore say by the first part of the *rule
of 3 composed*, if 74$\frac{1}{4}$ in 2$\frac{1}{3}$ months doe gain$\frac{12}{46}$ of a li. what
shall 100li. gain in 12 months after that rate? Work, &
you shall finde 78$\frac{172}{603}$ li. And so much doth he gain upon
100. pounds in 12 months.

*A Merchant hath bought 30.clothes at 6 li.the peice for
ready mony : Afterward he selleth 10 of them for 7 li.the
peice,for 3 months term : And the other 20. he selleth for
8li. the piece, for 4 months term,the question is now, what
he gaineth upon 100 pounds in 12 monsbs ?*

Ans. First find the value of the 30 clothes,which amount
to 180li. Secondly seek what the ten pieces come to at 7.
li.& what the 20 pieces come to at 80 li. the one comes to
70. and the other to 60, both which together make 230.
which is 50li. more thē they cost. Thirdly, as I have taught
you in the rule of payment, proportionate the first and 2^d
prices unto the proportiō they bear unto 230 the product
of their 2 prices,& you shall find $\frac{21}{23}$ for the first, & $\frac{16}{23}$ for the
latter. Then fourthly, multiply those parts by their times, &
you shall have $\frac{21}{23}$ and $\frac{64}{23}$ both which together maketh 3
whole months, & $\frac{16}{23}$ of a month, which is the just time that
both those payments ought to be paid at once.

Then say by the first part of the *Rule of* 3.composed, If
180li. in 3$\frac{16}{23}$ months doe gain 50li. what shall 100 gain in
12 mōths? Multiply & divide,& you shall find 90$\frac{10}{21}$ li. &
so much doth he gain upon 100 pounds in 12 months.

*A Merchāt hath bought Cinnamō which cost him 9s.the
li. ready money. The question is now, at what price he ought
to sell the 100 weight.To wit 112li. to be paid the $\frac{1}{4}$ at 2
months,& the residue at the end of 3 mōths,so that he may
gain after the rate of ten li. upon 100li.for 12 months.*

Ans. Seek first by the Rule of payment at what term
both the payments ought to be paid at once, where

the $\frac{1}{4}$ multiplyed by his term two months, maketh $\frac{4}{4}$ months.

Likewiſe the next payment, which is $\frac{3}{4}$ multiplied by his term three months, maketh $2\frac{1}{4}$ moneths, both which added together, maketh $2\frac{3}{4}$ months, which is the time that both the payments ought to be paid at once. Then ſay by the *Rule of Three*, If 12. months do give me ten pounds, what will $2\frac{3}{4}$ months give? Multiply and divide, and you ſhall finde $2\frac{7}{24}$ pounds. Then ſay again by the *Rule of Three*, If one pound coſt me 9s. what will 112. pounds coſt? Multiply and divide, and you ſhall finde 50 li. 8s. Then ſay once again: If 100. pound doe give $102\frac{7}{24}$, what will $50\frac{2}{5}$ pounds give? Multiply and divide, and you ſhall finde 51 li. 11 s. $1\frac{1}{5}$ d. and for that price ought I to ſell 112. pound of Cinnamon to be paid at the two ſeverall payments aforeſaid, to gain thereby after the rate of tenne pounds upon the hundred pound in twelve months.

Brief Rules for our hundred weight here at
London, which is after 112. pound
for the 160.

Item, *Who that multiplieth the pence that one pound weight is worth by* 7, *and divided the* Product *by* 15. *ſhall finde how many pounds in mony* 112 *pound weight is worth.*

And contrariwiſe, *he that multiplieth the pounds that* 112. *pounds weight is worth by* 15. *and divideth the* Product *by* 7. *ſhall finde how many pence in money the one pounds weight is worth.*

Example.

At 10 pence the pound weight, what is 112. pounds weight worth? *Anſwer,*

Anſwer, Multiply 10 by 7. and thereof commeth 70. the which divided by 15. and you finde $4\frac{2}{3}$ pounds. And thus the 112 pounds is worth 4 li. 13s. 4d, after the rate of 10. pence the pound aforeſaid.

At 6 pounds the 112 pounds weight, what is one pound worth?

Anſwer, Multiply 6 by 15, and thereof commeth 90. the which divide by 7. and you ſhall finde $12\frac{6}{7}$d. So much is one pound worth when the 112. pounds did coſt 6 pounds.

The eighth Chapter intreateth of Tares, and allowances of Merchandize ſold by weight, and of Loſſes and Gains therein, &c.

AT 16 pound the 100 Suttle, what ſhall 895 pound Suttle be worth, in giving 4 pound weight upon every 100. for Treat?

Anſwer, *Adde 4 unto 100, and you ſhall have 104. Then ſay by the Rule of Three, If 104. be worth 16. pounds, what are 895 pounds worth? Multiply and divide, and you ſhall finde 137li. 13s. $10\frac{2}{11}$d. and ſo much ſhall the 895. pounds weight be worth.*

Item, at 3s, 4d. the pound weight what ſhall $754\frac{1}{2}$ pound be worth, in giving 4 pounds weight upon every hundred for Treat?

Anſwer, See firſt by the *Rule of three* what the 100. pound is worth, ſaying, If one coſt $3\frac{1}{3}$s. what 100? Multiply and divide, and you ſhall finde $16\frac{2}{3}$ pounds. Then adde 4 unto one 100, and they are 104. Then

104. Then fay again by the *Rule of Three*, If 104 be fold for 16 $\frac{2}{3}$ pounds, for how much fhall 754 $\frac{1}{2}$ be fold for? Multiply and divide, and you fhall find 120 li. 18s. 3 $\frac{2}{13}$ d. And for fo much fhall the 754 $\frac{1}{2}$ pound be fold for at 3s. 4d. the pound, in giving 4 upon the 100.

Other neceffary briefe Rules there are for the finding of Treats, or cafting up of Chefts of Sugar, &c. which for that it is a myftery, I omit : if any lack inftruction that way they fhall finde me ready to pleafure them.

Item, If 100 pounds be worth 36s. 8d. what fhall 860 pounds be worth in rebating 4 pounds upon every hundred for tare and cloff?

Anfwer, Multiply 860 by 4, and thereof commeth 3440, the which divide by 100, and you fhall have 34 $\frac{2}{5}$ pounds, abate 34 $\frac{2}{5}$ from 860 and there will remain 825 $\frac{3}{5}$ pounds. Then fay, by the *Rule of 3.* If 100 li. coft 36 $\frac{2}{3}$ s. what will 825 $\frac{3}{5}$ coft after that rate? Multiply and divide and you fhall finde 15 li. 2s. 8 $\frac{16}{25}$ d. And fo much fhall the 860 coft in rebating 4 li. upon every 100, for tare and cloff.

Item, *whether doth he lofe more that giveth* 4 li. *upon the* 100, *or he that rebateth* 4 li. *upon the* 100 ?

Anfwer, Firft note, that he that giveth 4. pound on 100 giveth 104 for 100. And he which rebateth 4. pounds upon the 100. giveth the 100. for 96. Therefore fay by the *Rule of Three*, if 104 be delivered for 100, for how much fhall the 100 be delivered. Multiply and divide, and you fhall find 96 $\frac{2}{13}$, and he which rebateth 4 in the 100, maketh but 96 pounds of 100, fo that he lofeth 4 pounds in the 100, and the other which giveth 4 pounds unto the 100 lofeth but 3 $\frac{11}{13}$ pounds upon the 100.   Thus
you

you may ſee, that he which abateth 4 pounds in the 100, loſeth more by $\frac{15}{15}$ pound in the 100 pounds, then the other which gave 4 pounds upon the 100, for tare and cloſſe.

If 100 pounds of any thing coſt me 23s. 4d. the queſtion is, how I ſhall ſell the pound, to gain after the rate of ten pounds, upon the 100 pound.

Anſwer, Say by the *Rule of three,* if 100 pounds give 100 pounds, what ſhall $23\frac{1}{3}$ s. give, multiply and divide, and you ſhall finde $1\frac{17}{60}$ pounds, Then ſay again, if 100 pound be worth $1\frac{17}{60}$ pounds, what is one pound worth, multiply and divide and you ſhall finde 3 d. $\frac{2}{15}$ And ſo much is the pound worth in gaining ten pounds upon the 100.

Item, *A Grocer hath bought C. weight of commodity for 6 li. 10 s. The queſtion is now to know how many li. thereof he ſhall ſell for 33s. 4d. to gain 20s in C. weight*

Anſwer, Adde 20s. unto 6 li. 10s, and they make 7 li, 10s. Then ſay, if $7\frac{1}{2}$ pound, yeeld me 112 pound, what ſhall $1\frac{2}{3}$ pounds yeeld; multiply and divide, and you ſhall find $24\frac{8}{9}$ li. And ſo many pound ought he to ſell to gain 20s. in his C. weight.

Item, *If one pound weight coſt 3s. 4d. and I ſell the ſame again for 4s, what is gained in a hundred pound of mony laid out in that commodity;*

Anſwer, You may ſay, If $3\frac{1}{3}$ s, give 4 what will 100 pound gain; But then when you have found, you muſt ſubtract 100 pounds out of the Product, the reſt is your neat gain: or elſe to produce the neat gain in your work at the firſt, ſubtract the juſt price out of the overprice, as I taught before in the firſt beginning of Loſſe and Gain, and your concluſion ſhall be all one. Multiply and divide by which of

the

the two wayes you think good, and you shall finde that he gaineth 20 pounds in the 100 pound.

Item, *If the pound weight which cost* 4 s. *be sold again for* 3. s. 4 d. *I demand what is lost in the* 100 *pounds of money.*

Answer, Say if 4 s. lose ⅔ s. what shall 100 lose? Multiply and divide and you shall finde 16 li. 13. s. 4 d. and so much is lost upon the 100 of money.

Item, *If C. weight of any commodity cost* 45 li. *and the buyer repenting would lose* 5 *pounds in the* 100 *of money. I demand how the pounds may be sold, his losse to be neither more nor lesse then after the rate aforesaid of five by the hundred.*

Answer, By the *Rule of three*, if a 100 lose 5, what shall 45 lose? Work and you shall finde 2¼ pound, which rebated from the principall 45, resteth 42 li. 15 s. Lastly say if 112 yeeldeth but 4 li. 15 s. what one pound? Multiply and divide and you shall find 7 s. 7 d. $\frac{17}{28}$. And so much is the pound worth after that losse.

A Grocer hath bought three pieces of Raisins weighing 175 ½ *pounds,* 182 ¼ *pounds,* 191 *pounds: tare for each fraile* 2¼ *pounds at* 25 ⅕ s. *the C. weight. The question is, what they amount to in money.*

I answer, 6 li.————— 3 s.————— 4 d. $\frac{23}{28}$ d.

A Grocer hath bought three sacks of Almonds weighing 267 ½ *pound, tare two pound,* 257 ½ *pounds, tare* 2½ *pound,* 252 *pound, tare* 3 *pound, at* 2 s. 10½ d. *the pound, what amount they to in money.*

I answer, 110 li.————— 12 s.————— 3 ¾ d.

The

The ninth Chapter intreateth of lengths and breadths of Arras and other Cloths, with other questions incident unto length and breadth.

IF a peice of Arras be 7 Elles and $\frac{3}{4}$ long and 5 Elles and $\frac{2}{3}$ broad, how many Elles square doth the same peice containe?

Answer, Multiply the length by breadth, that is to say, 7 $\frac{3}{4}$ by 5 $\frac{2}{3}$ And thereof will come 43 $\frac{11}{12}$ elles: so many elles square doth the same piece contain.

Item, more, a piece of Arras doth contain 2 2 Elles square, and if the same were in length 3 $\frac{1}{2}$ elles, I demand how many elles in breadth the same peice doth contain.

Answer, Divide 22 elles by 3 $\frac{1}{2}$ and thereof commeth 6 $6\frac{2}{7}$ So many elles doth the same contain in breadth.

Item, more, a Merchant hath 3 $\frac{1}{4}$ elles of Arras, at 1 $\frac{2}{3}$ elles broad, which he will change with another man for a peice of Arras, that is $\frac{7}{8}$ elles square. The question is, how many elles of that squarenesse ought the first Merchant to have?

Answer, Multiply the first merchants peice his length by the breadth, and you shall finde it containeth 5 $\frac{-1}{12}$ elles, which $\frac{65}{12}$ elles, you shall divide by $\frac{7}{8}$ and you shall finde 6 $\frac{-4}{21}$ elles, and so many elles of that squarenesse ought the latter merchant to give the first.

Item, A Student hath bought 3 $\frac{1}{2}$ yards of broad Cloth, at 7 quarters broad, to make a gown, and should line the same throughout with Lamb at a foot square each skin, the question is now how many skins he ought to have.

Answer,

Anfwer, Seek firft the number of yards fquare that his cloth containeth, which to doe, multiply $3\frac{1}{2}$ his length, by $1\frac{3}{4}$ his breadth, and you fhall finde $6\frac{1}{8}$ yards fquare : then fay by the *Rule of three,* if one yard fquare give 9 foot, what fhall $6\frac{1}{8}$? Work, and you fhall find $55\frac{1}{8}$ skins.

Item, *more, a Lawyer hath a rich peece of feeling come home which is 24 foot and 3 inches long, and 7 foot and $2\frac{1}{2}$ inches high : the Joyner is to be paid by the yard fquare : the queftion is how many yards this containeth.*

Anfwer, Multiply his length by his breadth, that is to wit, $24\frac{1}{4}$ foot by $7\frac{5}{24}$ foot, and you fhall find $174\frac{77}{96}$ foot fquare, which 174 you fhall divide by 9 (for fo many foot make a yard fquare) and you fhall finde 19 yards 3 foot and $\frac{77}{96}$ of a foot, and fo many yards doth this piece hold.

Item, *I bought a piece of Holland cloth containing 36 Els $\frac{1}{3}$ Flemmifh. The queftion is how many Ells Englifh it makes.*

Anfwer, You muft note, that five ells *Flemmifh* doth make but three ells *Englifh.*

Therefore fay by the *Rule of three,* if five ells *Flemmifh* make but three ells of *Englifh,* how many ells *Englifh* will $36\frac{1}{3}$ ells *Flemmifh* make ? Multiply and divide and you fhall finde $21\frac{4}{5}$ and fo many *Englifh* doth $36\frac{1}{3}$ ells *Flemmifh* contein. The like is to be done of others.

Item, *more, I have bought 342 Ells Flemmifh, of Arras work, at two Ells broad Flemmifh, and I would line the fame with Ell broad Canvas of Englifh meafure. The queftion is, how many Ells Englifh will ferve my turn ?*

Anfwer,

Anſwer, For as much as three ells *Engliſh* are worth five ells *Flemmiſh*, therefore put three ells *Flemmiſh* into his ſquare, in multiplying three by himſelfe, which maketh nine. Likewiſe multiply the *Engliſh* ell, which is five quarters, every way into himſelfe ſquarely, and you ſhall finde 25. Then multiply 342 which is the length of the peice, by 2, which is the breadth, and thereof commeth 684, then ſay by the *Rule of three*, as before : if 25 ells ſquare of *Flemmiſh* meaſure, be worth nine ells ſquare of *Engliſh* meaſure, what are 684 of *Flemmiſh* meaſure? Multiply and divide, and you ſhall finde 246 $\frac{4}{25}$ ells *Engliſh*.

The ſame is alſo wrought by the Backer *Rule of three*, in ſeeking the ſquares contained in the *Flemmiſh* ell of two ells broad (which are 18) and alſo in ſeeking the ſquares contained in the *Engliſh* ell (which are 25) then ſay by the *Rule of three* backward, If 18 quarters require 342 ells, what ſhall 25 quarters give? Multiply and divide by the *Rule of three* Reverſe, and you ſhall find as before 246 $\frac{4}{25}$ ells *Engliſh*?

Item, *more, at three ſhillings foure pence the Flemmiſh Ell, what is the Engliſh Ell worth after the rate?*

Anſwer, Say, if three quarters give 3 $\frac{1}{3}$ s. what giveth five quarters? Multiply and divide, and you ſhall finde 5 s. 6 $\frac{2}{3}$ d.

Item, *more, at 8 s. 4 d. the Flemmiſh Ell ſquare, what is the Engliſh Ell worth after that rate?*

Anſwer, According to the reaſon of the laſt Queſtion, conſider that a *Flemmiſh* ell ſquare is equall to nine quarters of a yard *Engliſh*, and an *Engliſh* ell ſquare is equall to 25 quarters of a yard. There-

F f   fore

fore say by the *Rule of three*, if 9 quarters give 1.$\frac{2}{3}$ s. what 25 quarters? Work and find 23 s. 1 $\frac{7}{9}$ pence And so is the *English* ell worth.

Item, *more, at* 6 s. 8 d. *the ell square : what shall a piece of Cloth cost that is* 7 $\frac{1}{2}$ *ells long, and* 3 $\frac{1}{4}$ *ells broad ?*

Answer, Multiply the breadth by the length, and you shall find 24 $\frac{3}{8}$ ells square cost 6 $\frac{2}{3}$ s. what 24 $\frac{3}{8}$ els? Multiply and divide, and you shall find 8 pounds, 2 s. 6 pence, and so much the same piece of cloth cost.

Item, *more, a Mercer sold* 3 *pieces of Silk. To wit* 24 $\frac{1}{4}$, 13 $\frac{2}{3}$ *and* 25 *yards, at* 9 $\frac{3}{4}$ s. *the yard, and was glad to receive in part of payment again, a cloth containing* 34 $\frac{1}{3}$ *yards at* 7 $\frac{2}{3}$ *shillings the yard. The question is now, what the Debtor is in the Creditors debt ? Work, and you shall finde he oweth the Mercer* 17 l. 8 s. 11 d $\frac{1}{4}$.

The tenth Chapter intreateth of reducing of Pawns of Geanes into English yards.

 Ote that 100 *Pawnes doe make* 26 *yards, whereupon three Pawnes* $\frac{11}{13}$ *doe make one yard, and one Pawn after the rate and proportion is* $\frac{13}{50}$ *of a yard.*

In 4563 *Pawnes of Geanes, how many yards english ?*

Answer, Say by the *Rule of three,* if a hundred Pawns do make 26 yards, what will 4563 Pawns make? Multiply and divide, and you shall find 1186 yards $\frac{18}{50}$ So many yards doe 4563 Pawns make.

Otherwise take some other number at your pleasure;

fure, as ten pawns, which is the $\frac{1}{10}$ part of 100, then to find his proportion, take the $\frac{1}{10}$ part of 26, which is 2 $\frac{3}{5}$; & then say also by the *Rule of three*, if ten pawns give 2 $\frac{3}{5}$ yards, what will 4563 pawns give? Work and you shall finde 1186 $\frac{18}{50}$ yards, as before.

More, at 2 s. 6 d. *the Pawns of Geans, what will the English yard be worth after the rate?*

Answer, Say by the *Rule of three*, if $\frac{13}{50}$ of a yard cost 2 $\frac{1}{2}$ s. what one yard? Multiply and divide and you shall finde 9 s. 7 $\frac{9}{13}$ d.

More, if 346 $\frac{1}{2}$ *Pawnes cost* 30 li. 13 s. 4 d. *sterling, what is that the English yard after the rate?*

Answer, Say by the *Rule of three*, if 346 $\frac{1}{2}$ Pawns cost 30 $\frac{2}{3}$ pounds what are 3 $\frac{11}{13}$ pawns worth (for so many Pawns make a yard?) Multiply and divide, and you shall find $\frac{2300}{27027}$ parts of a pound, which in known numbers is worth 6 s. 9 d. $\frac{6271}{9009}$.

The eleventh Chapter intreateth of Rules of Loan and Interest, with certain necessary questions and proofs incident thereunto, &c.

Tem, *I lent my friend* 326 *pounds for* 5 $\frac{1}{2}$ *months simply without any Interest, upon condition, to have the like courtesie againe when I need. But when I came to borrow, he could spare me but* 149 li. 8 s. 4 d. *The question is now how long time I ought to hvea the use thereof, to countervaile my friendshid before time shewed him?*

Answer, Say by the backer *Rule of three*, if 326 pounds give 5 $\frac{1}{2}$ months, what time will 149 $\frac{5}{12}$ pound,

give ? Multiply and divide, and you fhall finde twelve
moneths, and fo long time ought I to ufe his money.

The Proofe.

Item, lent my friend 149 *li.* 8 *s.* 4 *d. for twelve
months. The queftion is now how much mony he ought
to lend me again for* 5 ½ *moneths to recompence my
friendfhip fhewed him ?*

Anfwer, Say by the Backer or Reverfe *Rule of
three.* If twelve months give 149 $\frac{4}{12}$ what fhall 5 ½
months give ? Work, and you fhall finde 326 pounds,
and fo much ought he to lend me to requite my gentle-
neffe or good turn.

Two other branches, yet more, for proofe out of the fame queftion.

Item, lent my friend 149 *li.* 8 *s.* 4 *d. for* 12 *moneths,
to have the like friendfhip again when I need. And
comming to borrow of him, he very courteoufly took
me* 326 *pounds (for that he could well then fpare the
fame) The queftion is now, how long I ought to occupy
it, not ufurping friendfhip, but in his due time to reftore it again.*

Anfwer, Say by the *Rule of three* reverfe, if 149
$\frac{1}{12}$ pounds give 12 moneths, what fhall 326 pounds
give ? Multiply and divide, and you fhall finde
that at 5 ½ moneths terme, I ought to reftore it again.

Proofe.

Item, Lent my friend 326 *pounds for* 5 ½ *moneths
The queftion is now, how many pounds he ought to lend
me for* 12 *moneths to recompence this pleafure again.*

Scholar. Work by the *Rule of three* reverfe, as you
have

have done before, and you ſhall finde 149 li.——8 s.
——4 d.

Again four other ſelected queſtions, of Loan and
Intereſt, all out of one branch, and each one alſo a
neceſſary queſtion, and a particular proofe to other.

ITem, *Lent my friend* 430 *pounds at Intereſt for
three moneths, to receive after the rate of* 8 *pounds
in the* 100 *pounds for* 12 *moneths. The queſtion is
what the Intereſt commeth to.*

You may, if you pleaſe, work it at two workings
by the *Rule of three* direct, in ſaying, if 12 moneths
give 8 pounds, what giveth three moneths ? Multiply
and divide ; and it giveth 2 pound.

Then for the ſecond work ſay : If a hundred pound
yeeld 2 pounds, what yeeldeth 430 li ? Multiply and
divide, and you ſhall find 8 li. 12 s. and ſo much
comes the loane of 430 li. to for 3 moneths after the
rate of 8 pound in the hundred pounds of 12 moneths.

Otherwiſe wrought thus by the rule of three at
twice alſo.

If 100 pounds give 8 pounds, what giveth 430
pounds ? Multiply and divide, & you ſhall find 34 li.
$\frac{2}{5}$. Then again for the ſecond work ſay : If 12 moneths
give 34 pounds ? what giveth three moneths ? Worke
and find 8 li, 12 s. as before.

Otherwiſe yet at one working : By the firſt part
of the rule of five numbers forward , in ſaying, if
100 pounds in twelve moneths gain 8 pounds,
what ſhall 430 pounds gain in three moneths ? Mul-

tiply

tiply the firſt by the ſecond for your Diviſor, and the other three, the one into the other for the dividend, and you ſhall finde eight pounds 1 2 ſhillings, as afore-ſaid.

Proofe.

Item, *A friend of mine received of me 8 pounds 12 ſhillings for the Intereſt and Vſe of 430 pounds for three moneths terme: The queſtion is now, what he tooke in the 100 pound for 12 months after that rate?*

Anſwer, For moſt brief, ſay by the firſt part or rule of five numbers forward : If 430 pounds in three moneths did pay 8 pound 12 ſhillings, what doth 100 pound in 12 moneths take after the rate? Work, and you ſhall finde 8 pounds, and ſo much he took upon the 100 pounds for 12 moneths.

A third Queſtion and proof alſo by the Backer Rule of five Numbers.

Item, *I lent my friend 430 pounds to receive for the intereſt thereof, after the rate of 8 pounds in the 100 for 12 moneths. The queſtion is now how long time my friend ought to give the uſe thereof, that it may be returned with 8 pounds 12 ſhillings gains.*

You may work it, if you pleaſe, by the Rule of three direct at twice, in ſaying : If 100 pounds, yeeld 8 pounds, what yeeldeth 430 pound? Multiply and divide, and find 34 pound and $\frac{2}{5}$.

Then again for the ſecond work ſay, if 34 $\frac{2}{5}$ pounds, give twelve moneths, what giveth 8 $\frac{3}{5}$ pounds? Multiply and divide, and you ſhall finde three moneths, and ſo long time ought my friend to uſe it to return with 8 li. 12 s. gain.   Other-

Otherwiſe at one working by the Backer Rule of 5 numbers, in ſaying, if 100 li. in 12 moneths doe gain 8 li. how much time ſhall 430 li. be a gaining of 8 li. 12 s. ? Multipliy the firſt and the ſecond into the laſt for your dividend, and the third and fourth multiply together for your Diviſor, and then divide, and you ſhall finde three moneths, the juſt time that my friend ought to uſe it to return it with 8 li. 12 s. gain.

A fourth derived queſtion out of this Branch, which is a proof of this laſt, and alſo of the other two going before.

ITem, *How much money, ought a Merchant to de-liver after* 8 li. *in the* 100 *for twelve moneths, that in three moneths he may gain* 8 *pounds twelve ſhillings.*

Anſwer, You may alſo if you pleaſe, work it by the Golden Rule of three at twice : firſt ſaying, If three moneths give 8 ⅖ pound, what 12 moneths gain? You ſhall finde 34 ⅔. Then ſay again, If 8 pounds be come of 100 pounds, what ſhall come of 34 pounds 8 ſhillings ? Work, and you ſhall finde the anſwer to the queſtion, which is 430 pounds, and ſo much ought the Merchant to deliver.

But moſt briefly it is anſwered by the Backer Rule of 5 numbers, where I argue thus, ſaying : If 100 pounds be 12 moneths a gaining of 8 pound, then but for three moneths terme onely to take 8 pounds 12 s. muſt needs be a good round ſumme to work it, ſet your numbers thus: 100 ——— 12 — 8 ——— 3 —— 8⅖ multiplying the firſt into the ſecond, and alſo by 43

the

the product of the fifth, for your dividend, and the third and fourth together with 5. the Denominator of your fraction for your Divisor: then divide, and you shall find as before 430 pounds : the true solution to your question.

The twelfth Chapter intreateth of the making of Factors, which is taken in two sorts.

He first is, when the estimation of the Factor is taken upon the sending of the Merchant, as if the estimation of his person be $\frac{1}{4}$ it is understood that he shall have $\frac{1}{4}$ of the gain, the Merchant the other $\frac{3}{4}$.

The other sort is, when the estimation of his making is out of the sending of the Merchant, as if the order and agreement between them were such, that the Merchant shall put in 800 li. and the Factor for his making shall have $\frac{1}{4}$: neverthelesse he shall have but $\frac{1}{5}$ of the gain or profit, for the $\frac{1}{4}$ of 800 is 200 (for the estimation of his making) which with the 800 pounds in all make 1000 pounds, whereof the 200 li. is $\frac{1}{5}$.

A Merchant doth put in 800 pound into the hands of his Factor, under such condition, that the said Factor shall have the $\frac{1}{4}$. And after certain time they raise in profit 124 li. 6 s. 8 d. I demand how much the Merchant shall have hereof, and how much ought the Factor to have?

Answer. *When the estimation of the Factor is out of the sending of the Merchant, it maketh,*

I

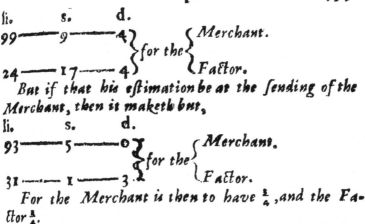

li.   s.   d.

99 —— 9 —— 4 } for the { Merchant.

24 —— 17 —— 4 }   { Factor.

But if that his eſtimation be at the ſending of the
Merchant, then it maketh but,

li.   s.   d.

93 —— 5 —— 0 } for the { Merchant.

31 —— 1 —— 3 }   { Factor.

For the Merchant is then to have $\frac{3}{4}$, and the Factor $\frac{1}{4}$.

A Merchant doth put into the hands of his Factor
800 pounds, and the Factor 400 li. to have the $\frac{1}{4}$ part
of the profit: I demand now for how much his per-
ſon is eſtemed, when the ſame is counted upon the ſending of the Merchant.

Anſwer. According to the Tenour and order before
preſcribed in the firſt Rule, that is, if his eſtimate be $\frac{1}{4}$
he ſhall have the $\frac{1}{4}$ of the gain. Therefore ſay by the Rule
of three direct: If $\frac{1}{4}$ taken put in 400 pound, what is the
eſtimate, or putting in of $\frac{1}{4}$ taking? Multiply and
divide, and you ſhall finde 320 pounds, and ſo much is
the perſon of the Factor eſtimated.

Otherwiſe.

To finde the eſtimation of the perſon of the Factor, you ſhall conſider, that ſeeing it was agreed
between them, that the Factor ſhould take the $\frac{1}{5}$,
then the Merchant ſhall have the reſidue, which
are $\frac{4}{5}$: wherefore the gain of the Merchant unto that
of the Factor is in ſuch proportion as 5 unto 4.
Then if you will know the eſtimation of the perſon
of

of the Factor, ſay, If 5 give 4, what will 400 give ? Multiply and divide and you ſhall finde 320 pound. And ſo much is the perſon of the Factor eſteemed to be worth.

Other conditions then theſe aforeſaid, may alſo be between Merchants and Factors, without reſpect eigher of ſending, or not ſending of the Merchant, where moſt commoly the eſtimation of the body of the Factor is in ſuch proportion of the ſtock which the Merchant layeth in, as the gain of the ſaid Factor is unto the gain of the merchant. As thus, if a merchant doe deliver into the hands of his Factor 400 pound, and he to have half the profit, the perſon of the ſaid Factor ſhall be eſteemed to be worth 400 pound : and if the Factor do take but $\frac{1}{3}$ of the gain, he ſhould have but $\frac{1}{2}$ ſo much of the gain as the merchant taketh, which muſt have $\frac{2}{3}$, wherefore the perſon of the Factor is eſteemed but the $\frac{1}{2}$ of that which the merchant layeth in, that is to ſay, two hundred pound.

And if the Factor did take the $\frac{2}{3}$ of the gain, then the merchant ſhall take the reſidue which are $\frac{1}{3}$, wherefore the gain of the merchants unto the Factor is then in ſuch proportion as 3 unto 2 : whereupon if you will then know the eſtimation of the perſon of the Factor, ſay, if 3 give 2, what ſhall 400 give ? Work, and you ſhall finde 266 $\frac{2}{3}$ pounds. And ſo much is the perſon of the Factor eſteemed to be worth.

And if the merchant ſhould deliver unto his Factor 400 pound, and the Factor would lay in 80, and his perſon, to the end he might have the $\frac{1}{2}$ of the gain, I demand how much ſhall his perſon be eſteemed ?

Anſwer, *Abate 80 from 400, and there will remain 320. And at ſo much ſhall his perſon be eſteemed.*

A

A Merchant hath delivered unto his Factor 900 pounds to govern in the Trade of Merchandize, upon condition that he ſhall have the $\frac{1}{3}$ of the gain if any thing be gained, and alſo to bear the $\frac{1}{3}$ of the loſſe, if any thing be loſt. Now I demand how much his perſon was eſteemed at ?

Anſwer, *Seeing that the Factor taketh the $\frac{1}{3}$ of the gain, his Perſon ought to be eſteemed as much as $\frac{1}{2}$ of the ſtocke which the Merchant layeth in : that is to ſay, the $\frac{1}{2}$ of 900 pound, which is 450. The reaſon is, becauſe $\frac{1}{3}$ of the gain that the Factor taketh is the $\frac{1}{2}$ of the $\frac{2}{3}$ of the gaine that the Merchant taketh, & ſo the Factor his perſon is eſteemed to be worth 450 pounds.*

A Merchant hath delivered unto his Factor 600 pounds, and the factor layeth in 250 pound, and his perſon. Now becauſe he layeth in 250 pounds, and his perſon, it is agreed between them, that he ſhall take the $\frac{2}{5}$ of the gain. I demand for how much his perſon was eſteemed ?

Anſwer, *For as much as the Factor taketh $\frac{2}{5}$ of the gain, he taketh $\frac{2}{3}$ of that which the Merchant taketh, for $\frac{2}{5}$ are the $\frac{2}{3}$ of $\frac{3}{5}$. And therefore the Factors laying in ought to be 400 pound, which is $\frac{2}{3}$ of 600 pound that the Merchant laid in. Then ſubtract 250, which the Factor did lay in, from 400 pound which ſhould have been his whole ſtock, and there remaineth 150 pound for the eſtimation of his perſon.*

More, a Merchant hath delivered unto his factor 800 li. upon condition that the factor ſhall have the gain of 160 li. as though he laid in ſo much ready mony : I demand what portion of the gain the factor ſhall take?

Anſwer, *See what part the 160 (which the Factor laid in) is of 960, which is the whole ſtock of their company,*

company, and you ſhall find ⅕ : And ſuch part of the
gain ſhall the Factor take.

But in caſe that in making their Covenants it were
ſo agreed between them, that the Factor ſhould have
the gain of 160 pound of the whole ſtock which the
Merchant layeth in, that is to ſay, of the 800 pound:
then ſhould the Factor take ⅕ of the gains, for 160 is ⅕
of 800 pound.

The thirteenth Chapter intreateth of Rules of Barter,
and exchanging Merchandize, which is diſtinct into ſeven Rules, with divers other neceſſary queſtions
incident thereunto.

The firſt Rule.

TWo Merchants, willing to change their Merchandize the one with the other : The one hath
24 broad clothes at 10 li. 10 s. the piece : The other
hath Mace at 12 ſhillings the pound. The queſtion
is, how many pounds of Mace he ought to give for
his Cloth, to ſave himſelfe harmleſſe, and be no loſer?

Anſwer, *Seek firſt by the* Rule *of three what the*
24 Cloths coſt at 10 pound 10 ſhillings the piece, and
you ſhall finde 252 pound : Then to finde the quantity
of Mace, ſay again by the Rule of three, If 12 ſhillings
buy one pound, what ſhall 252 pound buy me ? Works,
and you ſhall finde 420 pound of Mace : And ſo many pound ought he to give for his Clothes.

The Proof.

Two barter. The one hath 420 pounds of Mace.

a

at 12 s. the pound, to barter or change broad Clothes, at 10 pounds 10 shillings the piece. The question is, how many broad Clothes he ought to give for all his Mace?

Answer, *First say, If one cost* 12 *shillings, what* 420 *? you shall finde* 5040 *shillings. Then say again, If* 10½ *pounds give one cloth, what shall* 5040 *shillings give? Worke, and you shall finde* 24 *Clothes, your desire.*

The second Rule.

Two change merchandize for merchandize : The one hath Pepper at two shillings foure pence the pound, to sell for ready money. But in barter he will have no lesse then three shillings the pound. And the other hath Holland at five shillings six pence the Ell ready money. The question is now, at what price he ought to deliver the Ell in the barter to save himself harmlesse.

Answer, *Say by the* Rule of Three direct : *If* 2⅓ *ready money give* 3 *shillings in barter, what shall* 5½ *give in barter? You shall finde* 7 1/14 *shillings, and at that price ought the second merchant to sell his Holland in barter.*

The Proof.

Two barter. The one hath Holland at 5 s. 6 pence the Ell to sell for ready money. And in barter he will have 7 1/14 shillings. The other hath Pepper at 2 s. 4 pence the pound, to sell for ready money. The, question is now how he ought to sell in barter.

Answer, *Say by the* Rule of Three direct, *If* 5½ *ready mony give* 7 1/14 *shillings, in barter ; what ought* 2⅓ *to take in barter? multiply and divide and you shall finde* 3 *shillings your desire.*

The

Two barter. The one hath cloth of Arras at 30 s. the Ell ready money, but in barter he will have 35 ½ s. And the other hath white Wines which he delivered in barter for 16 pounds the Tun. The question is now, what his Wines cost the Tun in ready mony.

Answer, *Say by the* Rule of Three direct. *If* 35 ½ *shillings in barter, give but* 30 *shillings ready money, what did* 16 *pound in barter cost ? Worke, and you shall finde* 13 *pound* 10 *shillings* $\frac{30}{71}$*. And so much cost his Wines for a Tun ready money.*

The proofe.

Two barter Merchandize for Merchandize: The one hath white Wines at 13 pounds 10 s. $\frac{30}{71}$ s the Tun to sell for ready mony : But in barter he delivered it for 16 li. The other, to make his match good and save himself harmlesse, delivereth Arras at 35 ½ s. the ell. The question is now, what an ell of his Arras cost in ready mony ?

Answer, *Say by the* Rule of Three direct : *If* 16 *pounds in barter give but* 13 *pounds* 10 $\frac{30}{71}$ *shillings, in ready money, what shall* 35 ½ *shillings yeeld in barter Work, and you shall finde* 30 *shillings your desire.*

The fourth Rule.

Two barter : The one hath Kerseyes at 14 *poune the piece ready money: But in barter he will will hat* 18 *pounds : and yet he will have the* ⅓ *part of his over price in ready money. And the other hath Ginger eight groats the pound to sell for ready money. T question is, how he ought to deliver the Ginger by t pound in barter to save himselfe harmlesse, and ma the barter equall.* Answe

Answer, *Item*, for the working of this queſtion, and ſuch other the like, you muſt underſtand, if the party over-ſelling his wares, require to have alſo ſome portion in ready mony, as $\frac{1}{2}\frac{2}{3}\frac{3}{4}$. &c. Then ſhall you firſt rebate the ſame demanded part, whatſoever it be, from the over price, and alſo from the juſt price. And thoſe two numbers that ſhall remain after the ſubtraction is made, ſhall be the two firſt numbers in the *Rule of Three*. And the juſt price of the ſame Merchandize ſhall be the third number, which by the operation of the *Rule of Three direct*, ſhall yeeld you a true ſolution, how, and at what price you ſhall overſell that your merchandize, to have your ſelf harmleſſe, and make the barter equall.

Example.

Take the $\frac{1}{3}$ (of eighteen) which is the over-price of his Cloth, which $\frac{1}{3}$ of eighteen is ſix, which you muſt ſubtract from 18, there reſt 12. And alſo abate it from 14, which is the juſt price of the Cloth and there remaineth 8, which 8 and 12 are the two firſt numbers in the *Rule of Three*. Then take eight groats, or 2 $\frac{2}{3}$ ſhillings for the third number. Then ſay by the *Rule of Three direct*. If eight pounds give 12 pounds, what ſhall 2 $\frac{2}{3}$ s. give? Multiply and divide, and you ſhall finde 4 ſhillings. And for ſo much ſhall the ſecond Merchant ſell his Ginger, or his commodity in barter, to ballance the ſame equall.

$$14 \text{———} 18$$
$$6 \qquad\qquad 6$$
$$\text{———} \qquad \text{———}$$
$$8 \qquad\qquad 12$$

The Proof.

Two barter; the one hath fine Kerſeys, at 14 pounds the piece ready money: But in barter he will have 18 pounds:

*pounds : and yet he will have the ⅓ part of his overprice
in ready money. And the other hath Ginger, which he
having cunning enough to make the barter equall, delivered in barter for 4 shillings the pound. The question is now, what his Ginger cost him in ready money?*

Answer, After you have made the subtraction,
abating 6 the ⅓ part of 18, both from 18 and 14 (as
before was taught you :) then will there remain 8 and
12 for your two first numbers in the *Rule of three.* Then
say, If 12 give 8, what shall come of 4 the over-price
of the pound of Ginger? Multiply and divide, and
you shall find 2 s. 8 pence, your desire.

*Two Merchants barter Merchandize for Mer-
chandize. The one hath* Devonshire *whites at 7 pound
13 shillings 4 pence the peice ready mony : but in barter he doth them away for 8 pound, 3 shillings 4 pence,
and yet he will have the ⅓ part of his price in ready
money. And the other hath Cottens at three pounds the
piece ready money. The question is now, at what price
he ought to sell or exchange his Cottens in barter to
save himselfe harmlesse, and make the barter equall?*

li.	ſ.	d.	li.	ſ	d.
7	13	4	8	3	4
2	14	5⅓	2	14	5⅓
4	18	10⅔	5	8	10⅔

Answer, First seek the ⅓ part of 8 li. 3 s. 4 d.
which is 2 li. 14 s. 5 ⅓ d. which rebated from 8 li. 3 s.
4 d. there resteth as appeareth by the Example above-
said, 5 li. 8 s. 10 ⅔ d. which is ⅔ parts of 8 li. 3 s. 4 d.
also rebated from 7 lib. 13 s. 4 d. there resteth 4 li.
18 s. 10 ⅔ d. the two first numbers in the *Rule of
Three,* and the three pounds, which is the neat price
of

of the piece of Cotten, is the third number : Then say by the *Rule of Three direct*, as was taught before. If 4 li. 18 s. 10 $\frac{2}{3}$ d. give 5 li. 8 s. 10 $\frac{2}{3}$ d. what shall three pounds give ? Multiply and divide, and you shall find three pounds 6 $\frac{6}{8\cdot3}$ s. the just price that he ought to deliver his Cottens in barter.

The fifth Rule.

Two Merchants will change Merchandize for Merchandize. The one hath Kerseys at 40 s. *the piece to sell them for ready money. And in barter he will sell them for* 56 s. 8 d. *and he will gain after ten pound upon the* 100 *pound. And yet he will have the* $\frac{1}{2}$ *of his over-price in ready money. The other hath Flax at* 3 d. *the pound ready money, The question is now, how he shall sell the pound of his Flax in barter ?*

Answer, See first at 10 pound upon the 100 pounds what the 56 $\frac{2}{3}$ s. commeth to, in saying (by the *Rule of Three direct*) If 100 pounds give 110 pounds, what 56 $\frac{2}{3}$ s ? Multiply and divide, and you shall finde 3 pound 2 shillings 4 pence, of which the $\frac{1}{2}$ that he demandeth in ready money, is 1 pound 11 shillings 2 pence; the same 31 s. 2 d. abated from 40 shillings, and also from 56 s. 8 d. there will remaine 8 s. 10 d. and 25 s. 6 d. for the two first numbers in the *Rule of Three*, and 3 pence the price of the pound of Flax for the third number. Then multiply and divide, and you shall find 8 $\frac{3\cdot1}{5\cdot3}$ d. And for so much shall he sell the pound of Flax in barter.

The sixt Rule.

Two are willing to exchange Merchandize : The one hath Norwich *Grograns at* 25 s. *the piece ready money :*

G g money :

money: and in barter he wiſſ have 30 s. and he will have the ¼ part of his over-price in ready money. The other hath Norwich *Stockins*, at 40 s. the dozen to ſell for ready money. But in as much as the firſt *Mer-chants Grograns* are no better, he would deliver them ſo to ballance the barter, that he may gain 10 pounds in the 100 pounds. The queſtion is now, how he ſhall ſell his hoſe the dozen in barter, according to his requeſt.

Anſwer, Say, if 100 give 110 li. what ſhall 40 s. give, which is the juſt price of the dozen of ſtockins? Multiply and divide, and you ſhall finde 44 s. Then take the ¼ of 30 s. which is 7 s. 6 d. and ſubtract it from 25 s. and alſo from 30 s. and there will remaine 17 s. 6 d. and 22 s. 6 d. for the two firſt numbers in the *Rule of three,* and 44 ſhillings, which is the juſt price (with his gain in the dozen of Stockins) for the third number. Then multiply and divide, and you ſhall finde 56 s. 6 6⁄7 d. and for ſo much he is to ſell his dozen of ſtockins in barter.

The ſeventh Rule.

Two Merchants will change their Merchandize one with the other: The one hath 720 *Ells of Cam-bricke at* 5 s- *the Ell to ſell for ready money, but in barter he requireth* 6 s. 8 d. *And yet notwithſtand-ing, he loſeth by it after* 10 *pounds upon the hundred pounds, whereupon he requireth one halfe of his over-price in ready money: and the other Merchant ha-ving skill enough to make the barter equall, deliverd Engliſh Saffrons at* 30 s. *the pound, The queſtion is now, what his Saffrons coſt the pound in ready money.*

Anſ. You muſt firſt ſeeke what is loſt upon the 100 li. which to do, you may ſay (if you pleaſe) if 10 pou pound lose 10, what shall 6 $\frac{2}{3}$ lose? Work, and you shall finde $\frac{2}{3}$ s. (or 8 d.) which must be rebated from 6 s. 8 d. so resteth 6 s. still. Or you may say, If 100 pound give me but 9 pounds, what shall 6 s, 8 d. give? Work this way either, and you shall find also as before directly in your quotient 6 s. your desire. Then are you next to cast up what the 720 Ells of Cambrick commeth to at 6 s. 8 d. the Ell, and you shall find 240 pounds : the $\frac{1}{2}$ whereof the Cambrick Merchant will have in ready money (which is 120 pounds :) Nextly, you must cast what the Cambrick commeth to after his losse in the 100 pound, which as you found, is but 6 s. an Ell, and you shall find 216 pounds : Now must you subtract his ready money (which is 120 pounds in all) out of 240 pound, and also out of 216 pound, and there will remain 120 pounds, and 96 pounds for your two first numbers in the *Rule of three*, and 30 shillings is the over-price of your Saffron for the third number : Then multiply and divide, and you shall finde 24 shillings. And so much did his Saffron cost in ready money.

Two Merchants barter ; the one hath fifty Clothes to put away for ready money at 11 pounds the Cloth, and in barter putteth them away for 12 pounds, taking Holland Cloth at 20 d. the Flemmish Ell, which was worth no more but 18 d. The question is now, what Holland payeth for the Cloth, and what he winneth or loseth by the bargain?

Answer, Fifty Clothes at 11 pounds the Cloth commeth to 550 pounds, and put away at 12 pounds the piece, maketh 600 pound. Then to finde what Holland payeth for the Cloth, say by the *Rule of three direct*, If 20 d. buy one Ell, what 600 pounds ?

Gg2Worke,

Worke, and you shall finde 7200 Ells. Now to finde the Estate of his gain or losse, you must seek what his 7200 Ells commeth to at 18 d. the Ell : Work by the Rule of proportion direct, and you shall finde 540 pounds, which is not so much as his Clothes were worth in ready money by tenne pounds : and so much lost the first Merchant by his Exchange.

A Venetian hath in London 100 peices of silk, to put away for ready money at 3 li. the piece. But in barter he delivered them for 4 li. the piece, taking wools of a Felmonger at 7 li. 10 s. the C. weight, which was worth no more but six pounds the C. ready money. The question is now, what wools payeth for the silkes, & which of them winneth or loseth by the barter?

Ans. A hundred pieces of Silke at 3 li. is in all 300 li. and at 4 li. is 400 li. Then to find what Wools pay for the Silke, say by the *Rule of three direct* : If 7½ li. buy me 1 C. weight, what 400 ? Work, and finde 53 ⅓ C. weight of wooll. Now to find the Estate of their gaine and losse, cast up his wooll at 6 lib. the C. (for so much they were worth ready money) and you shall finde 320 pound which is 20 pound more then the Silks were to be sold for ready money, whereby the *Venetian* gained 20 pounds by the Barter.

A Merchant hath 53½ weight of Wooll at 6 pounds the C. to sell for ready money, but in barter he will have 7 pounds 10 s. and another doth barter with him for Silks, which are worth three pounds a piece ready money. The question is now, how he ought to deliver his Silks the peice in barter, and how many payeth for the wools.

Answer, Say by the Rule of proportion, (or by the *Rule of three direct*) If, 6 pounds for C. weight ready money,

money, yeeld me 7 li. 10 s. what will 3 li. yeeld, which is the juſt price of a piece of Silk in barter, to make the Truck equall? Work, and find 3li. 15 s, the price of a piece of Silk, in barter : then ſay, If 3 li. 15 s. require one piece of Silk, how many pieces of Silk are bought with 400 pound, which is the value of 53 ½ C. weight of wooll, at 7 li. 10 s? Worke by the *Rule of three direct*, and you ſhall finde 160 pieces of Silk and ⅔ of a piece, and ſo many of Silk pay for the wooll, and neither party hath advantage of other.

Two men will change Merchandize, the one with the other. The one of them hath beer at 6 s. 8 d. the barrell, to ſell for ready money, but in barter he will ſell the barrell for 8 s. and yet he will gain moreover after 10 pound upon the 100 pounds. And the other hath white Spaniſh *Wooll at 20 s. the Rove, to ſell for ready money. The queſtion is now, how he ſhall deliver the Rove of Wooll in barter to ſave himſelf harmleſſe.*

Anſwer, Say, if 6 ⅔ s. which is the juſt price of the barrell of Beer, be ſold in barter for 8 ſhillings : for how much ſhall 20 ſhillings (which is the juſt price of the Rove of Wooll) be ſold in barter? Work by the Rule of Three direct, and you ſhall finde 24 s. Then for becauſe the firſt Merchant will gain after 10 pounds upon the 100 pounds, he maketh his 100 pounds, 110 pounds. And therefore ſay by the Rule of Three, If the ſecond Merchant of 110 pounds doe make but 100 pounds, how much ſhall he make of 24 s? Multiply and divide, and you ſhall finde 21 s. 9 d. $\frac{2}{11}$ of a peny. And for ſo much ſhall he ſell the Rove of Wooll to be delivered in barter, to the end the firſt Merchant may give 10 in the 100.

Two Merchants will change their Commodities the

one

*one with the other. The one of them hath white Paper
at* 4 s. *the Reame, to ſell for ready money. And in barter he will doe it away for* 5 s. *and yet he will gain
moreover after the rate of* 10 *pounds upon the* 100
pounds. And the other hath Mace at 14 ſ. 6 d. *the
pound weight to ſell in barter. Now I demand what
the pound did coſt in ready money.*

Anſwer, Say, if 5 s. (which is the over-price of
the Paper in barter) be come of 4 s. the juſt price, of
how much ſhall come 14½ ſhillings, which is the
ſurprize of the pound of Mace in barter ? Multiply
and divide, and you ſhall find 11⅖ s. Then for becauſe the firſt Merchant of Paper will gain after 10
upon the 100, Say if 100 doe give 110, what ſhall
11⅖ ſhillings give ? Worke and you ſhall find 12 s.
9⅗ d. and ſo much did the pound of Mace coſt in
ready money.

The fourteenth Chapter intreateth of exchanging of
money from one place to the other.

Xchange is no other thing, then to take or
receive money in one City, to render or pay
the value thereof in another City, or elſe
to give money in one place, and receive the
value thereof in another, as term of cer-
tain dayes, moneths, or fairs, according to the diverſit
of the place.

But this practice chiefly conſiſteth in the knowledg
of the money or Coynes in divers places, of which for th
benefit (after a few examples given to the Introduct
ion of this work) I will ſet down certain notes of th
diverſit

diverfity of the common and ufuall Coynes in moft places in Chriftendome for trafficke.

And firft I will begin at *Antwerp*, where they ufe to make their accounts by *Deniers de groffe*, that is to fay, pence *Flemmifh*, whereof 12 doe make 1 s. *Flemmifh*, and 20 s. doe make one pound *de groffe*.

Item, *A Merchant delivered at* Antwerp, 400 pound Flemmifh *to receive in* London 20 s. *fterling*, for every 23 s.———4 d. Flemmifh : *The queftion is now how much fterling money is to be received at* London *for* 400 *pounds* Flemmifh ?

Anfwer, Say by the Rule of Three, If 23 ½ *Flemmifh* give 20 s. fterling, what 400 pounds *Flemmifh*? Worke, and you fhall finde 342 l———17 s.——1 $\frac{5}{7}$ pence, and fo much fterling fhall I receive in *London* for the faid 400 pounds *Flemmifh*.

Otherwife alfo wrought by Rules of Practice in taking the $\frac{4}{7}$ of the *Flemmifh* money delivered, and abating the fame from the principall, the reft is *Englifh* money, as before.

$$400 \text{ li} ———— 0 \text{ s.} ———— 0 \text{ d.}$$
$$57 ———— 2 ———— 10 \tfrac{2}{7}$$

$$342 ———— 17 ———— 1 \tfrac{5}{7} \text{ fterling.}$$

A Merchant at London *delivered* 200 lib. *fterling for* Antwerp, *at* 23 s.———5 d. Flemmifh *the pounds fterling : the queftion is, how much he muft receive at* Antwerp.

Anfwer, Say by the Rule of Three, if 1 pound fterling give 23 s. 5 d. *Flemmifh*, what 200 li. fterling? Work, and thou fhalt finde 234 li.———3 s.——4 d. So many pounds *Flemmifh* fhall he receive at *Antwerp* for the faid 200 pounds fterling.

O.her-

Otherwiſe by Practice.

```
1 ——— 3 ——— 5 ——— 200
3 s. 4 d.                    33 ——— 6 — 8
     1 d.                        ——— 16 — 8
maketh ſterling ——— ——— 234 li. — 3 s. 4 d
```

In London 20 *pound ſterling is delivered by Ex-*
change for Antwerp, *at* 23 s. 9 d. Flemmiſh *the*
pound ſterling: the queſtion is, at what rate the Flem-
miſh money ought to be returned to gain foure pounds
upon the 100 *pound ſterling at* London.

Anſwer, Firſt, ſay by the Rule of three direct :
If 1 pound ſterling give 23 ¾ Flemmiſh, what 200
pounds ſterling ? Multiply and divide and you ſhall
finde 237 pounds 10 ſhillings. The which to return
to gain 8 pounds ſterling in *London,* ſay by the backer
Rule, If 200 pounds ſterling require the exchange
23 s. 9 d. Flemmiſh, what the exchange to make 208
li. ſterling ? Work by the Rule, and finde 22 s. 10 $\frac{1}{26}$ d.
Flemmiſh, the effect in the queſtion required.

If I take up money at Antwerp *after* 19 s. 4 d.
Flemmiſh, *to pay for the ſame at* London, 20 s. *ſter-*
ling, and when the day of payment is come, I am
forced to return the ſame money again in London, *to*
pay my Bill of Exchange : ſo that for 20 *ſhillings*
which I take up here at London, *I muſt pay* 19 s. 6 d.
at Antwerp, *I demand whether I doe win or loſe, and*
how much in or upon the 100 *pounds of money ?*

Anſwer, Say by the Rule of three : If 19 ½ give 19⅜
what will 100 pounds give ? Multiply and divide, and
you ſhall finde 99 li. 2 $\frac{6}{117}$ s. which being abated
from 100 pounds, there will remain 17 s. $\frac{1}{7}$, and ſo
much I do loſe upon the 100 pounds of money.

If

If I take up at London *20 shillings sterling to pay at* Antwerp *22 s. 4d. and when the day of payment is come, my Factor is constrained to take up money again at* Antwerp, *wherewith to pay the aforesaid summe, and there he doth receive 23 s. 4d.* Flemmish, *for the which I must pay 20 s. at* London : *The question is now, whether I doe win or lose, and how much upon the 100 lib. of money after that rate.*

Answer, Say by the Rule of Proportion, If 22 $\frac{1}{3}$ s. give 23 $\frac{1}{3}$ s. what will 100 pounds give ? Multiply and divide, and you shall finde 104 pounds 9 shillings $\frac{12}{67}$, from the which abate 100 pounds, and there will remain 4 pounds 9 shillings $\frac{12}{67}$, and so much is there gained upon the 100 pounds of money.

In Antwerp *is delivered 200 pounds* Flemmish *by exchange for* London *at 20 shillings sterling for every 23 shillings 4d.* Flemmish, *The question is, at what rate the same is to be returned to gain 10 pounds upon the 100 pounds* Flemmish *in* Antwerp?

Answer, First, say by the Rule of three, if 23 $\frac{1}{3}$ Flemmish give 20 s. what shall 200 pounds gain ? Work, and you shall finde 171 pounds 8 s. 6 $\frac{6}{7}$ d. then say again by the Rule of three direct, if 171 pounds 8 s. 6 $\frac{6}{7}$ d. sterling, give me 210 pounds Flemmish, what shall 20 s. sterling give ? Work, and you shall finde 24 s. 6 d. Flemmish. And at the same rate ought the same to be returned at *Antwerp,* to gain 10 pounds upon the 100 Flemmish.

A Merchant of Antwerp *delivereth 234 lib. 3 s. 4 d.* Flemmish, *to receive at* London *200 li. sterling : The question is now, how the exchange goeth after this rate?*

Answer, Say by the Rule of three direct, if 200 give 20, what 234 $\frac{1}{6}$? Multiply and divide, and you shall

ſhall find 23 s. —— 5 d. And for ſo much goeth the exchange.

Item, *the exchange from* London *into* France, *is not like as it is to* Flanders, *but it is delivered by the* French *Crown, which is worth* 50 Soulx Turnois *the piece.*

Whereupon alſo you muſt note that in *France* they make their accounts by Franks, Soulx, and Deniers Turnois, whereof 12 Deniers make one Soulx Turnois, and 20 Soulx maketh one pound Turnois, which they call a Liure or Frank. But the Merchants to make their Accounts, doe uſe French Crowns, which is currant among them for 51 Soulx Turnois. But by exchange it is otherwiſe, for it is delivered but for 50 Soulx Turnois the Crown, or as the taker up of the money can agree with the deliverer. And note that this △ Character repreſenteth the Crown by exchange , and is ever 50 Soulx Turnois or *French* money.

A Merchant delivereth at London 240 *pounds ſterling, after five ſhillings ſixe pence the Crown, to receive at* Paris 50 *Soulx Turnois for every Crowne. I demand how much* Turnois *or French money payeth the Bills for the ſaid* 240 *pounds ſterling.*

Anſwer, Say by the Rule of Three. If 5½ s. ſterling give me 50 s. Turnois, what ſhall 240 pounds ſterling give? Reduce the pounds into ſhillings, then multiply and divide, and you ſhall finde 2181 Liures, 16 Soulx. 4 Deniers, and $\frac{4}{11}$ Turnois, and ſo much payeth the Bills at *Paris,* for the 240 pound ſterling.

A Merchant delivereth at Roan, *or elſewhere in* France, 1430 *pounds or franks, the which franke or pound is* 20 Soulx, *or a pound* Turnois *to receive in*
London

London 6 s. 4 d. *sterling for every* △ *of* 50 *Soulx* Turnois. *The question is, how much sterling money I ought to receive* at London *for my* 1430 pound Turnois.

Answer, Say, if 2 ½ pounds give me 6 ⅓ s. what will 1430 give me? Worke, and you shall find 3622 ⅓ shillings sterling, which maketh 181 li. 2 s. 8 d. and so much money is to be received at *London,* for the said 1430 Liure Turnois, after 6 s. 4 d. for every △ of 50 Soulx.

In London *is delivered* 200 *pound sterling by exchange for* Paris, *at* 5 s. 9 d. *the* △ *of* 50 *Soulx* Turnois. *The question is, at what price the said* △ *is to be returned to gain* 6 *pounds upon the* 100 *pounds sterling at* London.

Answer, First, say (by the Rule of Three direct) if 5 ¼ s. sterling give 50 Soulx Turnois; what shall 200 pound sterling give? Work, and you shall find 1739 Franks of Liures, 2 ¼⁴/₂₅ Soulx. Then the which to return and gaine 6 pounds upon the hundred pounds in *London,* say by the Rule of Three direct, if 1739 Franks 2 ¼⁴/₂₃ Soulx yeeld 212 pound, what the △ of 50 Soulx? work and finde 6 s. 1 ¹/₁₀ d. the effect required in the question.

A Merchant delivered in London 160 li. sterling *to receive in* Biskay *for every* 5 s. 6 d. *one Ducat of* 374 Marvides. *The question is, how many* Marvides *ought I to receive at* Biskay?

Answer, Say, if 5 ½ s. sterling give 374 Marvides: what shall 160 pounds sterling give? Multiply and divide, and you shall finde 217600 Marvides, and so many I ought to receive at *Biskay* for my 160 pounds sterling.

A

A Merchant delivered in Brion, 40000 Marvides *to receive in* London 5 s. 8 d. *ſterling for every Ducat of* 374 Marvides. *The queſtion is now, how much ſterling money payeth the Bills of Exchange for the ſaid* 40000 Marvides?

Anſwer, Say, if 374 Marvides make one Ducat, what 40000 Marvides? Multiply and divide, and finde, $106\frac{178}{187}$.

Then ſay again, if 1 Ducat give $5\frac{2}{3}$ s. what giveth $106\frac{178}{187}$ Ducats? Work, and find 30 l. 6 s. $\frac{24}{561}$ s. Otherwiſe it is wrought more briefe at one working, as in the laſt queſtion before, in conſidering that 5 s. 8 d. containeth one Ducat, or 374 Marvides. Therefore ſay by the Rule of Three, if 374 Marvides give $5\frac{2}{3}$ s. what 40000 Marvides? Work, and you ſhall alſo finde in your quotient 30 l. 6 s. $\frac{24}{561}$ s. And ſo many pounds ſterling is to be received for the 40000 Marvides.

In London 200 *pounds delivered by exchange for* Vigo, 374 Marvides *the Ducat of* 5 s. 10 d. *ſterling, maketh* $256457\frac{1}{7}$ Marvides: *the which to return and gain* 10 li. *upon the* 100 *pounds in* London, *ſay by the* Rule of three direct, *if* 220 li. *require* 256457 $\frac{1}{7}$ Marvides: *what* 5 s. 10 d? *Work, and finde* 340 Marvides, *the price of every Ducat in return, which is the effect in the queſtion required.*

Theſe may ſeem ſufficient for inſtructions.

Notwithſtanding *for thy further aid and benefit, hereafter follow ſix ſpeciall and moſt briefe* Rules of Practice, *for* Engliſh, French, *and* Flemmiſh *money.*

How

1) (How to turn Flemmish to English sterling.
2 / \ How to turn English sterling to Flemmish.
3 (teacheth) How to turn Flemmish to French.
4 (/ How to turn French into Flemmish.
5) / How to turn Sterling into French.
6) (How to turn French into Sterling.

The fifteenth Chapter intreateth of the said six Rules of brevity, and of valuation of *English*, *Flemmish* and *French* money, and how each of them may easily bee brought to others value.

How briefly to reduce pounds, shillings, and pence Flemmish into pounds, shillings, and pence English sterling.

IT is to be noted, that 7 pounds *Flemmish* maketh but 6 pounds sterling: 7 s *Flemmish* maketh 6 s sterling, and 7 d. *Flemmish* 6 d sterling so that 7 yeeldeth but 6. Wherein is evident that then is left $\frac{1}{7}$, (if it may be so called) when it is reduced into *English* money: wherefore to know how much 233 l.——13 s.—— 4 d. *Flemmish* maketh *English*, you must subtract from it $\frac{1}{7}$, beginning with the pounds, &c. and that which resteth after this subtraction, is the summe required: so that 233 li.——13 s.—— 4 d. *Flemmish*, maketh 200 li. 5 s. 8 $\frac{4}{7}$ d. sterling.

Example

Example.

li.	s.	d.
233 —— 13 —— 4		
$\frac{1}{7}$ 33 —— 7 —— $7\frac{1}{7}$		
200	5	$8\frac{4}{7}$ Ster.

Another Example

li.	s.	d.
311 —— 0 —— 0		
$\frac{1}{7}$ 44 —— 8 —— $6\frac{6}{7}$		
266	11	$5\frac{1}{7}$

Rule 2. *To reduce pounds, shillings, and pence sterling, into pounds, shillings, and pence* Flemmish.

Note that a pound sterling maketh 1 li. 3 s. 4 d. *Flemmish*: that is, 1 $\frac{1}{6}$ li : 1 s. sterling maketh 1 $\frac{1}{6}$ s. *Flemmish*, and 1 d. sterling maketh 1 $\frac{1}{6}$ d. *Flemmish*. So that there is gained (if it may be so called) $\frac{1}{6}$ of the summe being thus reduced to *Flemmish*, for of 1 $\frac{1}{6}$ is made $\frac{7}{6}$ which is one whole and $\frac{1}{6}$. Then to know how much 237 li. 7 s. 6 d. sterling maketh *Flemmish*, Subtract from your sterling the $\frac{1}{6}$ of the whole summe, & adde it to the same summe, & it maketh 276 li. 18 s. 4 d. which is the summe required.

Example.

li.	s.	d.
237 —— 7 —— 6 Ster.		
$\frac{1}{6}$ 39 —— 11 —— 3		
276 —— 18 —— 9 Flem.		

Another Example.

li.	s.	d.
337		
$\frac{1}{6}$ 56 — 3 —— 4		
393 —— 3 —— 4		

Rule 3. *To reduce pounds, shillings, and pence* Flemmish, *into pounds, shillings, and pence* French.

Ye shall note, that the equality of Flemmish and French money is this, that is to say, the pound Flemmish, maketh 7 pound $\frac{1}{5}$ French, or Turnois. 1 s. Flemmish maketh 7 $\frac{1}{5}$ s. French, and a groat Flemmish maketh 7 $\frac{1}{5}$ d. French.   Where-

Wherefore to know how much 143 li. 4 s. 9 d. Flemmiſh maketh French, ye muſt multiply the whole number twice by *6*, beginning at pence, and ſo forward, and the Product of your ſecond multiplication divide by 5, ſo the work is finiſhed. Or multiply the ſaid ſumme by 7, and take out of it $\frac{1}{5}$ adding it to the Product of your multiplication by 7, and that is your number required. So that as well by the one as by the other, 143 li. 4 s. 9 d. Flemmiſh, maketh 1031 li. 6 s. 2$\frac{1}{3}$ d. French or Turnois.

Example.		The ſame otherwiſe.	

li.	s.	d.		li.	s.	d.
143	4	9 Flem.		143	4	9
		6				7

859	8	6		1002	13	3
		6		$\frac{1}{5}$ 28	12	11$\frac{3}{5}$

5156	11	0 Fren.		1031	5	2$\frac{2}{5}$
$\frac{1}{5}$ 1031	6	2$\frac{2}{5}$ Fren.				

Another Example.		Or thus :	
143 l. Flem.		143	
6		7	

858		1001	
6		$\frac{1}{5}$ 28	12

$\frac{1}{5}$ 5148 French.		1029 li.	12

1029 li. $\frac{1}{5}$ or 12 s. French.

Rule 4.

To reduce pounds, shillings, and pence, French, into pounds, shillings, and pence, Flemmish.

Multiply 233 li.——8 s.——4 d French by 5, and divide the Product twice by 6, that is, the said number by 6, and the Product or Quotient again by 6, and the quotient of this second Division is the thing required. So that 233 li.--8 s.--4 d. French, maketh 32 li.-- 8s.--4 $\frac{5}{9}$ d. Flemmish.

Example.			Another Example.		
li.	s.	d.	li.	s.	d.
233——8——4 Fren.			753 French.		
	5		5		
1167——1——8			3765——		
$\frac{1}{6}$194——10——3$\frac{1}{3}$			$\frac{1}{6}$627——10——		
$\frac{1}{6}$32——8——4$\frac{5}{9}$ Flem.			$\frac{1}{6}$104——11-8 Flem.		

Rule 5.

To reduce pounds, shillings, and pence Sterling, into pounds, shillings, and pence, French, or Turnois.

The pound Sterling maketh 8 li. 8 s. *French*, that is to say, 8$\frac{2}{5}$ pounds: the shillings maketh 8$\frac{2}{5}$ shillings, and the peny 8$\frac{2}{5}$ d. *French*. Wherefore to know what 231 li. 13 s. 4 d. Sterling maketh *French*, ye must multiply the whole summe by 42, that is, by 7. and the Product of it by 6. and divide this second Product by 5. and that is the summe required.

Otherwise, multiply the summe Sterling by 8, and adde twice to the Product $\frac{2}{5}$. and it shall produce the summe required. So that both waies 231 li.--13 s. --4 d. Sterling, maketh 1945 pound *French*, as here under followeth.

Example

Example.			*Sier.*		*The same otherwise.*			Sterling:
li.	s.	d.			li.	s.	d.	
231 —— 13 —— 4					331 —— 13 —— 4			
		6					8	
————————————					————————————			
1390 —— 0 —— 0					1853 —— 6 —— 8			
		7			46 —— 6 —— 8			
————————————					46 —— 6 —— 8			
⅟₇ 9730 —— 0 —— 0					————————————			
⅟₆ 1946 —— 0 —— 0 *Fren.*					1946 —— 0 —— 0			French.

Another example.		*fter.*	The fame.	Sterling.
753			753	
6			8	
———————			———————	
4518			6024	
7		½	150 ——— 12	
———————			——————————	
31626		½	150 ——— 12	French.
⅟₅ 6325 —— 4 *Fren.*			6325 —— 4	Rule 6.

To reduce pounds, shillings, and pence, French, *into pounds, shillings, and pence, sterling.*

To know how much **1256** li. **12** s. **6** d. *French* maketh in sterling money : multiply the summe by 5, and divide the Product by 7 and 6 at twice, and the last Quotient shall be the thing required, that is to say, 1256 li. 12 s. 6 d. maketh 149 pounds, 11 s. 1 1⁵⁄₇ d. sterling.

H h   Example.

	Example,			Another example.		
li.	s.	d. French.	li.	s.	d.
1256	12	6	2531	0	0 French.
	5			5	
6283	2	6	12655		
$\frac{1}{6}$1047	3	9	$\frac{1}{6}$2109	3	4
$\frac{1}{7}$149	11	11$\frac{4}{7}$ Ster.	$\frac{1}{7}$301	6	2$\frac{2}{7}$Ster.

Note, that when any money is given by exchange at *London* for *Roan* at 71½ d. or rather 71$\frac{1}{7}$ for the Crown of 50 s. *French*, there is neither gain nor losse : for it is one money for another, accounting 8 li. 8 s. *French*, for one pound sterling. So the giver loseth the time of payment, which is about 15 dayes, and he that taketh it, hath the gain of the same.

They of *Roan*, that put forth or take money by exchange for *London*, ought to have like consideration.

Item, When any man giveth at *London* 64 pence$\frac{1}{3}$, or rather 65 $\frac{2}{7}$d. to have at one of the fairs of *Lions* a Crown *de Marc*, he that so giveth the money, loseth the time, and he that taketh it, gaineth the same : for 62 pence $\frac{3}{7}$ is equall in value to 45 s. *French*. He that putteth or taketh mony at *Lions* for *London*, ought to consider the same.

Item, when any deliver in *Antwerp* 75 pence, to receive at *Lions* a Crowne *de Marc*, he that putteth it forth, loseth the time, and he that take it gaineth the same. For 75 groats *Flemmish* is equall in value to 45 s. *French*,

Thus for this time I make an end of the Practice of
Exchange,

Exchange, and the instructions thereunto belonging, and according to my promise : yet further to gratifie such as are desirous to know the common Coynes used for trafficke among merchants in these Cities following, a briefe declaration of their moneyes, and the reckonings, and account of them I will here set downe.

The sixteenth Chapter containeth a Declaration of the valuation and diuersity of Coyns of most places of Christendome for traffick ; And the manner of exchange in those places from one City or Town to another : which known, is right necessary for Merchants, by means whereof they do finde the gain or losse upon the exchange.

*I*Tem, for as much as the greatest diversity of money of exchange is at *Lions* ; therefore I will begin duely of the money of that place.

At *Lions* they use Franks, Soulx, and Deniers Turnois. A Frank maketh 20 Soulx, and one Soulx 12 Deniers ; but the Merchants to keep their Bookes of accounts, doe use French crownes of the marke at 45 Soulx the piece, and doe divide it into 20 Soulx, one Suolx is 12 Deniers.

Item, a mark of gold maketh 65 △ of the mark, which serveth for exchange, and divide it into 8 ounces, the ounce into 24 pence or deniers, the denier into 24 grains, and so the summe or whole by imagination or guesse.

△ This mark standeth for a Crown.

Also at *Lions* there are four Fairs in a year, at the which they doe commonly exchange, which are

from

from 3 moneths to three moneths.

At *Geans* they uſe the Soulx : one Ducat maketh 3 pound.

At *Naples* they uſe Ducats, Taries, and Grains ; the Ducat maketh five Taries, and one Tary 20 Grains : but they take 6 Dueats which maketh 30 Taries for the ounce.

A Ducat maketh ten Carlins, and a Carlin ten grains, ſo that 2 Carlins make a Tary, and 100 Grains make a Ducat.

At *Rome* they uſe the Ducats of the Chamber ; one Ducat is worth 12 Guillis, and one Guillis ten Soulx.

At *Venice* they uſe Ducats currant at 124 Soulx a piece, or 24 Deniers, & one Denier maketh 32 Picolis.

At *Palerme* and *Meſſina* they write, after ounce, tary, and grains, and one ounce is worth 6 Ducats of 30 Taries, and 1 tary is 20 Grains, and 1 grain 6 Picolis, 1 Ducat is alſo worth 24 Carlins.

At *Millan* they uſe li. s d, of Ducat Imperials, and $\triangle$ of exchange is worth 4 li.

At *Lucques*, *Florence*, and *Ancone*, they uſe the $\triangle$ of Gold : in Gold the *French* Crown is worth 7 li. but at *Boloigne* 3 li. 10s.

At *Barcelone* they uſe the Soulx : the Ducat of exchange is worth 22 Soulx.

At *Valence* and *Saragoſſe* they uſe the Liver, Soulx, and Denier : the *French* Crowne of exchange is worth 20 Soulx, and one Soulx is 12 Deniers.

At the Fairs of *Caſtile* they uſe the Marvides, the Ducat is worth 375 Marvides.

At *Lisbone* they uſe the Rayes, one Ducat of exchange is worth 400 Rayes.

At *Norenburg*, *Frankeford*, and *Auguſt* in *Germany*

many, they ufe the *Krantzers*, whereof 60 make a Floren.

At *Antwerp* they ufe li. s. d. de Gros, and they exchange into the Denier de Gros, to wit, our Englifh peny.

At *London* they ufe the li. the s. and d. fterling, and they exchange in pence fterling.

The exchange of Lions *at fundry places.*

Item, at *Lions* there is exchange in three forts, at the Cities and towns following.

Firft, they deliver at Lions one mark to have or receive at Naples almoft 41 ½ Ducats, at Venice 70 ducats currant, at Rome 63 ducats of the Chamber, at Lucques and Florence 65 △ of Gold, at Millan 82 △.

And contrariwife, at the faid Cities aforefaid they doe give fo much of money, to have a mark of Lions.

Secondly, they give at Lisbone one of △ of Mark of 45 Soulx Turnois a piece, to have at Geans almoft 68 Soulx, at Palerme and Meffine almoft 24 Carlins, at Barcelone 22 Soulx, at Valence or Saragoffe 20 Soulx, at the Faire at Caftile 350 Marveides, at Lisbone 360 Rayes, in Antwerp 57 Deniers de Gros, and at London, 70 d. fterling.

And contrariwife, they give in the faid Cities almoft as much of their money to have a French Crown of the mark at *Lions*.

Thirdly, they do give at *Lions* a △ of the Sun to have almoft 93 Krentzars at *Frankeford*, *Augf-burg*, *Noremberge*, or other Cities in *Almain*.

Alfo at *Lions* onely they do pay, they change the ⅔ in Gold, and ⅓ in money, or elfe all in money, in giving 1 ½ for the hundreth.

Changes

Changes at Naples *and other Towns.*

Item, at *Naples* they give or deliver almoſt 112 Ducats, to receive at *Rome* 100 Ducats of the Chamber at the old value.

Through *Lucques* and *Florence* they deliver 100 Ducats Carlins, to receive there almoſt 86 △ of Gold.

Through *Palerm* and *Meſſine,* one Ducat of 5 Tary, to receive there almoſt 164 grains.

Through *Millain,* one Ducat to receive there almoſt 90 Soulx.

Through *Geans,* one ducat to receive there almoſt 65 Soulx. The whole ſumme to be paid within ten dayes after the ſight of the Bill of exchange.

Alſo at *Naples,* they deliver one ducat to receive in *Antwerp* almoſt 67 d. or Deniers *de Gros,* within two moneths. At *London* almoſt 60 d. ſterling in three moneths. At *Barſelone* almoſt 20 Soulx within two moneths.

At *Valence* almoſt 18 Soulx within two moneths At *Lisbone* 333 Rayes within three moneths : and at the Fair at *Caſtile* almoſt 340 Marvides.

Change of Venice *to other places.*

At *Venice* they deliver 100 Ducats currant to receive in *Almain* almoſt 140 Florens at 60 Krentze the piece.

At *Lucques* and *Florence* almoſt 108 △ of Go in 10 dayes.

Likewiſe at *Venice* they deliver a Ducat curra to receive at *Palerm* and *Meſſine* almoſt 21 Carlin at *Millan* almoſt 93 Soulx : at *Geans* almoſt Souls, the who'e at 10 dayes end.

Of the Pair or Pari.

As touching the exchange, it is necessary to under-stand or know the *Pair*, which the *Italians* call *Pari*, which is no other thing then to make the money of the change of one City or Town, to or with the money of another, by means whereof they doe finde the gains or losse upon the exchange.

Example.

Item, having received Letters of credit of one of *Antwerp*, that the △ of the sunne is there worth 7 Soulx : The question is, what the same is worth at *London*, when the *Pair* of exchange goeth for 23 shillings ?

Answer, *Say, if 23 give but 20, what giveth 7 ? Worke, and finde 81. 1.½ d. and so much is the △ of the Sun worth at London.*

The seventeenth Chapter containeth also a Declaration of the diversity of the Weights and Measures of most places of Christendome for traffique. At the end of which discourse are two Tables, the one for Weight, and the other for Measure, proportionate and reduced to an equality of our *English* Measure and Weight, by the aide whereof, the ingenious may easily by the *Rule of Three*, convert the one into the other at pleasure, &c.

T *London*, and so all *England* thorow, are used 2 kindes of Weights and Measures, as the *Troy* weight, & the *Haberdepoise*. From the *Troy* weight is derived the proportion and quantity of all kinde of dry and liquid Measures, as Pecks, Bushels, Quarters, &c.

H h 4 Where.

wherewith is bought and ſold all kinde of grain and
other Commodities meet by the Buſhell.   And in
liquids Ale, Beere, Wine, Oyle, Butter, Honey, &c,
Upon theſe grounds and Statutes is bread made, and
ſold by the *Troy* weight : and ſo is Gold, Silver,
Pearle , precious Stones , and Jewels. The leaſt
quantity of this *Troy* weight is a grain : twenty four
of theſe graines make a penny weight, twenty penny-
weights an Ounce, and twelve ounces a pound, two
pounds or pints of this weight maketh a quart. And
ſo aſcending into bigger quantities, is produced the
Meaſures whereby are ſold our other naturall ſuſte-
nance : *viz.* Ale or Beere, with all other neceſſary
commodities, as Butter, Hony, Herrings, Eeles, Sope,
&c. All which laſt before rehearſed, though their
Meaſures (wherein they are contained) bee framed
and derived from the *Troy* weight, yet are they in
traffique with divers Commodities, as Lead, Tinne,
Flax, Wax, with all other commodities, both of this
Realme, and of other forraine Countries whatſoever,
bought and ſold by the *Haberdepoiſe* weight after
ſixteen ounces to the pound, and 112 pound to the
hundred weight. And to every hundred is allowed
but 12 pound weight at the common-beam. From
hence is alſo derived the weight of *Suffolke* Cheeſe,
which containeth thirty two Cloves, 8 pound to a
Clove, and weigheth in all 256 pounds. And alſo
the Barrell of *Suffolke* Butter is, or ſhould be of like
weight with the weight of Cheeſe, *viz.* 256 pounds.
More 14 of theſe pounds make a Stone, and 26 Stone
containeth a Sacke of *Engliſh* Woll : Forraine Wool
to wit, *French*, *Spaniſh*, and *Eſtrich*, is alſo ſold by
the pound, or C. weight, but moſt Commonly by the
Rove

See further
of theſe
Weights
and mea-
ſures in
Reduction.
beginning
pag. 111.

Rove, 25 pound to a Rove : other commodities of
Tale, are bought and fold by the C fivefcore to the C.
except headed ware, to wit, Cattell, Nails, & Fifh, which
are fold after fixfcore to the C.   There are alfo two
other fort of meafures, to wit, the Ell and the yard.
By the Ell is ufually meete Linnen cloth, as Canvas,
&c. And by the yard Silks, woollen clothes, &c.

Antwerp.

At *Antwerp* are alfo two forts of weights, their
Gold and Silver weight and their common weight.
Gold and Silver is weighed by the mark, the mark is
8 ounces, the ounce 20 efterlings, and the efterling 32,
as our grains. The Goldfmiths divide that into
fmaller, but not the Merchants.   The proof of Gold
is made by Karects, whereof 24 maketh a marke of
fine Gold, the Karect is 24 grains : the proof of the
money is made by Deniers : 12 Deniers is 1 s. fine,
that is, a mark of fine filver : the Denier alfo is divided into 24 grains, & the grain into foure quarters.

Item, 100 marks in *Antwerp* Troy weight, maketh at *Lions* 103 marks, 2 ½ ounces, and 20 grains
23 d. At *Noremberg* 103 marks 2 ¼ ounces, 2 Quints,
3 Deniers; at *Frankeford* 105 marks : at *Ausburg*
104 marks, 3 ounces, 1 Quint : at *Venice* 103 marks, 1
ounce, 7. Deniers. 18 grains : at *London* 66 pounds.

The Mark of gold or filver at Antwerp, *Troy
weight, which is 8 ounces, maketh* 7 ½ *ounces common
weight, with which all other Merchandize is weighed.
So that the Troy weight is greater then the common
weight by* 6 ¼ *in the* C. *By this weight of Troy, they
also weigh* Musk, Amber, Pearl, &c.

All filks are bought at Antwerp, *by the Barges Ell,
which*

which is greater then the common meaſure, by which they retaile by 2 in the hundred. Their common Ell is $\frac{3}{4}$ of our yard, and $\frac{3}{5}$ of our Ell.

Lions.

At *Lions* is uſed 3 ſorts of weight, whereof the firſt is the common Towne weight, with which they weigh all kind of Spicery, and divers other Merchandize. The ſecond is called *Geneva* weight, which is 8 in the C. greater then the common weight, with which they weigh Silks, &c. The third is *French* weight, called commonly the mark weight, and 100 pounds thereof maketh 106 $\frac{1}{4}$ li. *Geneva*, and 114 $\frac{1}{2}$ of their common weight: with which *French* weight, is weighed all things that paid Cuſtome or Toll.

At *Lions* is alſo uſed two ſorts of Ells or Aulnes. The one wherewith they meaſure groſſe clothes, as Canvaſſe, and ſuch like. The other is called the *French* Ell or Aulne, with which they meaſure all other kind of Merchandize, whereof 7 common Towne Ells maketh 11 ordinary *French* ells.

Roan.

At *Roan*, 6 $\frac{1}{2}$ Muides of Salt, being the meaſure of the place, make a C. at *Armviden* in *Zeland*, and the C. of *Bronage* meaſure of *Armviden*, maketh a *Roan* 11 Muides. 30 Muides make a Laſt of Corn, and 16 a Laſt of Oats, 100 pound weight there, maketh at *London* 114 $\frac{1}{2}$, and 190 $\frac{1}{4}$ at *Antwerp*. And 20 Ells make at *London*, 115 $\frac{3}{4}$.

Noremburge.

A 100 pound weight at *Noremburge* maketh

Lord

London 111 ¾; at *Antwerp* 107 ⁴⁄₇. and 100 Ells at *Noremburge* make at *London* 75 ⅕, at *Antwerp* 95 ⅗ &c.

Lisbone

The C. weight at *Lisbone* maketh 4 Roves, every Rove 32 pounds, ſo that their C. weight is 128 pounds, and their pound containeth 14 Ounces, and 100 pounds of their weight maketh at *London* 113 ⅛.

Their Silke, Cloth of Gold, and Woollen is meaſured with a meaſure which they call a cubit, containing about ¾ of a Vare of *Caſtile*. Howbeit their common Meaſure is called a Varre, which maketh five Palms, and containeth 1 ¼ of a Varre of *Caſtile*, our Ell of *London* is equall with the Varre of *Lisbone*.

All kinde of Merchandize brought from *Flanders*, *Roan*, or *Brittain*, payeth at *Lisbone*, as a duty or cuſtome to the King, 20 in the C. which they call the tenth in Merchandize, and the other tenth in money.

Note alſo, that all kinde of Merchandize comming to *Lisbone* by land, payeth leſſe in cuſtome, then that that commeth by water.

Sivill.

The Rove of *Sivill* is 30 pound, 4 Roves make their C. weight, which is 120 pounds. The 100 pounds of *Sivill* maketh at *London* 102 pounds. Their other common meaſure is a Varre, whereof 100 maketh at *London* 74 Ells, and at *Rome* 40 Canes, &c.

Venice.

At *Venice* bee two ſorts of weight, the one called *la Groſſe*, the other *la Suttle*; with the groſſe is weighed

weighed all kinde of great wares, and with the ſmall all kind of ſpicery, and ſuch like : 96 pounds of groſſe weight there maketh at *London* 100 pound, and 100 pounds of ſpicery there, without any tare or allowance, make at *London* 94, and with tare 65.

Their owne common Meaſure are Braces, whereof 100 make at *London* 55 ¼ ells, at *Antwerp* 92 ½, &c.

Florence.

At *Florence* the 100 lib. weight maketh at *Aquila*, for Saffron, 110, and 145 pounds of *Florence*, make at *Roan* but 100 pounds ; the weight of *Florence*, and that of *Lucque* is all one.

Their other meaſures are Braces, whereof 100 maketh at *Antwerp*, *Burges* meaſure, 81 ⅖ els, 100 Braces there, make at *London* 49 els, &c.

Lucque.

The *Lucque* Sattens commonly ſold at *Lions* by weight, and 133 ⅓ pounds maketh at *Lions* 100 pound, ſo that 1 pound ⅓ maketh at *Lions* but one pound.

Their other meaſures are Braces whereof 100 of them make at *London* 50 ells, at *Antwerp* 83 ⅓ ells, &c.

Aquila.

At *Aquila*, their 100 pounds maketh at *London* 71 ¾, their 136 ⅔ pounds of ſaffron maketh at *Geneva* but 100, and 11 li. of *Geneva*, maketh 15 li. at *Aquila*.

Valentia.

At *Valentia* be two ſorts of weights, a great and

a

a small. The C. weight or great weight containeth four Roves, the Rove 36 li. so the C. great weight is 144 li. and the C. weight small containeth but 120 pounds, and is also parted into four Roves, which is 30 pounds to a Rove. By the small is sold the scarlet grain, with all other kind of spicery, and by the great is sold wooll, with all such like grosse wares. The 1⅕ pounds of Silke at *Valentia*, maketh at *Lions* one pound *Geneva* weight. The charge of great Merchandize at *Valentia* containeth 432 pounds, and in small wares 360 pounds.

The weight here and at *Barsellone* is all one.

Their 100 pound weight maketh at *London* 78 pound, at *Antwerp* 75.

Dansicke.

At *Dansick* or *Spruceland* the rule is, that whosoever buyeth any merchandize there, buyeth it by the ship-pound, which, is 320 li. 20 Lispounds make a ship-pound, and the Lispound containeth 16 pound, which ship-pound of *Dansick* maketh at *Antwerp* 266⅔ li. Their 100 li. weight maketh at *London* 86, &c.

Their other common measures are ells, whereof 100 maketh at *London* 72¼, and at *Antwerp* 120⅛ ells.

Toulhouse.

At *Toulhouse* 6 cabes of Woad maketh a charge, two cisterns of corn-measure, and all kinde of grain maketh a charge, the Cistern weigheth 160 l. weight of that place. Their 100 in weight maketh at *London* but 91¼ pound.

Geans.

At *Genua* or *Geans*, 100 li. of their weight maketh at *London* 71 ¼, and at *Antwerp* 68 ⅗. 100 li. weight at *Genua*, maketh at *Venice*, to Suttle wit, 106 li.

Their other common Measures are *Palmes*, whereof 100 make at *London* 20 ⅗ Ells, and at *Antwerp* 34 ⅗.

The rest are supplied in two Tables, which hereafter followeth: whereby the ingenious may gather his desire.

The Table of the agreement of the weights of divers Countries, the one with the other, being reduced to an equality, as followeth.

112 pounds weight at London, make at		112 pounds weight at London make at	
Antwerp	107 ⅕	Venice gross weight.	105 ⅜
Frankeford	099	Venice suttle weight.	166 2/7
Colen and Ausburgh	102 ¼	Aquina	157 ⅐
Noremburg.	100 ⅓	Vienna	089 ⅜
Roan	098	Preslaw	134 ⅔
Paris	102 ¼	Leipsig	101 ¼
Lions	118 ⅕	Dansick	129 ½
Deep	100 ¼	Lubeck	097 ½
Geneva	090 ⅓	Barcellon	144
Toulouse	122 ¼	Lisbone	099
Rochell	124 ¼	Geans	157
Marsellis	124 ¾		
Sivill, &c.	109 ¾		

The other Table of agreement of Measures of divers Countries reduced unto an equality, by the aid whereof you may with the use of the Rule of three convert either more or lesse of any one Measure unto the other.

Antwer

Antwerp	100	
Noremburg	104 $\frac{1}{2}$	
Frankeford	125	
Leipfig	125	Ells.
Preflaw	125	
Danfick	183	
Vienna in *Auftria*	87	
Lions in *France*	60 $\frac{3}{4}$	Aulnes.
Paris in *France*	57	
Roan in *Normandy*	52	
Lisbone	60	
Sivil in *Spain*	81	
Caftile in *Spain*	81	Varres.
Methera Ifles	62	
Venice	108	
Lucques	120	Braccs.
Florence	122 $\frac{1}{2}$	
Millain	138	
Rome	90	Canes.
Geans	288 $\frac{6}{13}$	Palms.

60 Ells, or 75 Yards at London make at

The eighteenth Chapter treateth of Sports, and Paſtimes, done by number.

IF you would know the number that any man doth think or imagine in his minde, as though you could divine, bid them triple it, or put twice ſo much more to it as it is, which done, aske him whether it be even or odd; if he ſay odde, bid him take one to it, to make it even, and for that one, keepe one in your minde. Now after he hath taken one to it, to make it

even,

even, bid him give away half, and keep the other half
for himſelf, which when he hath done, bid him triple
that half, and again, after he hath tripled it ; aske him
whether it be even or odde : if hee ſay odde, then bid
him take one to make it even again, and for that laſt
one, keep two in your minde : now after hee hath made
his number even, bid him caſt away the one halfe, and
keep the other ſtill, from which halfe that he keepeth,
cauſe him ſubtilly to put away or give you nine out of
his number, as oft as he can, and for each 9 that he
giveth you, keep in mind, and thereunto joyn the 3
which I bad you keep, and you ſhall have your
deſire.

Example.

*Imagine he thought 7, the triple whereof is 21, and
becauſe it is odde; he is to take 1 to make it even,
which firſt 1 given, is for you to keep in mind. Then
the halfe of his 22 being caſt away, he reſerveth ſtill
11, which after you have bid him triple, it maketh 33 :
then in giving of him one again to make it even, upon
that laſt 1 reſerve 2 in your minde, then his halfe of
34 maketh 17 ; from whence he can give you 9
but once. Therefore that yeelding to you 4, and the 3
that you keep, make 7 your deſire.*

Another kind of Divination, to tell your friend
how many pence of ſingle pieces, reckoning them one
with another, he hath in his purſe, or ſhould think in
his mind.

*Which to doe, firſt bid him double the pieces he hath
in his purſe, or the number he thinketh (if he partici-
pate his number or ſecrecie unto ſome one friend that
ſitteth by him that can but multiplie, and adde never*

so

fo little, if their number be great, then fhall they worke
as you bid them fo much the furer.)

Now after he hath doubled his number, bid him
adde thereunto 5 more ; which done, bid him multiply that his number by 5 alfo ; which done, bid him
tell you the juft fumme of his laft multiplication,
which fumme the giver thinketh it nothing available,
becaufe it is fo great above his pretended imagination :
yet thereby fhall you prefently with the help of Subtraction tell his propofed mumber.

The Rule is this.

Imagine he thought 17, double 17, and it   17
maketh 34, whereunto if you adde 5, it mak-   2
eth 39 : which multiplied by 5, as here is   ——
practifed, it yieldeth 195, which 195 is the   34
fumme delivered you in the worke : then for   5
a generall Rule you fhall evermore cut off   ——
the laft figure toward your right hand, with   39
a dafh of your pen, as here is performed, as   5
a figure nothing available unto your work,   ——
and then rebate 2 from your firft figure, after   195
5 is cut off, and the reft fhall evermore be   2
your defire, as by this example doth ap-   ——
peare.   17

Another of a Ring.

If in any company you are difpofed to make them
merry by manner of divining, in delivering a Ring
unto any one of them, which after you have delivered
It unto them, that you will abfent your felfe from
them, and they to devife after you are gone,
which of them fhall have the keeping thereof, and
that you at your return will tell them what

I i   perfon

person hath it, upon what hand, upon what finger, and what joynt: Which to doe, cause the persons to sit down all in a row, and to keep likewise an order of their fingers: now, after ye are gone out from them to some other place, say unto one of the lookers on, that he double the numbers of him that hath the Ring, and unto the double bid him adde 5. and then cause him to multiply the Addition by 5, and unto the product bid him adde the number of the finger of the person that hath the Ring. And lastly, to end the work, beyond that number, towards his right hand, let him set down a figure signifiying upon which of the joynts he hath the Ring, as if it be upon the second joynt, let him put down 2. Then demand of him what number he keepeth, from the which you shall abate 250, and you shall have three figures remaining at the least. The first towards your left hand, shall signifie the number of the person which hath the Ring, the second or middle number shall declare the number of the finger, and the last figure towards your right hand shall betoken the number of the joynt.

Example.

Imagine the seventh person is determined to keep the Ring upon the fifth finger, and the third joynt: first double 7, it maketh 14, thereto adde 5, it maketh 19, which multiplyed by 5, yeeldeth 95, unto which 95, adde the number of the finger, and it maketh 100: and beyond 100 toward the right hand, I set down 3 the number of the joynt, all maketh 1003, which is the number that is to be delivered you, from which abating 205, there resteth 753, which prefigureth unto you the seventh person, the fifth finger, and the third joynt.

But

But note, that when you have made your ſubtraŝi-
on, if there doe remain 0, *in the place of tens, that is to*
ſay, in the ſecond place, you muſt then abate 1, *from*
the figure which is in the place of the hundreds, that
is to wit, from the figure which is next your left hand,
and that ſhall be worth 10 *tenths, ſignifying the tenth*
finger, as if there ſhould remain 803, *you muſt ſay,*
that the ſeventh perſon upon his tenth finger, and upon
his third joynt, hath the Ring.

Another of three Dice.

If a man doe caſt 3 Dice, you may know the
points of one of every of them. For if you cauſe him
to double the points of one Die, and to the double to
adde 5, and the ſame ſumme to multiply by 5, and
unto the produĉt adde the points of one of the other
Dice ; and behind the number towards the right hand,
to put the figure which ſignifieth the points of the laſt
Die, and then aske what number he keepeth, from
which abate 250, and there will remaine 3 figures,
which doe note unto you the points of every Die.

Another of things hidden.

If three divers things are to be hidden of three
divers perſons, and you to divine which of the three
perſons hath the three divers things, doe thus : imagine the three things to be repreſented. *A,B,C.* Then
ſecondly, keep well in your mind which of the perſons you mean to be the firſt, ſecond, and third. Then
take 24 Counters or Stones, and your three things,
and give *A* to the party whom you imagine to be
your firſt man, and therewithall give him one of your
24 Counters in his hand, & *B,* unto your ſecond man,

I i 2   and

and therewithall 2 Counters, and *C*, unto your third
man, and therewithall 3 Counters : and leave the reſt,
which are 18, ſtill among them : which done, ſeparate
your ſelfe from them, and afterwards bid them change
the things among them as they ſhall think good :
which done, after they are agreed, bid him that hath
ſuch a thing, as before you have repreſented by *A*,
for every Counter that he hath in his hand, to take up
as many more.   And for him that hath *B*, for every
one in his hand to take up 2. And for him that hath *C*,
for every one in his hand to take up 4, and the reſt of
them to leave ſtill upon the board. Theſe 3 things,
and the three perſons being fully printed in your
minde , come to the Table , and you ſhall evermore finde out of theſe 6 numbers, 1, 2, 3, 4, 5, 6, or
7. If therefore one remaine ſtill upon the board, then
have they made no exchange, but keep them ſtill as
they were delivered unto them. So that the firſt man
hath *A*, the ſecond *B*, and the third man *C*. But if 2
remain, then the firſt man hath *B*, your ſecond man
A, and your third man *C*. The reſt of the work and
the order thereof are here apparent by the Table following.

	1	A			1	B
1	2	B		5	2	C
	3	C			3	A
	1	A			1	C
2	2	B		6	2	A
	3				3	B
	1	C			1	C
2	2	B		7	2	B
	3				3	A

A

Another divination of a number upon the Cafting of two Dice.

Firft let the Cafter caft both the Dice, and mark well the number : then let him take up one of them, it maketh no matter which, and look what number it hath in the bottom, and adde all together : then caft the Die again, and keep in his mind what all toge-ther maketh : then let the Dice ftand, and bring feven with you, and thereunto adde the reft of the pits that you fee upon the upper fide of the Dice, and fo many did the Cafter caft in all.

FINIS.

An Appendix concerning the Refolution of the *Square* and *Cube* in Numbers, to the finding of their fides by *Ro. Hartwel.*

Figurate Number is a number made by the multiplication of one number or more by another.

The fides of a figurative number, are the numbers by whofe multiplication it is made.

A figurate number what.

The fides of a figu-rate num-ber what.

A Figurate number is two-fold, as { Plain.
{ Solid.

And it is { Of one Multiplica-tion.
{ Or confequently of many. } as a { Plain.
{ Solid.

And in each { Both Æquilater.
{ And Inæquilater.

A

A figurate number made of one multiplication by two sides or numbers multiplyed together, is called a plain figurate number.

A plain figurate number.

For every number made by the mutuall multiplication of two numbers, may be called a *Plaine*, because it bringeth forth a rightangled parallelogramme, according to his unites disposed in *length* and *beadth*, the *sides* whereof are the two multiplying numbers. As the number 20, made by the mutuall multiplication of 4 and 5 is called a *Plaine*, and the *sides* thereof are 4 and 5 as here

* * * * *
* * * * *
* * * * *
* * * * *

Because the unites thereof disposed in *length* and *breadth*, as the *sides* do expresse, do bring forth an *inæquilater Parallelogram*, for that the numbers, or *sides* are inequall.

By like reason 36 made by multiplication of 6 by 6, is called an *Æquilater plain*, for the *sides* thereof 6 and 6 are equall.

Moreover one and the same *plain number* may have many *sides*, as the *plain number* 24, hath sides and six: 3 and eight: 2 and 12. For it is produced from the mutuall multiplication of these numbers whereupon for the invention of the sides, to wit, inæquilater *Plains*, it is needfull to give one of the *sides* by which, the *plain* it selfe divided, the other *side* made known. As the *plain* 48 being divided by the *side* 8, the *quotient* 6 is the remaining *side*. Notwithstanding another resolution and inquisition do happen in the *sides* of the *Æquilater plains*.

An Equilater Plain or quadrat what.

An Æquilater plain is a number made by two equall sides, or by any number multiplyed by it selfe is vulgarly called a square or quadrat; by the
Arabia

Arabians Zenſus, *it is commonly expreſſed by this note z, by us q.*

A quadrat or ſquare in *Geometry* is called a right lined *plain figure,* made by four equall right lines, and ſo many right Angles, & every one of the lines is called the ſide of the quadrat, as this *figure* a b c d, whoſe ſide is *a b,* or b c, as alſo c d, and *a d.*

To the ſimilitude hereof, that number is called a *Qua-drate,* which is made by the multiplication of two equall numbers, or of one in it ſelf, of which manner 36 is made, by 6 multiplyed in it ſelfe, or by the mutuall multiplication

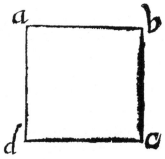

of 6 and 6. For if 36 unites be placed in plain forme, it bringeth forth a perfect *Geometricall Quadrat,* having in every ſide ſixe unities as here.

The number whereof the Quadrat is produced by multiplication in it ſelfe, is called the ſide or root of the Quadrat.

The ſide or root of a number what.

Concerning the extraction of the Qua-drat or ſquare Root.

THerefore to find the *Quadrat Root,* or the *ſide* of any *Quadrat number,* is to ſearch a number, which brought or multiplyed in it ſelf, maketh the number propounded : concerning the finding whereof, as it is requiſite that the *ſides* (being leſſer then 10) of

the

the *squares* under an hundred should be gathered by the Table of Multiplication: so the *sides* of the *greater squares* are to be sought out by *Art*. First, the *squares* whose *sides* are simple numbers, are here set down as you see.

1	2	3	4	5	6	7	8	9
1	4	9	16	25	36	49	64	81

The knowledge of a square is by finding out his side expressed by a whole number.

Although the finding out of the *side* of a *square* be applyed to each number given, as to a *square*, yet *square numbers* only have a *side* to be expressed by a certain number of unites, or by rationall numbers, the other are to be expressed but onely in power. The *sides* are commonly called *Roots* by a *Metaphoricall phrase*.

The Root or side of a square is to be found by the Theorem following.

If the odde degrees of a *square number* being marked from the right toward the left hand with points you subduct from the number given, the particular *square* of the last *period*, setting the *side* thereof alone by it selfe.

Then going on, if you divide the remainder (if there be any) with the figure going before it, by the double of the side set alone by it selfe.

And multiply the *quotient*. found out (being placed by the side, which was first set alone by it selfe, and also before the doubled number on the right hand by both the numbers (namely by the double number, & the Figure set by it selfe) being counted as of divisor, subducting the products from the given number her, and then renue this laſt work of diviſion ſo many times as there are pricks remaining, the *ſide* of the *ſquare* ſhall be found out.

This artificiall device is taken out of the 4 Pro. 2 l. of Euclide. Where by demonſtration it is proved, that if a right line be cut into two ſegments, *howſoever the* ſquare *of the whole line is* equall *to the* ſquares *of the* ſegments, *and the two right-angled figures made of the* ſegments *as in the figure annexed, the two Diagonals, k g, and b f, are the* ſquares *of the* ſegments, *a b, and b c. Alſo the complements b k, and f g, are the right-angled figures made by multiplying the line a b, by b c.*

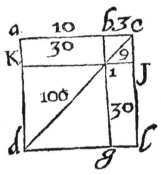

To extract the ſquare roo

The ſelfe ſame parts are to be found in any *ſquare number.* As for *example,* let the *number* be 169, whoſe *ſide* is 13. This *ſide* being divided into two pieces, 10 and 3, multiplying each pieces, by it ſelfe once, namely, 10 by 10, and 3 by 3: then multiply one by another, as 10 by 3, and 3 by 10, ſo ſhall you have 4 *plain numbers,* whereof 2 are *ſquare,* as here you ſee.

The firſt example.

Therefore as the *ſquare* 169 is made by adding together of theſe 4 *plain numbers,* ſo by ſubducting them ſeverally it is reſolved.

10	3
10	3
100	
30	
30	
9	

169

First therefore I mark each odde place with points, becauſe the *particular ſquares* are to be found in the odde places. Then for ſo much as the unity ſtanding under the firſt point next the left hand, and repreſenting the laſt *period,* is both a *ſquare* and the *ſide* of a *ſquare:*

square: that *figure* therefore being set alone in the *quotient*, and being subducted from the unity standing over the point, there remaineth nothing.

This unity set alone by it self in the *quotient*, shall signify 10, when another *figure* is set by it, representing the side of some other particular *square*. Whereupon I say, that the greater *Diagonall* k g; is now subducted from the whole *square*, and the side of it k i, or a b, (for they are equall one to another) and also the *side* of the *complement* is found out.

This is the first step to this Resolution.

Moreover, I double the *figure* found out, because being doubled, it is the side of both the *complements* taken jointly together, namely, k i, and g i. Then setting 2 the doubled number under 6, I divide 6 (which in this place is as much as 60, and representeth both the *complements*) by 2, the *quotient* is 3, representing the other *side* remaining of the *complement*; namely, i f for b c, which number I set in the *quotient*, & count it for the *segment remaining* of the right line given. Wherefore because this number 3 is the side of the remaining *Diagonall*, that is to say, of the lesser *square* b f, therefore being set by the *divisor* on the right hand, and multiplyed by it selfe, and also by the *divisor*, it bringeth forth three *plain numbers*, namely the *square* b f, and the two *complements* a i, and i l, which being subducted from the numbers standing over them, there remaineth nothing.

The

The example is thus.

$$169 \; (12 \qquad \qquad 169$$

Which is all one, as if you had put downe the numbers found out in this manner.

169
100 *The greater Diagonall.*
60 *The complements twofold.*
9 *The lessor Diagonall.*

169

The subtilty of this invention is illustrated by many examples.

Let the *square* given be 1764. This number being marked with two points, telleth us, that the side thereof is to be written with two Figures.

First therefore beginning at the point on the left hand, I seek the side of the last period, namely, 17. But for so much as it is no *square* number, I take 4 the side of the next lesser *square*, which I set alone by it self in the quotient, and then multiply it by it self, the *Product* is 16, which being subducted from 17, there resteth 1. Moreover I double the side found out, the product is 8, I place this doubled number under 6, and by it divide 16 standing above it, the *quotient* is 2, which must be set by 4. This *quotient* 2, must be set before the *Divisor* 8, on the right hand under the point, and then must it be multiplyed both by it selfe, and into 8, the product is 164, which being subducted from the figures standing over them, there remaineth nothing : whereby I gather that the number given is a just *square*.

.The

The Example standeth thus.

$$
\begin{array}{r}
\text{r} \\
1764\ (42 \\
1682 \\
2 \\
\hline
164 \\
\hline
1764 \quad \text{The Collection.}
\end{array}
$$

The same manner of working is to be followed in greater square numbers given, saving that the former part of the works is to be used but once, but the later part is to be followed so many times as there are points remaining excepting the last.

The third example.

As in 5 47 56, I say, that the *side* of the *square* next unto 5 is 2 : therefore 2 being set in the *quotient*, and multiplyed by it self, makes 4, and taken from 5, the remainder is 1. Moreover I double the *quotient* the product is 4, which I set under the next *figure* toward the right hand, and thereby divide 14, the *quotient* is 3. which three being set both in the *quotient*, and also before the *Divisor* toward the right hand, I multiply both the numbers by it, the *product* is 129: this being subducted from 147 standing above it, the remainder is 18. But because there is yet one point remaining, with which I have not medled ; I therefore again double all the whole *quotient*, for in this case I must take 23 for the side of one former *square*,

Note.

and generally in great numbers, when I light upon more particular *squares* then two, I must esteeme them but as two, and take the sides which are first found out , but as the *sides* of one only *square*. Therefore twice 23 is 46 : by this I divide 185, the number to be set in the *quotient* is foure , which

number

number alfo muft be fet before the *Divifor* on the right hand : then muft 464 be multiplyed by 4 : the *product* is 1856, this *product* being fubtracted from the numbers ftanding over it, there remaineth nothing. The example ftandeth thus.

```
118
54756 (234.
44364
129
```

```
    4
  1856
54756        The Collection.
```

See alfo the Example following.

10942864 (3308.

Therefore out of this invention is this *confectarie*. *The number whofe fide cannot be expreffed by whole numbers, is not a fquare number.*

Such are all *prime numbers*, and (the *fquares* themfelves excepted) all other *compound numbers*. For if in them you defire to find out the *fquare fide*, you fhall labour in vain, becaufe they are not *fquares*, for to the whole numbers arifing in the *quotient*, there will be fome *fraction* adjoyned, whereby it commeth to paffe. that the number of the *fide* is not to be expreffed by a true number, and it is commonly called a *furd number*.

Notwithftanding, if you adjoyne to the *fide* found out, the number remaining, taking his *denomination* from the double of the *fide augmented* by an unity, you

4 Example of a furd number.

you shall finde the next side that may be like to the *side of a square.*

As if from 40 you take the neerest *square*, to wit, 36, the remainder is 4. Here therefore the side sought for of the *square* exceedeth not the *side* found out by an unity, but either by one, or more parts of some whole number : wherefore I double 6, the side found out, and adde an unity to it being doubled, the totall is 13, this number I set under 4 the remainder, and say that the side of 40 demanded as neere as may be is 6 $\frac{4}{13}$: the *Denominator* of the *Fraction* being added to the greatest *square* in the number given, namely unto 36, maketh the next greatest *square* above it, namely 49, whose side is 7. But this surd side, to wit, 6 $\frac{4}{13}$, multiplyed by it selfe, maketh 39 $\frac{111}{169}$, which are not just equall to 40, the given number.

Judge the like concerning the rest which are not *squares.*

Thus much concerning plain figurate numbers, but especially such as are square numbers.

Concerning solid figurate Numbers.

A *Solid figurate Number is made of two multiplications by three numbers, or sides, multiplyed together, admitting length, breadth, and thicknesse.* Therefore every *number* made by the mutuall multiplication of three numbers, may be called a *solid,* because it bringeth forth *a right angled Parallelepipedon,* disposed according to his unities in *length, breadth, and thicknesse,* the *sides* whereof are the three multiplying *numbers.* As the *number* 30 made by the mutuall

A solid figurate number.

tuall multiplication of 2, 3, and 5, is called an *Inequilater solid number*, and the sides thereof are 2, 3, and 5; because the unities thereof disposed by a certain distance one from another, in *length*, *breadth*, and *depth*, as the sides doe expresse, do bring forth in resemblance an *Inequilater Parallelepipedon*, for that the *numbers* or sides are inequall.

By like reason 216 made by multiplication of 6 by 6, and the product thereof by 6, is called an *Equaliter solid*, for the sides thereof 6, 6, and 6 are equall.

An Equilater, is a number made by three equall sides, or by any number multiplyed by it selfe, and that product againe by the foresaid number. And it is called an Equilater and Equiangled Parallelepipedon or Cube, and is commonly represented by us thus C. An Equilater Solid or Cube.

A Cube in Geometry is a right-angled Parallelepipedon, having six equall *surfaces* : and 8 solid *angles*,

and 12 sides, as this figure *a. b. c. d. e. f. g. h.* whose side is *a b*, or *a d*, also *b c*, or *c d*, either *c e*, or *e f*, likewise *e h*, or *h g*, also *g f*, or *d f*, or *d a*, and *g a*.

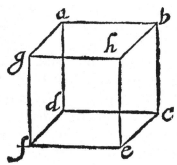

The number whereof the Cube is produced by Multiplication in it selfe twice, is called the side or root of the Cube, which being found out in whole numbers, the Cube is known. The side or root of the Cube.

Con-

Concerning the extraction of the
Cubick Root.

THerefore every *Cube* in numbers hath such a
side as may be expressed in whole numbers, but
in magnitudes it is not alwayes so, as indeed in magnitudes there are many things not to be expessed in *whole
number*. Now for as much as the side of any *Cube*
under 1000, is a simple figure, it is necessary, before
we undertake to finde out the side of any great number, to know what *Cube* is made of each simple figure, and what is the side of any *Cube* lesser then
1000, as I have here set them down.

1	2	3	4	5	6	7	8	9
1	4	9	16	25	36	49	64	81
1	8	27	64	125	216	343	512	729

But in searching out the *sides* of greater *Cubes* we
are to proceed as the *Theorem* following teacheth us.

If you distinguish with *points* as it were in o
periods, the given *Cube*, beginning at the first figure
on the right hand, and omitting each two figures
continually, and first of all subduct the *particular
Cube* of the last *period* from the given number, setting
the *side* thereof in the *quotient* : and then set triple of
the *quotient* under the *figure* next following the former *point*, on the right hand, and the *square* of the
quotient being tripled beneath it one *degree* more
toward the left hand : and afterward divide the number
above written by the triple of the *square*, setting the
quotient by it self, and then multiply the *divisor* by
the *quotient* found out, and the triple *square* by the
square

square of the *quotient*, and the *quotient cubically*, subducting the *products* (so orderly added together, that each *figure* may answer the *numbers* whereof it was multiplied) from the *number* given, and renue this last manner of division so many times as there are *points* remayning, the *side* of the *Cube* shal be found out.

This artificiall device is drawne out of that *theorem*, which *Ramus* made, imitating that of *Euclide* concerning *square numbers* in this manner.

If a right line be cut into two segments, the Cube of the whole line shall be equall to the Cubes of the segments, and the two solid figures comprehended three times under the square of his segment, and the segment remayning. The extraction of the Cubick side or root.

As the line *c l.* which is **13**, is cut into two *segments*, **10** and **3**, therefore the *Cube* of the whole line, mamely **2197**, is equall unto the *Cubes* of the *Segments*, namely unto **100**, and **27** also to *the two-fold Solids* or *Parallelepipedons thrice* taken whereof three have like *soliditie*, the *soliditie* of each of the three

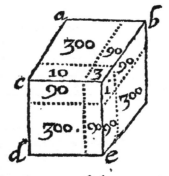

lesser is **90**, being made of the *Square of the Segment* **3**, that is to say of **9**. multiplied by the other *Segment* **10**. These three *Parallelepipedons* joyntly taken together, make **270**. But of the three greater *Parallelepipedons* each containing **300**, being made of **100**, the *Square* of the greater *Segment* **10**, multiplyed by the *lesser Segment* **3**, and they being taken joyntly together, make **900**.

K k   *The*

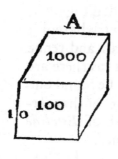

The Cube of
the leſſer ſeg-
ment 3.

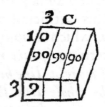

The Cube of the
greater ſegment 10.

The 3 leſſer Paral-
lelepipedons.

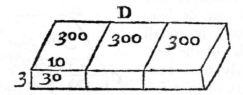

The three greater Parallelepipedons.

TheCube therefore hath eight *particular*   10   3
ſolids in number, which are made of the   10   3
parts of the number given, namely of 10
and 3 in this manner. Firſt, let there be   9
foure *plain numbers* made, each part be-   30
ing multiplied by it ſelf, and one by   30
another.   100

If

If againe I multiply the *Plaines* by the same parts, there will arise 8 *solids*, as you see here,

9	9
30	30
30	30
100	100
3	10
27	90
90	300
90	300
300	1000

All these being added together, are equall to the Cube of the whole, to wit, 2197.

Therefore the same way that is kept in making the Cube, *is also to be followed in resolving the* Cube.

As for example, I marke the *Cube* given with points in this manner, 2197.

Then I subduct the particular *Cube* of the number set under the last point : but for so much as that number is no *Cube*, I take the neerest to it, namely, an unity, which also I set in the *quotient*. This unity in the number given, is 1000, but in the *quotient* it is but 10, the unity subducted from 2, the remainder is 1, which must be written over the number given. So that the greater *Cube A*, is to be supposed to be subducted from the number given.

This is the first step of this work,

After I triple the *quotient* found out (that is to say, I multiply it by 3) this triple representeth the three *sides* (joyntly taken together) of the three lesser *solids* marked with *C*, I place the tripled number under 9, Again, I multiply the *quotient* square wise, and triple the *product*, which

The first example to extract the Cubick root.

```
     1
2197 (1
  3
```

K k 2   maketh

maketh likewife 3. This *product* refembleth the three *fquare fides* (taken joyntly together) of the three *greater folids*, marked with *D*, I place the *product* on a degree lower toward the left hand underneath 1. With it I divide 11, which written above it, the quotient is 3. This *fegment* or *quotient* 3, being multiplyed by 3, the *Divifor* maketh 9, which in refpect of the place wherein it ftandeth, is 900, and reprefenteth the three greater *folids* marked with *D*, taken joyntly together. Furthermore, the fame *quotient* being multiplyed fquare wife, maketh 9. and multiplyed afterward by the *triple number* ftanding under 9, it maketh 27, which in refpect of the place wherein it ftandeth, is 270, and reprefenteth the 3 *leffer folids* marked with *C*. Laft of all, the fame *quotient* multiplyed *cubically*, breedeth the leffer *Cube B*. Thefe 3 *products* therefore being added together, and the totall fubducted from the *numbers* ftanding over it, there remayneth nothing, which importeth the given number is a *Cube*.

The example is as you fee.

```
1                    2197 (13
2197 (13             1000 The greater Cube.
1 3                    3
  3                    3
 ──                   ──
  9 Or thus : 9 : 0 The 3 greater Parallelepipe.
 27                  270 The leffer Parallelepipedons,
 27                   27 The leffer Cube.
 ────                ────
 2197                2197
```

The fecond example of the Cubick.

The matter may be explained by many examples.

　　Let the fide of the given *Cube* 16387064, be fough out, contrive it therefore (as it were) into certai period

periods with points. Then firſt of all, ſearch out the
ſide of the *Cube* next to the left hand : But for as much
as 16 is no *Cube*, take 2 the *ſide* of the next *Cube* under it, that is to ſay, of 8, and ſet in the *quotient*, and
ſubduct 8 the *Cube* thereof from 16, there remaineth
8. The firſt work is not to be renued throughout the
whole *number*, but the *rules* following muſt be repeated as often as there are *points* remaining.

   The firſt ſtep to finde out the root,   8
is in this manner,   16387064 (2
   Moreover, triple the *quotient* now   8
found out and the *product* is 6, which is to be placed
under 8, namely, under the figure following the next
prick toward the right hand. Then multiply the *quotient* by this tripled *number* (or which is all to one pur-
poſe, ſquare the *quotient* and then triple the *product*)
it maketh 12, ſet that number in a lower place one degree neerer the left hand, & make it the *diviſor* : divide
83 by 12, obſerving this rule in chooſing your *quotient*, that it be no greater, then that the numbers afterward produced by multiplication may not exceed the
numbers ſtanding over it. So that here you ſhall take
1 in 8, but 5 times. Afterward by this number 5, multiply the *diviſor* 12, & by the ſquare of 5, multiply the
tripled number 6, & laſt of all multiply 5 cubically :
ſo ſhall you produce three numbers, namely, 60,150,
125, to be deſcribed in ſuch ſort as you ſee. Theſe numbers added together, and ſubducted from 8387, the
remainder is 762.

The second step to find out the root, in this manner.

```
    8762
  16387064 (25
     6
    12
      60
     150
     125
     7625
```

And because there is yet one point remayning, this last manner of Division must be wrought again.

First, therefore I triple the *quotient*, the product is 75. which must be so placed, that the first figure thereof, namely 5, may stand under 6, the second under the 0. Again, multiply the *quotient* by this *tripled number* (or which is all one, square the quotient, and triple the product) it maketh 1875, which must be the *Divisor*, whose first figure namely 5, must be placed under 7, the *last figure* of the *tripled number*. Then see that 1 may be contained in 7, many times, but I can take it but 4 times, I set 4 in the quotient and multiply the *Divisor* by 4, the product is 7500, afterward I square 4, it maketh 16, which I multiply by the tripled number 75, the product is 1200. Last of all I multiply 4 *cubically*, it maketh 64, these products added all together, make 762064, which number being subducted from the *Cube* given, there remayneth nothing, whereby I gather that the number given is exactly *cubicall*.

The

The third step to finde out the side is in this manner.

```
      762
  1·6·8·8·7·0·6·4 (254
        75
       1875
       7500
       1200
         64
      762064
```

Behold also the example following.

614125000 (850

Another manner of working.

HItherto the Princely high-way to find out the side of
the Cube hath been declared.

But there are moreover certain other wayes also
bending thereto, and leaning to the same principles,
whereof this is one.

Having found out in the Table of simple *Cubes*, the
first figure representing the side of the *cube* contained
in the number standing under the first *point* on the left
hand, set it in the *quotient*, and subduct the particular
Cube of that figure as you did before : then *square*
that *figure*, and triple that *square*, the *product* shall
be the *Divisor*, the first *figure* whereof shall be set
under that *figure* which is on the right hand next of
all to the point (now examined) before going.

See how many times the *Divisor* is contained in
the number written over it, and multiply the *Divisor*
in the *quotient*, and subduct the *product* from the *dividend*: yet here you must take heed, that you take
not a greater *quotient* then that the *products* made

afterward

afterward thereby may be subducted from the *number* given.

The subduction being done, triple the first *figure* which was set in the *quotient*, and adde to the triple the last *number* which was set in the *quotient* on the right hand of the *product*.

This totall multiplyed by the square of the *figure* last found out, and set down the *product* so, that the first *figure* thereof toward the right hand may stand under the *point* next before going on the same hand, and finaly, subduct the same from the *number* given.

The fourth example of the Cubick root.

As in 804357. The particular *Cube*, namely, 729 being taken from the number standing under the last *period* upon the left hand, there remaineth 75357, the side of that particular *Cube* being 9, I set in the *quotient*. Then I *square* that *side*, it maketh 81, and triple the *square*, the *product* 243 is my *Divisor*, which I set under the given number, so that 3 may stand under 3 with this *Divisor*, divide the number standing over it, you shall find 2 to be contained in 7 three times. Therefore I set 3 in the quotient, and multiply the *Divisor* by it, the *product* is 729, which being subducted from 753, the remainder is 24.

The Induction is thus:

2

754.

804357 (93

243

729

Moreover I triple 9, the product is 27, by which on the right hand I set 3 the quotient last found out, the totall is 273.

This

This *Number* I multiply by 9 the square of the quotient last found out, the product shall be 2457, which being subducted from the superiour number, there remaineth nothing.

The Induction is thus.

$$\begin{array}{l} \overset{\cdot\cdot}{24} \\ 8\overset{\cdot\cdot}{0}4\overset{\cdot}{3}5\overset{\cdot}{7}\,(93 \\ \quad 732 \\ \qquad 9 \\ \hline \quad 2457 \end{array}$$

Another manner.

THE selfe-same work may be dispatched another way, a little differing from the former in this manner.

The figure in the *quotient*, being found out by subducting the particular *Cube*, and also the second figure in the quotient being found by division, let the totall quotient be tripled, and let the tripled number be multiplyed by the former figure in the quotient. Then let the product be multiplyed againe, by the latter figure found out, and let a *cipher* be set on the right hand of that *product*. Last of all let the *Cube* of the latter figure found out, be added to this *product*, and let the totall summe be subducted from the number given. As in 373248.

The third forme.

The

The first induction is in this manner.

$$\overset{\overset{3\,0}{\bullet\;\;\bullet}}{373\,248}\;(7$$

$$\overline{}$$

$$343$$

The fifth
example

Moreover I *square* the fide found out, it maketh
49, and triple the fquare, the *product* is 147, which
fhall be the *divifor*, by this I divide 302, the number written over it, the *quotient* is 2. Now I triple
the *totall quotient* 72, it maketh 216, and multiply
this triple by 7, the former *figure* in the *quotient*, the
product is 1512. I multiply this *product* alfo by 2,
the latter *figure* of the *quotient*, and fet a *cipher* on the
right hand of it, fo as it maketh 30240, unto this
number laft of all I adde 8, the *Cube* of the latter
figure found out, the total is 30248, which being
fubducted from this *figure* above it, there remaineth
nothing.

The fecond Induction is thus.

$$\overset{\overset{\;\;\bullet\;3\,0}{\bullet\;\;\bullet}}{373\,248}\;(72$$

$$\overline{}$$

$$\cdot 147$$

$$\overline{}$$

$$30248$$

All the *points* of the *number* given being examined,
if any thing remaine, it fignifieth the *number* given
is no *Cube* : wherefore the true *fide* of it cannot be
exactly given in *numbers*. Yet if it pleafe you to fift
out the neereft *fide* that may be, by the firft kinde of
reduction of *mixt numbers*, you fhall reduce the *num-*
ber

ber given unto a *cubicall* fraction of a greater *denomination*, and afterward feeke out the *cubicall fide* of that *fraction*.

For example fake, becaufe 120 is no *Cube*, therefore let it be reduced into fixty *cubicall* parts, after this manner. Multiply 60 *cubically* in it felf, it maketh 216000, by this being taken for the *denominator* of the *fraction*, multiply 120 the *number* given, the product is 25920000 whofe *cubicall fide* is $\frac{325}{60}$ that is $4\frac{11}{12}$ the neareft to the true *fide* that can be.

To finde the neereft Cubick root in a furd number

For the extraction of all forts of roots, the tables of Logarithmes fet forth by Mr. Briggs *are moft excellent, and ready.*

FINIS.

A Table of Board and Timber meaſure more perfect then ever hath been made; ſhewing alſo the Squares between 4 and 37 from quarter to quarter, calculated by Robert Hartwell.

Board meaſure.	Inches & quarters.	Squares.	Timber meaſure.	Board meaſure.	Inches & quarters.	Squares.	Timber meaſure.
36. 0. 0	4	16	108.0.0	16. 0. 0	9	81	21. 3. 3
33.8.8	1	18	96. 0. 0	15. 5. 7	1	85	20. 3. 3
32.0.0	2	20	86. 4. 0	15. 1. 6	2	90	19. 2. 0
30. 3. 1	3	22	78. 5. 4	14. 7. 7	3	95	18. 1. 9
28. 8. 0	5	25	69. 1. 2	14. 4. 0	10	100	17. 2. 8
27.4.3	1	27	64. 0. 0	14. 0. 2	1	105	16. 4. 6
26.1.8	2	30	57. 6. 0	13. 7. 1	2	110	15. 7. 1
25. 0. 4	3	33	52. 3. 6	13. 3. 9	3	115	15. 0. 3
24.0.0	6	36	48. 0. 0	13. 0. 9	11	121	14. 2. 8
23.0.4	1	39	44 3. 0	12. 8. 0	1	126	13. 7. 1
22. 1. 5	2	42	41. 1. 4	12. 5. 2	2	132	13. 0. 9
21. 3. 3	3	45	38. 4. 0	12. 2. 5	3	138	12. 5. 2
20. 5. 7	7	49	35. 2. 6	12. 0. 0	12	144	12. 0. 0
19.8.6	1	52	33.2.3	11. 7. 5	1	150	11. 5. 2
19. 2. 0	2	56	30. 8. 6	11. 5. 2	2	156	11. 0. 8
18. 5. 8	3	60	28. 8. 0	11. 2. 9	3	162	10. 6. 7
18. 0. 0	8	64	27. 0. 0	11. 0. 7	13	169	10. 2. 2
17.4.5	1	68	25. 4. 1	10. 8. 7	1	175	9. 8. 7
16.9.4	2	72	24. 0. 0	10. 6. 7	2	182	9. 4. 9
16.4.6	3	76	22. 7. 4	10. 4. 7	3	189	9. 1. 4

Board

Board measure	Inches & quarters	Squares	Timber measure	Board measure	Inches & quarters	Squares	Timber measure
10.2.8	14	169	8.8.1	6.8.6	21	441	3.9.2
10.1.0	1	203	8.5.1	6.7.7	1	451	3.8.3
9.9.3	2	210	8.2.3	6.6 9	2	462	3.7.4
9.7.6	3	217	7.9.6	6.6.2	3	473	3.6.5
9.6 0	15	225	7.6.8	6.5.4	22	484	3.5.7
9.4.4	1	232	7.4.4	6.4.7	1	495	3.4 9
9.2.9	2	240	7.20	6.4.0	2	506	3.4.1
9.1.4	3	248	6.9.7	6.3.3	3	517	3.3.4
9.0.0	16	256	6.7.5	6.2.6	23	529	3.2.7
8.8.6	1	264	6.5 4	6.1.9	1	540	3.2.0
8.7.3	2	272	6.3.5	6 1.2	2	552	3.1.3
8.6.0	3	280	6.1.6	6.0.6	3	564	3.0.6
8.4.7	17	289	5.9.8	6.0.0	24	576	3.0.0
8.3.5	1	297	5.8.1	5.9.4	1	588	2 9.4
8.2.3	2	306	5.6.4	5.8.8	2	600	3.8.8
8 1.1	3	315	5.4.8	5.8.2	3	612	2.8.2
8.0.0	18	324	5.3.3	5.7.6	25	625	2.7.6
7.8.9	1	333	5.1.9	5.7.0	1	637	2.7.1
7.7.8	2	342	5.0.5	5.6.5	2	650	2.6.5
7.6.8	3	351	4.9.2	5.5.9	3	662	2.6.1
7.5.8	19	361	4 7.9	5.5.4	26	676	2.5.5
7.4.8	1	270	4.6.7	5.4.8	1	689	2.5.1
7.3.9	2	380	4.5.5	5.4.3	2	702	2.4.6
7.2.9	3	390	4.4.3	5.3.8	3	715	2.4.2
7.2.0	20	400	4.3.2	5.3.3	27	729	2.3.7
7.1.1	1	410	4.2.1	5.2.8	1	749	2.3.2
7.0.2	2	420	4.1.1	5.2.3	2	756	2.2.8
6.9.4	3	431	4.0 1	5.1.9	3	767	2.2 5

Board

Board measure	Inches & quarters	Squares	Timber measure	Board measure	Inches & quarters	Squares	Timber measure
5.1.4	28	784	2.2.0	4.3.6	33	1089	1.5.9
5.0.9	1	798	2.1.6	4.3.3	1	1104	1.5.6
5.0.5	2	812	2.1.2	4.3.0	2	1122	1.5.4
5.0.0	3	826	2.0.9	4.2.7	3	1139	1.5.2
4.9.6	29	841	2.0.5	4.2.3	34	1156	1.4.9
4.9.2	1	855	2.0.2	4.2.0	1	1173	1.4.7
4.8.8	2	870	1.9.8	4.1.8	2	1190	1.4.5
4.8.4	3	885	1.9.5	4.1.4	3	1208	1.4.3
4.8.0	30	900	1.9.2	4.1.1	35	1225	1.4.1
4.7.6	1	915	1.8.9	4.0.8	1	1242	1.3.9
4.7.2	2	930	1.8.6	4.0.5	2	1260	1.3.7
4.6.8	3	945	1.8.3	4.0.3	3	1278	1.3.5
4.6.4	31	961	1.7.9	4.0.0	36	1296	1.3.3
4.6.1	1	976	1.7.7	3.9.8	1	1313	1.3.1
4.5.7	2	992	1.7.4	3.9.4	2	1331	1.2.9
4.5.3	3	1008	1.7.1	3.9.1	3	1350	1.2.8
4.5.0	32	1024	1.6.9	3.8.9	37	1369	1.2.6
4.4.6	1	1040	1.6.6	3.8.7	1	1388	1.2.4
4.4.3	2	1056	1.6.4	3.8.4	2	1406	1.2.2
4.4.0	3	1072	1.6.1	3.8.1	3	1425	1.2.1

The use of this former Table.

IF upon a *Scale* or *Ruler* you divide one *inch* into ten *equall parts* or *primes*, and againe by *diagonals*, and *parallell-lines*, you fubdivide each of them into ten *equall parts* or *feconds*, with your *compaffes*, you may take a more exact running meafure for *board and timber*, then by any other meanes whatfoever, and ſo place the ſame, or this *Table* if you will, upon any *Ruler.*

Alſo by meanes of the *columnes of ſquares*, you may readily finde a *ſquare* equall to any *Parallelepipedon*, or piece of *timber*, which is thicker then it is broad. As for example, ſuppoſe a piece of *timber* to be ten *inches thicke*, and *9 inches broad* : if I multiply thoſe *ſides* one by another, they will produce 290, then ſeeking the *columne of ſquares* for 290, which I finde not, but I finde 289 the neereſt number to 290, to ſtand againſt **17** : thererore I ſay **17** *inches ferè*, will make a *ſquare* equall to ſuch an unlike *ſquared piece*, then looking in the *columne of timber meaſure* againſt 17, you ſhall finde that **5** *inches*, **9** *primes*, or $\frac{9}{10}$ and **8** *ſeconds*, or $\frac{28}{100}$ of an *inch in length*, of that piece will make a *foot of timber.*

Likewiſe for *board meaſure*, you may finde how much in *length* or *breadth* of *board* muſt be in one *foot.*

By the like meanes, ſuppoſe for example that a *board*, appointed to be meaſured, is **15** *inches* $\frac{2}{4}$ *broad*, if I deſire to know how much in *length* thereof will make a foot ; I ſeeke in the *columns* that ſtand under *unites* and *quarters*, for 15 $\frac{2}{4}$, and alſo againſt the ſame in the *columne* under the title of *board meaſure*, where I finde *9 inches*, **1** *prime*, or *tenth* of an *inch*, and 4 *ſeconds*, or *hundreds* of an *inch* will make a *foot* at that *breadth* : The like may be practiſed for any other *breadth of board* whatſoever.

Cer-

Certain *Tables* shewing the Interest of any *summe of mony* whatsoever unto 40 years; how much *Annuities* respited or forborn commeth unto. And for *buying* or *selling* of *Annuities* for the said time; and also the same in *reversion* after any *number of yeares* unto 30. What they may be worth in present ready money, by *R. C.* and now diligently corrected and amended by *Robert Hartwell*.

Definition of Interest.

PRincipall, *is the summe from which the Interest is reckoned.*

2 Interest *is the summe reckoned for the lending or forbearance of the Principall for any termes or time.*

3 Interest simple *is that which is counted from the* Principall onely.

4 Interest compound *is that which is counted for the* Principall, *together with the* Arrerage.

5 Interest profitable *is that which is added to the* Principall.

6 Interest Damageable *is that which is to be subtracted from the* Principall.

The use
per annum
of
{
1 li.
10 s.
5 s.
2 s. 6 d.
1 s.
}
is
{
2 s.
12 d.
6 d.
3 d.
1 , $\frac{1}{5}$ of a peny.
}

A

A Table shewing what 1 li. with increst, and interest upon interest after 10 in the 100 comes to every yeer under 41 yeers. As followeth.

yeers	l.	s.	d.	li.	s	d.	yeers.
1	1	2	0	7	8	0	21
2	1	4	2	8		9	22
3	1	6	7	8	19	1	23
4	1	9	3	9	16	11	24
5	1	12	2	10	16	8	25
6	1	15	5	11	18	4	26
7	1	18	11	13	2	2	27
8	2	2	10	14	8	5	28
9	2	7	1	15	17	3	29
10	2	11	10	17	8	11	30
11	2	17	1	19	3	10	31
12	3	2	9	21	2	3	32
13	3	9	0	23	4	6	33
14	3	15	11	25	10	11	34
15	4	3	6	28	2	0	35
16	4	11	10	30	18	0	36
17	5	1	1	34	0	0	37
18	5	11	2	37	8	1	38
19	6	2	3	41	2	10	39
20	6	14	6	45	5	2	40

L l   By

By the former Table, if you defire to know what 1 li. commeth to with _intereſt_, and _intereſt_ upon _intereſt_ after 10 in the 100, for any _number_ of years unto 40. Look in the _row_, or _margent_ (over which is written _years_) and againſt it on the right hand cloſe unto it in the _row or margent of pounds, ſhillings, and pence_ (which is titled thus, li. s. d.) you ſhall finde your deſire.

Example.

I would know what 1 li. _with intereſt, and intereſt upon intereſt commeth to in_ 7 _yeares ?_

I looke in the _row_ of yeares for the _number_ 7. And againſt it on the right hand I finde 1 li. 18 s. 11 d. Alſo what it commeth unto in 13 years. I ſeeke among the years for 13, and againſt it I finde 3 li. 9 s.

Again, for 21 years. I look for 21 among the years, and I finde 7 li. 8 s. 0 d. But if you would know for a greater ſumme then 1 li. Then multiply your ſumme by that ſumme of 1 li. in the Table for any of thoſe years, and you ſhall eaſily finde it. As thus, I would know what 10 li. commeth to for 7 years with intereſt, &c. I ſee that 1 li. commeth to 1 li. 18 s. 11 d. in that time. Then ſay I that 10 li. muſt bee 10 times as much in that ſpace, which is 19 li. 9 s. 2 d. Alſo of 10 li. in 13 years, I ſee that 1 li. in that time commeth unto 3 li. 9 s. Then muſt 10 li. be ten times as much in that ſpace, which is 34 li. 10 s. Alſo what 10 li. commeth to in 21 years. I finde firſt that 1 li. in that ſpace commeth to 7 li. 8 s. Then I ſay 10 muſt be 10 times as much, which is 74 li. Laſtly, I would know what 100 li. commeth to in 7 years, I ſee it muſt bee 100 times as much as 1 li. commeth to in

that

that space, which is 194 li. 11 s. 8. d Hereby you see the common saying is not true, that 100 li. doth double it selfe in 7 years, for it wants thereof 5 li. 8 s. 4 d. But in 8 years 100 l. commeth to 210 li. 8 s. 4 d. which you see is more then double it selfe by 10 li. 8 s. 4 d. And in this sort may any that can but cast with Counters, or indeed by memory finde the increase of any summe whatsoever for any of the number of years in the foresaid Table, after they have found what 1 li. commeth unto for that time, as before is specified.

A Table shewing if 1 li. annuity to endure for any number of years, under 41, be all respited or forborne, untill the last payment grow due, and then all be received together; with interest, and interest upon interest after 10 in the 100 per annum, what they will amount unto by any of the said number of yeers, As followeth.

Ll 2　　　　　years

yeers.	li.	s.	d.	li.	s.	d.	yeers.
1	1	0	0	64	0	0	21
2	2	2	0	71	8	0	22
3	3	6	2	79	10	10	23
4	4	10	0	88	9	11	24
5	6	2	1	98	6	11	25
6	7	14	3	109	3	7	26
7	9	9	8	121	1	11	27
8	11	8	8	134	4	2	28
9	13	11	7	148	12	7	29
10	15	18	8	164	9	10	30
11	18	10	7	181	18	10	31
12	21	7	8	201	2	9	32
13	24	10	5	221	5	0	33
14	27	19	5	245	9	6	34
15	31	5	6	271	0	5	35
16	35	18	11	299	2	6	36
17	40	10	10	330	0	9	37
18	45	1	11	364	0	10	38
19	51	3	2	401	8	11	39
20	57	0	6	442	11	10	40

By this Table you may know what any Annuity being respited or forborn for any number of yeers unto 41. with interest upon interest, after 10 in the 100 will come unto : first seeking in the *Table*, what 1 li. will come unto, in that time, and that being found to multiply it by the summe you desire to know.

Example.

Example.

First, I would know what 1 li. *Annuity being forborn or respited for* 14 *yeers commeth unto.*

I look in this last Table (which is for the purpose) and I finde 27 li . 19 s. 5 d.

Again, what 1 li. *Annuity* respited for 21 *yeers* commeth to, I looke in the said Table for 21 *years,* and I find 64 li. Also the like for 1 li. for 30 *yeers* respited. I look, and find it to be 164 li. 9 s. 10 d. as by the said Table may appeare. Now for greater *Annuities,* as 30 li. *per annum,* respited or forborn, what it amounteth to in 16 *yeers.* I seek first for 1 li. in this last *Table* before for 16 *yeers,* and against it I find 35 li. 18 s. 11 d. Then say I, that 30 li. *per annum* being respited for that time, will come to 30 times as much, which is 1078 li. 7 s. 6 d. Also if there be an *Annuity* of 45 li. due and unpayed for 12 *yeers,* I look in the said Table what 1 li. commeth to, 12 *yeers* being respited, and I find it is 21 li. 7 s. 8 d. Then I conclude that 5 li. must be 45 times as much, which is 962 li. 5 s.

Lastly, I have an Annuity of 50 li. per annum, *which hath been behind for* 16 yeers, *and must be answered unto me with interest, and interest upon interest, all at one payment, what shall or ought I to receive in all, at the* 16 *yeers end?*

I seek what 1 li. comes unto in that time (as before taught) and I finde 35 li. 18 s. 11 d. Then must my 50 li. *per annum* forborne for that time, come to 50 times as much, which is 1797 li. 5 s. 10 d. And thus may you find any other *summe* great or small, for any *number of yeers* contained in the aforesaid *Table,* without the helpe of *Arithmeticke,* if

you

you can but uſe your Counters, or by memory count well.

A Table ſhewing if 1 li. Annuity (to indure for any number of years unto 41) bee to be ſold for preſent ready money, how much ought that ready money to bee, reckoning 10 per 100 per annum abating intereſt, and intereſt upon intereſt. As followeth.

yeers.	li.	s.	d.	li.	s.	d.	yeers.
1	0	18	2	8	12	11	21
2	1	14	8	8	15	5	22
3	2	9	8	8	17	7	23
4	3	3	4	8	19	8	24
5	3	15	9	9	1	6	25
6	4	7	1	9	3	2	26
7	4	18	4	9	4	8	27
8	5	6	8	9	6	1	28
9	5	15	2	9	7	4	29
10	6	2	10	9	8	6	30
11	6	9	9	9	9	7	31
12	6	16	3	9	10	6	32
13	7	2	0	9	11	4	33
14	7	7	4	9	12	2	34
15	7	12	1	9	12	10	35
16	7	16	5	9	13	6	36
17	8	0	5	9	14	1	37
18	8	4	0	9	14	7	38
19	8	7	3	9	15	1	39
20	8	10	3	9	15	6	40

T

This Table before laſt ſpecified is very neceſſary and commodious for all *Gentlemen* or others, that ſhall have cauſe to buy or ſell *Annuities* or ſuch like, for by this they ſhall know what they doe, whether they demand, or take too little or too much, after the rate of ten in the 100, by which proportion all theſe *Tables* are ruled.

As for example, I am to buy an Annuity of 16 li. per annum, for 12 yeers, and am demanded for it ready mony 120 li. I would know, if I give this rate, whether I give too much or too little, according to the proportion of ten in the 100 per annum, &c.

I look in the *Table* laſt before what 1 li. is worth for 12 *yeers*, and I finde againſt 12 this *ſumme* 6 li. 16 s. 3 d. Now I ſay that 16 li. *Annuity* for that time, and after that proportion commeth to 16 times as much, which is 109 li. So that I ſee the party demanded of me 11 li. too much after the rate of ten in the 100 *per annum*, and therefore I muſt draw him to a lower price or leave it.

Again, I am offered an Annuity of 20 li. per annum of 14 yeers for 130 li. I would know if I give it whether I give too much or too little, according to the proportion aforeſaid.

I ſeek firſt what 1 li. *Annuity* is worth for 14 *yeers*, and I find in the ſaid laſt *Table* 7 li. 7 s. 4 d. Then ſay that the *Annuity* of 20 li. *per annum*, will come to 20 times as much, and will be worth 147 li. 6 s. 8 d. according to the proportion before mentioned: and is more then his demand by 17 li. 6 s. 8 d. So that I ſee if I accept of it, I ſhall have a good bargaine. And thus may you know readily by looking in your Table, and finding what 1 li. is worth for

L l 4   any

any time therein contained, how much any greater summe will come unto, if you multiply it by that summe of 1 li. as before is sufficiently shewed.

But suppose this, I have 300 li. ready mony, and would bestow the same for a valuable Annuity answerable thereunto according to the proportion aforesaid. I would know what Annuity to endure 21 yeers this 300 li. will buy?

I look in the former Table what 1 li. Annuity will cost for that time, and I find 8 li. 12 s. 11 d. Then I say by the *Rule of proportion.* If 8 li. 12. s. 11 d. will buy 1 li. Annuity for 21 yeeres : what Annuity shall 300 li. buy or be worth for that time ? I reduce the *summes* to the least denomination (which is pence) and I find 34 li. 13 s. 11 d. And after this manner (by the help of this *rule*) may you find all other summes for any time contained in the foresaid last *Table.*

A

A Table ſhewing what 1 li in reverſion for any number of years under 31 is worth in ready money, the buyer ſtaying untill the thing be falne in hand.

yeers.	li.	s.	d.	li.	s.	d.	yeers.
1	0	18	2	0	4	4	16
2	0	16	6	0	3	11	17
3	0	15	0	0	3	7	18
4	0	13	7	0	3	3	19
5	0	12	5	0	2	11	20
6	0	11	3	0	2	8	21
7	0	10	3	0	2	5	22
8	0	9	3	0	2	2	23
9	0	8	5	0	2	0	24
10	0	7	8	0	1	10	25
11	0	7	0	0	1	8	26
12	0	6	4	0	1	6	27
13	0	5	9	0	1	4	28
14	0	5	3	0	1	3	29
15	0	4	9	0	1	1	30

This laſt *Table* differeth, and is contrary to the other three before mentioned : For whereas the others increaſed more and more according to the *number of yeers* ſpecified, this doth grow and diminiſh leſſe and leſſe, as the *number of yeers* increaſeth. As for example.

Ther

There is a Tenement, the fee ſimple whereof after 7 *yeers will be worth* 40 li. *what am I to give for it in ready mony, now ſtaying untill it fall in hand?*

To know this I look in this laſt Table for 7 years, and againſt it I find 10s. 3d. So that a thing that after 7 years will be worth 1 li. is worth now in ready mony but 10 s. 3 d. Then ſay I, that the foreſaid Tenement (which after 7 years will be worth 40 li.) is now worth 40 times 10s. 3d. which is 20li. 10s.

Again there is a Farm which after 9 *yeers will be worth the Fee-ſimple* 420 li. *what is it now worth in ready mony, ſtaying untill it fall in hand?*

I look in the ſaid Table what 1 li. is worth in reverſion after 9 years, and I find 8s. 5d. Then ſay I, that the *Farm* of 420 li. ſo long in Reverſion, will be now worth in ready mony, 420 times as much, which is 176li. 15s.

Laſtly there is a Lordſhip to be ſold, the Feeſimple whereof after 14 *yeers will be worth* 7500 li. *I would know what the ſame is now worth in ready mony for the Reverſion.*

I look in this laſt *Table* for 14 yeers, and againſt it I find 5s. 3d. ſo much 1 li. is worth in reverſion after 14yeers. Then ſay I, that 7500 li. is worth no more in reverſion for that time then 7500 times 5s. 3d. which is 1968 li. 15s. And after this manner may you finde out any other ſumme whatſoever. And though ſome men of their own experience can ayme (as they think) neer enough the mark to ſerve their own turns : yet I dare undertake they ſhall never ſo exactly doe it, nor juſtify what they doe, as if they did it by Art.

FINIS.

New *Tables of Interest* at 8 *per centum per annum*, exactly calculated for 30 yeers by *Robert Hartwell*, with neceſſary *queſtions* for the uſe of them.

The firſt Table expreſſing the increaſe of one pound principall, put out and forborn for any number of yeers under 31 at 8 *per centum, per annum.*

yeers	li.	s.	d.	q.	li.	s.	d.	q.	yeers
1	1	1	7	0	3	8	6	0	16
2	1	3	3	3	3	14	0	0	17
3	1	5	2	1	3	19	6	0	18
4	1	7	2	2	4	6	3	3	19
5	1	9	4	2	4	13	2	2	20
6	1	11	8	3	5	0	8	0	21
7	1	14	3	1	5	8	8	3	22
8	1	17	0	0	5	17	0	0	23
9	1	19	11	3	6	6	9	3	24
10	2	3	2	0	6	16	1	2	25
11	2	6	7	2	7	7	11	0	26
12	2	10	4	1	7	19	9	0	27
13	2	14	4	2	8	12	6	2	28
14	2	18	8	3	9	6	4	0	29
15	3	3	5	1	10	1	3	0	30

The

The deſcription and uſe of the Tables of Intereſt at 8 per annum, being profitable.

The firſt of them.

THeſe Tables conſiſt of foure *Columns*, in the firſt and fourth, whereof is written over the head, yeeres, and under the firſt number of yeeres deſcending from 1 to 15, likewiſe in the fourth the number of yeeres deſcending from 16 to 30. And againſt every yeere in the ſecond Columne, toward the right hand the pounds, ſhillings, pence and farthings, which one pound, or 20s. principall will amount unto, being put forth and forborn for the number of yeeres ſet againſt it ; (but the pounds, ſhillings, pence, &c. in the third Columne belong to the yeeres ſet in the laſt Columne.)

1. Example.

Let it be required what one pound or 20 ſhillings, being put forth and forborn for 12 yeers, ariſeth to at 8 per centum, per annum, intereſt upon intereſt.

Seek in the firſt Columne under the title of yeeres, for 12 the number of yeers propoſed in the queſtion, & right againſt it toward the right hand in the ſecond Columne, you ſhall find 2li——10s——4d——1q. which is the principall and increaſe thereof due for the time required.

2 Example.

If 100 li. be put forth for 17 yeeres according to the ſame intereſt, I demand what it will amount to in that time?

Look in the Column under the title of yeers for 17, and right againſt it towards the left hand in the Table is found 3li——14s——0d——0q. which is the increaſe

increaſe of 1li. by which you may thus gather the increaſe of 100li. or any other *ſumme* ; a hundred times 3li. is 300 li. then 100 times **14** *ſhillings* is 70. li. both which added together doe make 370li

li	s	d.	q
300	0	0	0
70	0	0	0
370	0	0	0

make 370li——os——od.which is the *increaſe* of 100li. put forth and forborn 17. *yeeres*, the ſolution to the queſtion.

3 Example.

Suppoſe 60li. *be put forth for* 19 *yeeres according to that rate, what will it increaſe to in that time ?*

Seek 19 under the *title* of yeeres, and againſt it toward the left hand is found 4li——6s——3d ——3q. now ſay 60 times 4li. is 240,and 60 times 6 *ſhillings* is 360 *ſhillings*, or 18li. and 60 times 3d. is 180 d. or 15 *ſhillings*,& 60 times 3 *farthings* is 3 *ſhillings* 9d. all which

li	s	d	q
240	0	0	0
18	0	0	0
0	15	0	0
	3	9	0
258	18	9	0

added together make 258li. 18 s. 9d. the increaſe thereof demanded.

The

The second Table *shewing what one pound annuity or yeerly rent is worth at the end of any number of yeers under* 31, *being forborn, at* 8 per centum per annum.

yeers.	li.	s.	d.	q.		li.	s.	d.	q.	yeers.
1	1	0	0	0		30	6	5	3	16
2	2	1	7	0		33	15	0	0	17
3	3	4	6	0		37	9	0	0	18
4	4	10	1	1		41	8	11	0	19
5	5	12	2	3		45	15	2	3	20
6	7	6	8	0		50	8	5	2	21
7	8	18	5	1		55	9	1	2	22
8	10	12	8	3		60	17	10	1	23
9	12	9	9	0		66	15	3	2	24
10	14	9	8	3		73	2	1	1	25
11	16	12	10	3		79	19	1	0	26
12	18	19	0	2		87	7	0	0	27
13	21	9	10	3		95	6	1	2	28
14	24	4	3	2		103	19	3	3	29
15	27	3	0	2		113	5	7	3	30

The use of the second Table, *whose disposition is altogether like the former according to the title thereof, being profitable.*

1 Example

1 Example.

There is a Leafe worth 28. li. *per annum, to endure* 14 *years, I demand what it will rife unto at the end of thofe yeares being all forborne with the intereft upon intereft at the rate prefcribed in this* Table.

Look in the third Table for 14 yeares, againft which toward the right hand, you fhall find 24li———4f———3d———2q. Now Multiply 28 li. by 24, there arifeth 672li. then 28li. by 4 s. yeeldeth 112s. or 5 li. 12 s. Again 28li. by 3d. produceth 84d. or 7s. finally, 28. by 2 farthings yeeldeth 56 farthings or 1s. 2d. Al which added together make 678li. 0s. 2d. to be received at the end of 14 yeers, the fame rent or annuity being refpited.

li.	s	d	q
672	0	0	0
5	12	0	0
	7	0	0
	1	2	0
678	0	2	0

2 Example.

If 60 li. *yeerly rent or annuity be forborn* 20 *years: I demand how much it will increafe at the end of the faid term?*

In the Table I find that 1 pound in 20 yeers will arife to 45li———15s———2d———3q. therefore 60li. in the like terme wil yeeld 60 times as much; which I will reckon thus: 60 times 45 li. is 2700 li, 60 times 15 s.is 900 s. or 45l. 60 times

li.	s	d	q
2700	0	0	0
45	0	0	0
10	0	0	0
3	9	0	0
2745	13	9	0

times 2 d. is 120 d. or 10 s. laſt of all, 60
times 3 q. is 180 farthings, or 3 s.——9 d. all which
together amount unto 2745 li——13 s——9 d. the
value thereof to be received at the end of the term.

3 Example.

*The yeerly rent of 6 li.——13 s——4. being behind
and unpaid the ſpace of 7 yeeres at the end of which
term the Tenant is compelled to pay the ſame with the
intereſt thereof according to the above named rate ; I
demand what the payment ought to be.*

The increaſe of 1 li. *yeerly rent* anſwering to 7
yeeres, is 8 li. 18 s. 5 d. d which for 6 li. *rent* taken
6 times, ariſeth to

	li.	s.	d.	q.
53 li. 10 s. 7 d. 2 q.	53	10	7	2
now becauſe 13 s.	5	18	11	2
4 d. is two third				
parts of 1 li. there-	59	9	7	0

fore I take $\frac{2}{3}$ of 8 li. 18 s. 5 d. 1 q. which is the increaſe of 1 l. forborn for 7 *yeers*, that is 5 li. 18.
11 d. 1 q. which together make 59 li. 9 s. 7 d. 0 q.
the *ſumme* to be received, as was required.

The

The third Table declaring what one pound due at the end of any number of yeers under 31 is worth ready mony at 8 per centum, per annum.

yeers.	li.	s.	d.	q.	li.	s.	d.	q.	yeers.
1	0	18	7	0	0	5	10	0	16
2	0	17	10	3	0	5	4	3	17
3	0	15	1	2	0	5	0	0	18
4	0	14	8	1	0	4	7	2	19
5	0	13	7	1	0	4	3	1	20
6	0	12	7	0	0	3	11	2	21
7	0	11	8	1	0	3	8	0	22
8	0	10	9	2	0	3	4	3	23
9	0	10	0	0	0	3	1	3	24
10	0	9	3	0	0	2	11	0	25
11	0	8	6	3	0	2	8	1	26
12	0	7	11	1	0	2	5	0	27
13	0	7	4	0	0	2	2	3	28
14	0	6	9	2	0	2	1	3	29
15	0	6	3	3	0	1	11	3	30

M m

The

The third Table is diſpoſed as the firſt, the uſe according to the Title thereof, being damageable.

1 Example.

Suppoſe there is 750 li. due to be payed at the end of 9 yeeres, the Creditor would ſell this debt for preſent money, what ought that mony to be at the rate deſcribed in the Table?

Seek in this third Table for 9 yeers at the left ſide of the Table, and right againſt it toward the right hand, you ſhall find 10 ſhillings, which multiplyed or taken 750 times, yeeldeth 7500 ſhillings, which is 375 li. the value of that debt in preſent mony.

2 Example.

There is a Leaſe worth 500 li. after the end of 7 yeers; what is it worth preſent mony, according to the rate deſcribed in the Table ſtaying till it fall?

I ſeek in the Table for the 7 yeers, and right againſt it I find 11 s——8 d; now I multiply 500 by 11, it yeeldeth 5500 ſhillings, or 275 li. then 500 times 8 d. maketh 4000 d, which is 16 li. 13 s. 4 d. which added together is 291 li. 13 s. 4 d. the value of the Leaſe to be paid before it fall in hand.

li	ſ	d	q
275	0	0	0
16	13	4	0
291	13	4	0

The

The fourth Table *expressing what one pound yeerly rent or annuity for any number of yeeres not exceeding* 30 *is worth ready mony at* 8 *per centum per annum.*

yeers	li.	s.	d.	q.		li.	s.	d.	q.	yeers
1	0	18	6	0		8	17	0	1	16
2	1	15	7	3		9	2	5	0	17
3	2	11	6	2		9	7	5	1	18
4	3	6	2	3		9	12	0	3	19
5	3	19	10	1		9	16	4	1	20
6	4	12	5	1		10	0	4	0	21
7	5	4	1	2		10	4	0	0	22
8	5	14	11	0		10	7	5	0	23
9	6	4	11	1		10	10	6	3	24
10	6	14	2	1		10	13	5	3	25
11	7	2	9	1		10	16	2	1	26
12	7	10	8	2		10	18	8	1	27
13	7	18	0	3		11	1	0	0	28
14	8	4	10	2		11	3	2	0	29
15	8	11	2	1		11	5	1	3	30

The

The fourth Table is difpofed altogether as the for-mer, and the ufe thereof in like fort being damage-able.

1 Example.

There is an annuity or rent of 20 s: per annum *to endure* 25 *yeers, it is required what it is worth ready mony ?*

Look in the Table for 25 *yeers*, and right againft it you fhall finde 10 li. 13 s-5 d. 3 q. which is the folution.

2 Example.

What is the Leafe of certaine Land valued at 140 li. per annum, *to begin prefently and endure* 18 *yeers, worth ready mony ?*

Search in the *Table* for 18 *yeeres*, the *term* named in the *queftion*, and right againft it toward the left hand you fhall find 9 li-7 s 5 d-1 q. which expreffeth that *one pound rent* to be bought for that *terme* is worth fo much ; therefore that fumme 140 times is the value required. Now 140 times 9 li. is 1260, and 140 times 7 s. is 980 s. or 49 li ; likewife 140 times 5 d. is 700 d. or 2 li-18 4d. and 140 *farthings* is 2 s-11 d. all which added together make 1312 li--1 s--3 d. for the value of the faid *Leafe* paying no *rent*.

li.	s.	d.
1260	0	0
49	0	0
2	18	4
2		11
1312	1	3

3 Example

3 Example.

A Leafe taken for 21 *yeers at* 13 li. 6 s. 8 d. per annum, *which after* 5 *yeers expired, the Tenant is defirous to give a fine, and bring the rent down to* 8 li. per annum, *for the reft of the term, the demand is what fine is to be payed ?*

Subtract 5 *yeers* from 21, the remain 16, is the time unexpired ; likewife from the prefent *rent* abate 8 li. the reft will be 5 li-6 s- 8 d. now the drift of the *queftion* is, what 5 li-6 s-8 d. *yeerly rent* or *annuity* to indure 16 yeeres is worth prefent money.

The value of 1 li. *rent* or *annuity* anfwering to 16 *yeers* is, 8 li-17 s-0 d. 1 q. Now 5 times 8 li. is 40 li. and 5 times 17 s. 4 li 5 s. and 5 times one farthing, is 1 d-1 q. and becaufe 6 s-8 d. is ⅓ of 1 li. I take ⅓ of 8 li-17 s-0 d. 1 q. which is 2 li-19 s- 0 d. all which added together, make 47 li-4 s-0 d- 1 q. which is the *fine* that ought to be paid to bring the *rent* to 8 l. per annum.

li.	s.	d.	q.
40	0	0	0
4	5	0	0
2	19	0	1
47	4	0	1

The

The fifth Table declaring what yeerly rent or annuity of one pound ready mony will purchase for any
number of yeers under 31 at 8 per centum per annum.

yeers.	li.	s.	d.	q.	li.	s.	d.	q.	yeers.
1	0	18	7	0	0	2	9	3	16
2	0	14	0	2	0	2	8	3	17
3	0	9	8	2	0	2	8	0	18
4	0	7	6	0	0	2	7	0	19
5	0	6	3	0	0	2	6	2	20
6	0	5	4	3	0	2	5	3	21
7	0	4	9	2	0	2	5	1	22
8	0	4	4	0	0	2	4	3	23
9	0	4	0	0	0	2	4	1	24
10	0	3	8	2	0	2	4	0	25
11	0	3	6	0	0	2	0	3	26
12	0	3	3	3	0	2	3	1	27
13	0	3	1	2	0	2	3	0	28
14	0	3	0	1	0	2	2	3	29
15	0	2	11	0	0	2	2	2	30

In

In the fifth Table the Numbers and Columns are all difpofed as the former Tables, and needeth no further explanation but only Examples.

1 Example.

The Table declareth at firft fight what yearly rent or annuity one Pound ready mony will purchafe for any terme in the Table expreffed.

But if the *ready mony* be above one pound, then if any *value* or *rent* fet down in this Table, be multiplyed by the *number* belonging to the *yeers in queftion*, the *product* will fhew what *yearly rent* or *annuity* that *ready mony* will purchafe for the *time propofed*.

2 Example.

A certain man hath 750 li. *to purchafe an Annuity to endure* 27 *yeers, fo as it may yeeld him the like profit, as if it were put out according to the rate in the Table expreffed, it is required what that annuity ought to be?*

Becaufe the *annuity* is to endure **27** yeeres, feek out the value or rent fet againft **27** yeers, in this fifth Table, which is 2s———3d———1q. now this being the *Annuity* with 20s. ready

li	s	d	q
75	0	0	0
9	7	6	0
	15	7	2
85	3	1	2

mony will purchafe for that terme, it muft be multiplyed by 750 li. as followeth, becaufe 2s. is the tenth part of 20 s. therefore take the tenth part of 750 li. which is 75 li.

Mm 4   which

which set first down, then 750 times 3d, is 9l, 7s, 6d. which set under the former, last of all 750 *farthings* is 15s, 7d, 2q. All which added together, produce 85 li. 3s. 1d. 2q. the *yeerly Annuity* required.

Deo soli laus, omnis honor & gloria tribuatur.
A M E N.

FINIS.

Compendious **Tables** *of Interest money forborn any number of Dayes Weeks, Moneths, or Years under* 22 ; *exactly calculated at* 6 l. per Cent. per Annum.

Days.	Numbers.	Yeares.	Numbers.
I	100016.	I	106000.
II	100032.	II	112360.
III	100048.	III	119102.
IIII	100064.	IIII	126248.
V	100080.	V	133822.
VI	100096.	VI	141852.
Weekes.		VII	150363.
I	100112.	VIII	159385.
II	100224.	IX	168948.
III	100336.	X	179085.
Moneths.		XI	189830.
I	100487.	XII	201220.
II	100976.	XIII	213293.
III	101467.	XIV	226090.
IIII	101961.	XV	239656.
V	102458.	XVI	254035.
VI	102956.	XVII	269277.
VII	103457.	XVIII	285434.
VIII	103961.	XIX	302560.
IX	104467.	XX	320714.
X	104976.	XXI	339956.
XI	105486.		

The

The use of this Table *of proportionall numbers;*
the Radius 100000.

This Table is divided into 4 columnes, and the first in 3 parts afcending from 1 day unto 6. fecondly from 1 weeke to three, thirdly from one moneth proceeding to 11. the third columne comprehends the yeares inclufive from 1 to 21 as by their numerall letters doe appear : the fecond and fourth columnes are proportionall numbers, in arithmeticall characters, refpectively anfwering the time or times compounded of Days, Weeks, Moneths and Yeers to 21, as by examples fhall be evidenc'd.

An explanation.

What will 50 L. amount unto, if forborne 21 years, the Intereft allow'd at 6 L. *per Cent. per An.*

Look in the columne of years for the terme of forbearance propounded, in this 21, whofe *Decimall* is 339956, which multiplied by the principall forborn *viz.* 50 L. the product will be 169|97800, to be divided by the *Radius* 100000, therefore cut off 5 places (as in the margent) from the right hand, and

		33956
		5,0
L	169	97800
S	-19	56000
		112000
D	-6	72000
Q	-2	88|000
		100

there will appeare 169 L. the remainer 97800 multiply by 20 s. or by 2, and annex a cipher at the right hand, the product is 19 56000, cutt off 5 places as before ; on the left hand you will find 19 S. the remainder 56000, which encreaf'd by 12 D. produceth

ceth 6 72000, ftrike off 5 places, you will difcover the 6. remainder 72000 multiplied by 4 Q. and worke as before, you fhall find 2 $\frac{22}{25}$ Q. fo 50 L. forborn 21 yeares at 6 L. per Cent. per An. will amount unto the fumme of 169 L. 19 S. 6 D 2 $\frac{22}{25}$ Q. the demand folv'd.

If a queftion confifts of feverall denominations viz. Pounds, Shillings, Pence, fee the rules of Practice in the propofitions of Intereft ; if mixt, as in refpect of time, viz. Moneths, Weeks, Days &c. find firft the encreafe for the terme of years, and multiplie all that by the proportionall number found in the former Table, for the parts of a yeare the refult will anfwer gripple expectation ; for Decimall tables, (or what thefe Authors have not treated off) I referre the Reader unto my bookes of *Naturall* and *Artificiall* Arithmetick, or my Scales of commerce and Trade, this place being not convenient to enlarge my felfe, or build upon anothers ground neither will I leffen their workes (by me now corrected) whereby to magnifie my owne, but have inferted this Table, and here I fubfcribe my name, hoping my labours may give you eafe I reft

Your friend although unknown

Thomas Willsford.

To every young Arithme-
tician, or practicioner in Num-
bers, who shall peruse
these Bookes.

Candid Lector,

Ou will here receive an old Arithmeticke *from the authoritie of*
Record, entaild upon the People,
ratified and sign'd by the appro-
bation of Time, with multitudes
of surviving witnesses; so expect not from me
to confer Encomiums, when I had written no-
thing here, but that through mistakes and over-
sights of former Correctors, Errors have ap-
pear'd like infirmities incident to decrepit
Age, involv'd within the sheets, as if prepar'd
for a funerall, the Authors Senses departed,
or in a trance confused and ambiguous, their
names remaining like inscriptions on a Tombe,
where corruption of one have demonstrated
generations of others.
    Some places I found obliterated, other
parts dislocated, or false numbers have usurp'd
their

their roomes, and those established by sundrie
impressions, may animate some (ignorantly)
to plead Prescription for them ; although
multiplied into a numerous and adulterous
offspring, which deterred many young beginners from their progresse in these Numbers,
and invoked me to their assistance in the restitution of the Authors, notwithstanding I
have published some bookes of this subiect
alreadie, differing in forme, the scope I aimed
at being both Speculation and Practice, my
intentions dedicating that and this to the
Publicke good, whereby purblinde Suspicion,
or fond affection (the Parents of Partialitie)
are expulsed and vanished, and I elected as
an impartiall Corrector of this Treatise where
many of the Tables and rules of direction
were direct divided from the first Composers
meaning, the grounds of Truth, or wayes of
Art; some of which deviations I have rectifi-
ed, subtracted others, and totally cancelled
many; adding numbers in their places accom-
modated and reduced to the Authors sence, the
Questions stated, and the pristin Copies ;
usher'd from the Presse again by M^r. Mellys
attended by Hartwell, and all these expose to
publicke view, drest (without disguise) in their
old attires; otherwise it would seeme absurd, as to
see grave Antiquitie vested in French habite.

Humanum

Humanum eſt errare ; *ſo I will not promiſe that all the old errors are corrected (although above* 1000*) nor yet engage the Preſs ſhall commence no new, but ſo faithfully as I could, theſe Authors recovered are here preſented to your view* ; *which is all that was required of me, and from you (Courteous Reader) a friendly acceptance (in recompence of my labours) deſiring to be numberd amongſt the coadjutors of my Country-men* ; *in teſtimonie whereof I here ſubſcribe my name*

Thomas Willsford,

FINIS.

Lightning Source UK Ltd.
Milton Keynes UK
UKOW07f1140100417
298773UK00002BB/119/P